Megacities

International Year of Planet Earth

The book series is dedicated to the United Nations International Year of Planet Earth. The aim of the Year is to raise worldwide public and political awareness of the vast (but often under-used) potential of Earth sciences for improving the quality of life and safeguarding the planet. Geoscientific knowledge can save lives and protect property if threatened by natural disasters. Such knowledge is also needed to sustainably satisfy the growing need for Earth's resources by more people. Earth scientists are ready to contribute to a safer, healthier and more prosperous society. IYPE aims to develop a new generation of such experts to find new resources and to develop land more sustainably.

For further volumes:
http://www.springer.com/series/8096

Frauke Kraas • Surinder Aggarwal
Martin Coy • Günter Mertins
Editors

Megacities

Our Global Urban Future

Springer

Editors
Frauke Kraas
Institute of Geography
Cologne University
Cologne, Germany

Martin Coy
Institute of Geography
Innsbruck University
Innsbruck, Austria

Surinder Aggarwal
Mansara Housing Society
Delhi, India

Günter Mertins
Faculty of Geography
Marburg University
Marburg, Germany

ISBN 978-90-481-3416-8 ISBN 978-90-481-3417-5 (eBook)
DOI 10.1007/978-90-481-3417-5
Springer Dordrecht Heidelberg New York London

Library of Congress Control Number: 2013934967

Printed on acid-free paper

Springer is part of Springer Science+Business Media (www.springer.com)

Editorial: Megacities: Our Global Urban Future

Since the year 2007, according to the United Nations, more than half of the world's population live in cities – an increase from 30% in 1950 to 51.6% in 2000 – and the increase will probably reach 60% in 2030 (UN 2011). Urbanisation is proceeding rapidly, currently particularly in Asia and Africa. In the industrial North, urbanisation has slowed down and urban transitions have matured, except in the transition countries of industrialising Eastern Europe. The Asian region (in particular China and India with urban population weight of 700 million) sharing 75% of the developing countries' and 53% of the global urban population will remain the major contributor to future urban growth. Megacities as cities with more than five million inhabitants are most significant in this worldwide process of urbanisation. Almost 60 of them, with together more than 600 million people, are expected to exist by 2015. They are characterised by new development dynamics as well as intense and complex interaction of different demographic, social, political, economic and ecological processes. Many urban processes take place simultaneously, thereby often reinforcing themselves. In the developing world, megacities, population grows faster than the infrastructure. A fragmented and uncontrolled urban sprawl can foster high traffic volumes, high concentrations of industrial production, ecological overload, environmental degradation, informality and unregulated land development and property/housing markets. Both convergence and divergence forces are operating simultaneously to produce concentrated and highly differentiated and fragmented urban landscapes. Megacities often contain a mix of coexisting people: there are commonly groups with their own distinctive ethnic, community, cultural roots, lifestyles and social *milieux*. Differences in economic development, social polarisation, quality of infrastructure and governance must be recognised and taken into account.

Against this background, this book intends to take a closer look at selected mega-urban questions and case studies. It is directed to important dynamics and challenges of mega-urban development worldwide by focusing on three main topics, i.e. (1) physical space, land and resources; (2) economic, social and infrastructure transformation; and (3) governance and management of and in megacities.

At first, a general overview of major trends and global considerations aims at summarising important processes and challenges in mega-urban areas worldwide (contribution of Frauke Kraas and Günter Mertins). It points out major developments and trends and emphasises the complexity of interwoven processes of mega-urban regions.

In the *first part* of the book, focusing on *Physical Space, Land and Resources*, Bernd Hansjürgen and Dirk Heinrichs give a broad overview over the key challenges in respect to climate change – similarly as Eduardo de Mulder, Jacques Besner and Brian Marker summarise problems and options of using additional spaces in underground cities. In regard to the situation of rapid mega-urban development in China, Rafiq Azzam and co-authors in their research contribution focus on the problems of water quality and socio-economic vulnerability that arise from the massive influx of migrants into the emerging megacities. Forms and problems of ensuring the mega-urban food supply in Dhaka/Bangladesh are looked at closely in the paper from Markus Keck and co-authors. The contribution of Babette Wehrmann points out the important issue of land development strategies in the context of urban sprawl and informality, while the influence of foreign direct investment on land use changes and regional planning is elaborated in the chapter by Margareth Pugh O'Mara and Karen Seto. All chapters in this part of the book open the view for a critical perspective on the shortages of land and resources under the high pace of mega-urbanisation.

The deep and far-reaching *Economic, Social and Infrastructure Transformation of Mega-Urban Regions* is focused at in the *second part* of the book. Martin Coy and Tobias Töpfer investigate the diverse inner-city development in Latin American megacities in context to degradation and renewal. Agile firm organisation and upgrading processes in the Pearl River Delta, China, are dealt with in the contribution from Wan-Hsien Liu and co-authors. Tabea Bork-Hüffer and co-authors focus on the transformation of international migration patterns in Guangzhou and Foshan. A critical assessment of growing violence, fear and fragmentation is undertaken by Marcelo Lopes de Souza for the case of Rio de Janeiro, Brazil. Alexandra Hill and co-authors focus on informal processes of urban expansion and technical infrastructure in Dar es Salaam, and Jan Marco Müller assesses closely the bus rapid transit system of the TransMilenio in Bogotá as possible example for infrastructural problems in megacities worldwide.

The *third part* of the book directs the reader to questions of *Governance and Management* in mega-urban regions. Shipra Narang-Suri and Günter Taube elaborate on experiences, challenges and implications for international cooperation in megacities. Werner Gamerith focuses on planning processes in New York City, and Christoph Dittrich compares e-Governance initiatives in India, taking up the examples of Hyderabad and Bangalore.

All contributions show the necessity of international, interdisciplinary collaboration in addressing complex research questions of mega-urban development. They also prove the imperative to include and connect major stakeholders, ranging from local governments, private enterprises, non-governmental organisations to the civil societies, in order to understand and analyse current developments and achieve good practices for a more sustainable performance of mega-urban governance.

This book combines contributions which originated from several initiatives within the growing community of researchers working on megacity issues. Among the first was the conference series on megacities in 2002, organised by the Konrad Adenauer Foundation (documents of this conference can be downloaded at http://www.kas.de/wf/de/21.116/). Furthermore, the International Geographical Union's MegaCity TaskForce brought together researchers from all over the world, forming a transglobal, multidisciplinary network of scholars and practitioners exchanging knowledge, expertise and solution-oriented good practices worldwide (www.megacities.uni-koeln.de). The later development of three complementing research programmes which focus on key issues of megacity development has contributed strongly to deeper knowledge on and comparative analysis of mega-urban realities. These are, namely, (1) the German Ministry of Education and Research (BMBF) programme's "Research for the Sustainable Development of the Megacities of Tomorrow" (2005–2007), which later was renamed "Future Megacities – Energy- and climate-efficient structures in urban growth centers" (2008–2013; www.future-megacities.org); (2) the Priority Programme of the German Research Foundation (DFG) on "Megacities – Megachallenge: Informal Dynamics of Global Change" (2006–2013; www.megacities-megachallenge.org); and (3) "Risk Habitat Megacity" (2007–2011; www.risk-habitat-megacity.ufz.de) of the Helmholtz Association, which later transformed to "ClimateAdaptationSantiago" (2009–2012; www.climate-adaptation-santiago.ufz.de). Last but not least, over the International Year of Planet Earth (IUGS & UNESCO), one of the key topics was "Megacities – our global urban future" – this volume is a direct product of the activities during this global initiative.

The editors would like to sincerely thank first and foremost all the authors of this volume for their strong engagement, commitment and patience over its development period. Special thanks go to Dr. Günter Dill and Ulrich Nitschke for leading the first megacity conferences, to the members of the MegaCity TaskForce of the International Geographical Union (IGU) for their continuous support in the scientific development of the topic and in approaching several of the authors of this volume and to the chairs of the International Year of Planet Earth, in particular Eduardo de Mulder, as well as the publisher Springer for their always generous and very helpful support of the publication. We also reserve special thanks to Ursula Dörken, Rebeca Niemann and the staff at Springer for the reliable help in proofreading of the contributions.

Frauke Kraas
Surinder Aggarwal
Martin Coy
Günter Mertins

Contents

Contributors

Rafig Azzam Department of Engineering Geology and Hydrogeology, RWTH Aachen University, Aachen, Germany

Klaus Baier Department of Engineering Geology and Hydrogeology, RWTH Aachen University, Aachen, Germany

Anna Lena Bercht Department of Geography, University of Kiel, Kiel, Germany

Jacques Besner Associated Research Centers for the Urban Underground Space, ACUUS Secretariat, D.D.O. (Montreal) Quebec, Canada

F. Bickenbach Kiel Institute for the World Economy, Kiel, Germany

Hans-Georg Bohle Geography Department, University of Bonn, Bonn, Germany

Tabea Bork-Hüffer Asia Research Institute, National University of Singapore, Singapore

Martin Coy Institute of Geography, University of Innsbruck, Innsbruck, Austria

Eduardo F.J. de Mulder Geoscientists, Executive Director Earth Science Matters Foundation, Wageningen, The Netherlands

Marcelo Lopes de Souza Department of Geography, Federal University of Rio de Janeiro (UFRJ), Rio de Janeiro (RJ), Brazil

Christoph Dittrich Department of Geography, University of Göttingen, Göttingen, Germany

Benjamin Etzold Geography Department, University of Bonn, Bonn, Germany

W. Fu Institute of Economic and Cultural Geography, Leibniz University Hanover, Hanover, Germany

Werner Gamerith Fach Geographie der Universität Passau, Regionale Geographie, Passau, Germany

Bernd Hansjürgens Helmholtz Centre for Environmental Research - UFZ, Department of Economics, Leipzig, Germany

P. Hartmann Institute of Geography, University of Cologne, Cologne, Germany

Dirk Heinrichs Institute of Transport Research, German Aerospace Center, Berlin

S. Hennemann Department of Geography, Justus Liebig University Giessen, Giessen, Germany

Alexandra Hill Institute of Spatial Planning, Technical University Dortmund, Germany

Tanja Hühner Institute of Spatial Planning, Technical University Dortmund, Germany

Markus Keck Department of Economics, South Asia Institute, University of Heidelberg, Heidelberg, Germany

Frauke Kraas Institute of Geography, University of Cologne, Cologne, Germany

Volker Kreibich Institute of Spatial Planning, Technical University Dortmund, Germany

Zhigang Li School of Geography and Planning, Sun Yat-sen University, Guangzhou, China

I. Liefner Department of Geography, Justus Liebig University Giessen, Giessen, Germany

Christian Lindner Institute of Spatial Planning, Technical University Dortmund, Germany

W.-H. Liu Kiel Institute for the World Economy, Kiel, Germany

Lin Lu Department of Engineering Geology and Hydrogeology, RWTH Aachen University, Aachen, Germany

Brian Marker Independent Consultant, London, United Kingdom

Günter Mertins Faculty of Geography, Marburg University, Marburg, Germany

S. Meyer Joanneum Research Forschungsgesellschaft mbH, Graz and Vienna, Graz, Austria

Jan Marco Müller Bureau of European Policy Advisers, European Commission, Brussels, Belgium

S. Ohm Department of Geography, Justus Liebig University Giessen, Giessen, Germany

Margaret Pugh O'Mara Department of History, University of Washington, Seattle, WA, USA

Birte Rafflenbeul Department of Geography, University of Cologne, Albertus-Magnus Platz, Cologne, Germany

J. Revilla-Diez Institute of Economic and Cultural Geography, Leibniz University Hanover, Hanover, Germany

D. Schiller Institute of Economic and Cultural Geography, Leibniz University Hanover, Hanover, Germany

Karen C. Seto School of Forestry and Environmental Studies, Yale University, New Haven, CT, USA

R. Soltwedel Kiel Institute for the World Economy, Kiel, Germany

Ramona Strohschön Department of Engineering Geology and Hydrogeology, RWTH Aachen University, Aachen, Germany

Shipra Narang Suri Department of Politics, University of York, UK

Günther Taube InWEnt – Capacity Building International, Germany

Tobias Töpfer Institute of Geography, University of Innsbruck, Innsbruck, Austria

Rainer Wehrhahn Department of Geography, University of Kiel, Kiel, Germany

Babette Wehrmann Freelance consultant, Munich, Germany

Katharina Wiethoff Institute of Landscape Architecture, RWTH Aachen University, Aachen, Germany

Wolfgang-Peter Zingel Department of Economics, South Asia Institute, University of Heidelberg, Heidelberg, Germany

Megacities and Global Change

1

Frauke Kraas and Günter Mertins

Abstract

Megacities, due to their enormous population concentrations and rapid development dynamics, are gaining more and more importance as junctions of globalisation processes and governance centres in a world increasingly dominated by cities. Megacities are not just large cities. Their scale creates new dynamics, new complexity and new simultaneity of events and processes – physical, social, political and economic. A multitude of drivers, driving forces and actors as well as interacting and partially self-enhancing acceleration and feedback effects is further contributing to the complexity of their development dynamics.

Keywords

Megacities • Global change • Governance • Informality

Megacities are influenced by the manifold processes of global change, and they also co-determine it vice versa (Kraas 2003, 2007; Gurjar and Lelieveld 2005; Borsdorf and Coy 2009). In general, megacities are globally viewed more as risk areas (Mitchell 1999; Kraas 2003), where environmental pollution, symptoms of capacity overloads and stress, resource consumption, natural and human-made risks (e.g. inundation, earthquakes, storms, water shortage, economic crises, ethnic-religious conflicts and industrial accidents) endanger the functioning of mega-urban economies and societies (McGee 1991; Mertins 1992; McGee and Robinson 1995). In particular, disadvantaged population groups in megacities are subject to increasing poverty and vulnerability (Mertins 2006; Bohle and Sakdapolrak 2008; Bohle et al. 2008) and socio-spatial segregation and fragmentation processes (Mertins and Müller 2010); in addition, socioeconomic disparities and disintegration are worsening. However, the positive development chances inherent in megacities as global junctions should also be perceived (Ehlers 2006): There is substantial potential due to the wide range of available financial and human resources as well as widely networked and interacting stakeholders, especially

F. Kraas (✉)
Institute of Geography, Cologne University,
Albertus-Magnus-Platz, 50923 Cologne, Germany
e-mail: f.kraas@uni-koeln.de

G. Mertins
Faculty of Geography, Marburg University,
Deutschhausstraße 10 10, 35032 Marburg, Germany
e-mail: mertins@staff.uniunimarburg.de

F. Kraas et al. (eds.), *Megacities: Our Global Urban Future*, International Year of Planet Earth,
DOI 10.1007/978-90-481-3417-5_1, © Springer Science+Business Media B.V. 2014

as pioneers on the way to sustainable development, for instance, through decreasing spatial consumption per capita, efficient resource use or improved education and health care (Krämer et al. 2011; Bork et al. 2011a, b). Furthermore, technical innovations in megacities can be realised cost-efficiently and integrated in the existing structures (e.g. transport systems, networks, process innovations; Herrle et al. 2006).

1.1 Megacities: Definitions and Dimensions

Megacities are usually defined according to quantitative criteria and are thus – in accordance with various definitions – metropolises with five, eight or ten million inhabitants (Mertins 1992; Fuchs et al. 1994). In general, these cities have a mono-centric structure (Bronger 2004). However, polycentric mega-urban regions as functionally integrated, interacting agglomerations are also included, e.g. the Pearl River Delta in Southern China or the Rhine-Ruhr Area, Germany.

While seven megacities (with more than five million inhabitants) existed in 1950 and 24 in 1990, by 2010 there were 55 and by 2025 there will be – according to estimations – 87 megacities (UN 2012; Fig. 1). Currently, megacities in Asia are facing the world's most dynamic changes. Due to high urbanisation rates (up to about 5%/a), the degrees of urban population compared to the total national population have clearly climbed in the last three decades (UN 2012). Highest growth rates are due to natural population growth, especially however due to migration and administrative incorporation of urban territory. International and/or national migration modifies this, as do temporary, circular or seasonal migration with respect to underemployment and landlessness in rural areas.

However, purely quantitative delimitations based on population numbers remain unsatisfactory, since the statistical values are based on nonuniform surveys from different spatial administrative regional divisions. More essential are the qualitative characteristics of the megacities, among which – with considerable individual differences between megacities in industrial, emerging, transition and developing countries – are a series of frequently observed similarities. Among these are intensive expansion, suburbanisation and inner-urban consolidation processes, functional primate city dominance as well as symptoms of ecological and infrastructural capacity overloads and stress, diversification of inner-urban centre structures, development of polarised or respectively, fragmented societies and increasing loss of governance and control with increasing informality (Roy and Alsayad 2004; Qi et al. 2007; Kraas and Mertins 2008; Mertins and Müller 2010). Some megacities are also global economic governance centres and global cities with high-ranking services as well as headquarters of transnational businesses that produce for world, national and regional markets.

1.2 Megacities: Contributing to and Affected by Global Change

Above all, globalisation processes were and are the motors that drive these enormous changes and are also the driving forces, together with transformation and liberalisation policies, behind the economic developments of the last ca. 25 years (in China, especially the so-called socialism with Chinese characteristics that started under Deng Xiaoping in 1978/1979, in India essentially during the course of the economic reform policies of the so-called New Economic Policy as of 1991; Cartier 2001; Nissel 1999). Especially in megacities, these reforms led to enormous influx of foreign direct investments, to intensive industrialisation processes through international relocation of production locations and depending upon the location, partially to considerable expansion of the services sector with increasing demand for office space as well as to a reorientation of national support policies – with a not to be mistaken influence of transnationally acting conglomerates but also considerable transfer payments from overseas communities. In turn, these processes are flanked and intensified through, at times, massive migration movements of national and international migrants into the megacities (Baur et al. 2006).

Globalisation processes also cause or amplify three central developments, which are characteristic for many megacities with current high dynamics: (1) How vulnerable the population of a megacity, as well as its social and economical system, with respect to extreme or shock events is, is considerably dependent upon how much extensive parts of the population rely on the public supply systems (food, water and energy, but also health-care and education facilities) or if they are neglected by the systems (Bohle et al. 2008; Bork et al. 2011a, b). (2) Strength and measures of public-administrative governance and steering, on the one hand, and capability and flexibility of self-organisation of civil society, on the other hand, cause to what extent informal, self-organised structures, and negotiation processes are or must be responsible for the development. Both stakeholder groups act all the more strongly the more intensively they are integrated in the superordinated globalisation processes: State and administration are strengthened, e.g. through foreign investments or development cooperation; local population and NGOs are supported through foreign connections in their capacities and actions and thus have enhanced legitimation and negotiation margins. (3) Problems of governance and steering are also directly connected to globalisation processes: For example, the financial resources of a megacity accrue from the economic capacity and international competitiveness of its urban economy; the political direction of the governing powers contributes to international legitimation and financial support of the steering instruments.

Thus, the dynamics and complexity of the processes that take place in megacities as well as their impact upon the reorganisation of global spatial, sociocultural, economical and political-institutional relationships belong to the greatest challenges of our time. Historically, the processes connected with today's mega-urbanisation can only be compared with the profound changes that occurred as a result of the Middle European and North American industrialisation of the 19th and 20th century – if their importance is not clearly exceeded, especially with respect to the current, enormous mega-urbanisation dynamics in the populous countries of South, East and Southeast Asia and the global shift of production and services.

1.3 Governance and Informality

Some of the most critical deficits in megacities include overall weak governability and steering structures in the administrations with a prevalence of, for the most part, vertical governance, low inter-institutional horizontal networking and coordination as well as weak steering competency of the administrations (UN-Habitat 2002, Kreibich 2010). In addition, the dominance of, for the most part, sectoral planning with a simultaneous lack of integrated urban or regional planning and effective land-use planning and controls as well as binding construction plans with regulations for building and construction standards (e.g. storey heights, infrastructure equipment or safety measures) is a problem. An additional aggravating factor for all administration procedures is that there are different borderlines and thus competencies of administrative units within the megacities. Correspondingly, extensive areas of many megacities are characterised by uncontrolled, very heterogeneous land-use mosaics that are directly next to each other. Furthermore, there are at times striking bottlenecks in the provision of everyday basics, especially with respect to drinking water and electricity supplies, basic health care and emergency and catastrophe precautions.

While urban policy – seen against modern concepts of "governance" and "government" – is more and more viewed as a multilevel system, "since the borders of the state have been dissolved with respect to society as also international surroundings" (Benz 2004: 5) and new stakeholders are influencing development processes, a new complexity of political processes and structures in international system can be observed.

Due to strong economic dynamics and high immigration figures in many megacities, housing, infrastructure, jobs, supply and disposal systems as well as health care and education must be provided for hundreds of thousands, often within a few years. Naturally, the struggle for urban livelihood and reduced steering capacities supports regulation regimes where informality prevails (Hackenbroch et al. 2009). Insufficient or a lack of developmental and land-use planning and control as well as increasing loss of governance and administration capacities have an effect on the administration, organisation, planning, control and management of municipal tasks – with the result that many processes are unregulated, informal or illegal (Roy and Alsayad 2004; Kreibich 2010). Thus, a substantial and, in part, increasing amount of informal structures and processes can be observed beyond state-registered and state-regulated activities. Within the wide range of the informal economy, these include, e.g. domestic help, street hawkers and cook shop operators, as well as unregistered workers in transport and repair sectors, travelling hawkers, waste collectors and informal vendors (Kulke and Staffeld 2009). Earlier perceptions of the informal sector, as a sector that absorbs the masses, are being questioned because of the informal sector's adaptation capabilities and flexibility. Current discussions address the question as to whether and how much the informal sector in megacities will be able to realise their necessary survival absorption functions due to erosion of local supply cycles due to internationalisation of the markets (Bohle et al. 2008). Moreover, phenomena like conceded or experimental informality can be identified.

Beyond this essentially economically focussed perception, present concepts of informality include aspects such as informal construction, personal arrangements in personal networks as well as unregulated, semi-legal and illegal activities. The transitions of different socioculturally interpreted understandings of legitimacy, legality and illegality are not necessarily sharp, especially since at times competing legal systems can coincide as anchors of informal organisations. In addition, the polarising paired concept of formal and informal, in which the distinguishing criterion is state participation, has been found inadequate because it displaces the realities of the varied stakeholder interconnections. In addition to stakeholders in formal political-administrative systems and the private sector, self-organised networks and institutions are establishing themselves increasingly, and their complex governance mechanisms, negotiation processes and discourse influence the development dynamics of megacities. Especially a look at administrative governance capacities shows that conventional concepts, standards, strategies, tools and priorities of megacity development neither do justice to conditions of current urbanisation dynamics nor are suitable for adequately integrating the complexity of the stakeholders – also they have to be reconsidered under the auspices of global change.

1.4 Key Priorities for Research and Practice

The general perception of mega-urban regions, the international megacity research as well as the priorities in planning and governance are currently undergoing substantial changes: First, megacities are perceived more and more as areas of global importance, affected by and affecting themselves manifold levels of global change over wide distances and long periods of time. Consequently, their performance falls no longer just in the responsibility of local actors, but as they are embedded at least in transnational, if not global development processes, the responsibility for their sustainable development lays in the hands of numerous, more or less directly or indirectly responsible, internationally connected actors. Second, the general perception of megacities shifts from a predominantly negative view ("moloch", "global sink") to a more positive perception of mega-urban areas as priority areas and drivers of change, with at least often undiscovered potential of improved sustainability and quality of life for many, at least more, if not all inhabitants. Third, the complex reality of phenomena, processes and actors as well as the high pace of development in mega-urban areas inevitably deserve international, inter- and transdisciplinary, intercultural as well as multi-stakeholder-oriented research and action – including stakeholders from research, administration, the private sector and the general public and civil society. This necessarily implies a more engaged and committed interaction among all responsible levels. Fourth, as to the role and direction of research, the generation of not only knowledge based on fundamental descriptions, analyses and explanations but, moreover, the creation of knowledge for prediction, orientation and decision making are needed. Fifth and finally, for many megacities, particularly in the emerging economies, major shifts from a predominantly globalisation-driven, competitiveness-seeking top-down development to alternative priorities, such as of region-based and spatial transformations (Pain 2010) are regarded important. Beyond current priorities on structure-, pattern-, land-use-, infrastructure- and housing-based planning, more problem-, process- and people-oriented approaches are emerging.

References

Baur M et al (2006) Labour mobility in urban china. An integrated labour market in the making? Berliner China-Studien 46. Lit, Berlin

Benz A (2004) Governance – Modebegriff oder nützliches sozialwissenschaftliches Konzept? In: Benz A (ed) Governance – Regieren im komplexen Regelsystem. Eine Einführung, Wiesbaden, pp 12–28

Bohle HG, Sakdapolrak P (2008) Leben mit der Krise Vertreibung von Slumbewohnern in der Megastadt Chennai. Geogr Rundsch 60(4):12–20

Bohle HG et al (2008) Reis für die Megacity. Nahrungsversorgung von Dhaka zwischen globalen Risiken und lokalen Verwundbarkeiten. Geogr Rundsch 60(11):28–37

Bork T, Kraas F, Li Z, Xue D (2011a) Urban environmental health challenges in China's villages-in-the-city. Geogr Z 99(1):16–35

Bork T, Kraas F, Yuan Y (2011b) Governance challenges in China's urban health care system – the role of stakeholders. Erdkunde 65(2):121–135

Borsdorf A, Coy M (2009) Megacities and global change: case studies from Latin America. Die Erde 140(4):341–353

Bronger D (2004) Metropolen, Megastädte Global Cities. Wissenschaftliche Buchgesellschaft, Darmstadt

Cartier C (2001) Globalizing South China. Blackwell, Oxford

Ehlers E (2006) Stadtgeographie und Megastadt-Forschung. In: Gans P, Priebs A, Wehrhahn R (eds) Kulturgeographie der Stadt. Kieler Geographische Schriften 111., pp 51–62

Fuchs RJ et al (eds) (1994) Mega-city growth and the future. United Nations University Press, Tokyo

Gurjar BR, Lelieveld J (2005) New directions: megacities and global change. Atmos Environ 39:391–393

Hackenbroch K, Baumgart S, Kreibich V (2009) The spatiality of livelihoods: urban public space as an asset for the livelihoods of the urban poor in Dhaka, Bangladesh. Die Erde 140(1):47–68

Herrle P, Jachnow A, Ley A (2006) Die Metropolen des Südens: Labor für Innovationen? Mit neuen Allianzen zu besserem Stadtmanagement. Stiftung Entwicklung und Frieden, Policy paper 25. Bonn

Kraas F (2003) Megacities as global risk areas. Petermanns Geographische Mitteilungen 147(4):6–15

Kraas F (2007) Megacities and global change: key priorities. Geogr J 173(1):79–82

Kraas F, Mertins G (2008) Megastädte in Entwicklungsländern: Vulnerabilität, Informalität, Regier- und Steuerbarkeit. Geogr Rundsch 60(11):4–10

Krämer A, Khan MMH, Kraas F (eds) (2011) Health in megacities and urban areas. Springer, Heidelberg

Kreibich V (2010) The invisible hand: informal urbanisation in major cities of Tanzania. Geographische Rundschau International 6(2):38–43

Kulke E, Staffeld R (2009) Informal production systems – the role of the informal economy in the plastic recycling and processing industry in Dhaka. Die Erde 140(1):25–43

McGee TG (1991) The emergence of desakota regions in Asia: expanding a hypothesis. In: Ginsburg N, Koppel B, McGee TG (eds) The extended metropolis: settlement transition in Asia. University of Hawaii Press, Honolulu, pp 3–25

McGee TG, Robinson IM (eds) (1995) The mega-urban regions of Southeast Asia. UBC Press, Vancouver

Mertins G (1992) Urbanisierung, Metropolisierung und Megastädte. Ursachen der Stadt "explosion" in der Dritten Welt. In: Deutsche Gesellschaft für die Vereinten Nationen (ed): Mega-Städte – Zeitbombe mit globalen Folgen. Bonn, pp 7–21

Mertins G (2006) Wachsende Marginalisierung und Marginalviertel in Großstädten der Dritten Welt. Kieler Geographische Schriften 111:63–77

Mertins G, Müller U (2010) Gewalt und Unsicherheit in lateinamerikanischen Megastädten. Auswirkungen auf politische Fragmentierung, sozialräumliche Segregation und Regierbarkeit. Geogr Rundsch 60(11):48–55

Mitchell JK (ed) (1999) Crucibles of hazard: mega-cities and disasters in transition. United Nations University Press, Tokyo

Nissel H (1999) Megastadtentwicklung, Globalisierung und Migration. Fallstudie Bombay. In: Husa K, Wohlschlägl H (eds) Megastädte der Dritten Welt im Globalisierungsprozeß, Abhandlungen zur Geographie

und Regionalentwicklung, 6. Inst. für Geographie der Univ. Wien, Wien, pp 347–432
Pain K (2010) Spatial transformations of cities: global city-region? Mega-city region? www.lboro.ac.uk/gawc/rb/rb353.html
Qi C, Kreibich V, Baumgart S (2007) Informal elements in urban growth regulation in China – urban villages in Ningbo. Asien 103:23–44
Roy A, Alsayad N (eds) (2004) Urban informality. Transnational perspectives from the Middle East Latin America and South Asia. Lexington Books, Lanham
UN-Habitat (2002) United Nations human settlements programme. The challange of slums, Nairobi
United Nations (2012) World urbanization prospects. The 2011 revision. United Nations, New York

Part I

Physical Space, Land and Resources

Megacities and Climate Change: Early Adapters, Mainstream Adapters and Capacities

2

Bernd Hansjürgens and Dirk Heinrichs

Abstract

In this chapter, the relationship between the necessity of making urban adaptation responses to climate change and the competences and capacities in megacities is analysed. The experience of early adapters (cities that have already initiated local adaptation strategies and plans) reveals that these cities are able to seize several opportunities. These include the political recognition as local champions, innovation of local administration (e.g. by creating cross-departmental working groups and units), the development of new networks, the creation of new knowledge and the formation of strong and broad political consensus on development priorities that support and confirm existing strategies. However, these determining factors do not necessarily work for potential later adapters and, in particular, for megacities. They are either case-specific or exclusively available for early movers only and will no longer apply to mainstream adapters. Some other factors are of instructive value to mainstream adapters as good practice, for example, those relating to accommodating the integrative cross-cutting nature of climate change in the implementation arrangements. The diffusion of adaptation action to megacities will rely to a great extent on mobilising potentials and overcoming obstacles with respect to a range of political, financial and administrative competences and capacities. However, many megacities around the world suffer from significant deficits with respect to these capacities and competences. These shortcomings are probably among the causes for the fact that initiatives towards strengthening adaptation action are still the exception rather than the rule.

B. Hansjürgens (✉)
Helmholtz Centre for Environmental Research - UFZ,
Department of Economics,
Permoserstraße 15, 04318 Leipzig, Germany
e-mail: bernd.hansjuergens@ufz.de

D. Heinrichs
Institute of Transport Research,
German Aerospace Center, Rutherfordstr. 2, 12489 Berlin

F. Kraas et al. (eds.), *Megacities: Our Global Urban Future*, International Year of Planet Earth,
DOI 10.1007/978-90-481-3417-5_2, © Springer Science+Business Media B.V. 2014

Keywords
Megacities • Climate change • Adaptation • Early adapters • Mainstream adapters • Capacities

2.1 Introduction

Since the release of the Fourth Assessment Report of the Intergovernmental Panel on Climate Change (IPCC-AR4), we know that the change of the globally averaged temperature is undisputable. We also know that it is *very likely* (90–99% chance) that global warming is caused primarily by anthropogenic activities. It is *very likely* that temperature extremes, warm spells and heat waves will continue to become more frequent in most areas; it is *likely* that the areas affected by droughts will increase and that tropical cyclones will become more intense, with higher peaks of wind speed and heavy precipitation (IPCC 2007a; Fragkias 2009). The IPCC Fourth Assessment Report does not contain any new or surprising results. Instead, it confirms the statements of its predecessors and underlines the fact that there is now more scientific evidence and a higher likelihood that negative climate change impacts will occur.

Megacities are part of the story as a cause of global climate change. Because they concentrate large portions of the urban population, major shares of their countries' economic activities as well as the consumption of food and natural resources, megacities are a major source of anthropogenic greenhouse gas emissions. Large cities are responsible for a significant share of the greenhouse gases released into the atmosphere. As 80–90% of future population growth is expected to occur in megacities, especially in the agglomerations of the global South, megacities' greenhouse gas emissions into the atmosphere will contribute considerably to the increase of total greenhouse gas concentrations. Therefore, the success of greenhouse gas mitigation strategies will depend crucially on whether cities are included in the response to climate change issues by developing and applying greenhouse gas mitigation strategies, in particular, in the fields of urban housing and buildings, energy supply and demand and urban traffic.

However, megacities are also affected by global climate change. Many megacities are located in climate-sensitive areas, and their growth is in particular predicted to occur in these areas. Therefore, risks associated with climate change will rise in areas where megacities are located, for example, coastal areas, flood-prone areas or areas suffering from water scarcity and droughts. What is more important is that many megacities are not only exposed to climate change risks but also concentrate large numbers of the most vulnerable parts of the population. The world's urban population is likely to reach more than eight billion by 2020, and the urban slum population is expected to increase to 1.4 billion by this time. This means that one out of three persons living in cities will live in slums under impoverished, overcrowded and insecure living conditions. This also means that these people comprise individuals and groups which are extremely vulnerable to the impacts of global climate change, such as the poor, women, old-aged people, children, migrants or other groups suffering from segregation (Bartlett et al. 2009).

As climate change is to a large degree unavoidable, at least in the short and middle term, there is no chance for urban areas and their residents to escape from the predicted adverse climate change impacts. The retention period of greenhouse gas concentrations in the atmosphere is decades or even centuries. The world's current climate is influenced by the greenhouse gas emissions of the past, mainly by emissions of the industrialised countries since the beginning of industrialisation. Thus, climate change risks will increase even if the world manages to reduce the release of greenhouse gases into the atmosphere, for example, in the follow-up process of the Kyoto Protocol (which can be doubted). If mitigation measures have long time horizons before becoming effective, cities do not have any other choice besides developing strategies and measures to adapt to the impacts of climate change.

The relationship between the urgency to act and the adaptation capacities in megacities is the focus of this chapter. We will explore this relationship with respect to several questions. What climate change adaptation challenges do megacities face? What opportunities but also obstacles do exist for meeting these challenges? What are key competences and capacities?

To address these questions, the chapter draws primarily on a review of experiences of early adapters. These are cities – not necessarily megacities – that have initiated local adaptation strategies and plans. To develop the arguments, Sect. 2.2 establishes the relationship between global climate change and urban vulnerability. We will point out that vulnerability is not only determined by the exposure to climate change alone but also by the capacity of cities and their residents to cope with climate change impacts. Based on this, Sect. 2.3 puts forward reasons why adaptation with its central objective to reduce vulnerability deserves greater attention as an unavoidable action and complement to mitigation. In Sect. 2.4, we deepen the understanding of the adaptation challenges in megacities. We argue that they face a particular combination of climate change impacts, levels of already existing vulnerability and distinct challenges to coping capacities. Section 2.5 examines the experience on how cities have started to confront these combined challenges, their strategies and the factors that drive and motivate adaptation action. The lessons from early adaptation are discussed in Sect. 2.6, where we discuss whether 'mainstream adapters' will be able to follow a similar model. Section 2.7 provides a summary of results and a conclusion.

2.2 Climate Change Is Associated with Vulnerability of Cities

Working Group II of the IPCC, in its Fourth Assessment Report, deals with the impacts of global climate change (IPCC 2007b). With respect to cities, there are four major climate change impacts which will lead to severe risks and which will increase over time: temperature increase, changing precipitation rates, sea-level rise and extreme events (storms, floods and droughts) (World Bank 2008; McCarney et al. 2009).

Temperature change is very likely to occur. Atmospheric concentrations of carbon dioxide (CO_2), methane (CH_4) and nitrous oxide (N_2O) have increased markedly as a result of human activities in the last 250 years and now exceed by far preindustrial values. Consequently, the IPCC-AR 4 (2007a) states a global increase in temperatures for the past 100 years (1906–2005) of 0.74°C. The last 12 years have seen the warmest 11 years since 1850. According to the IPCC scenarios, the mean temperature on Earth will further increase between 2 and 6°C by the year 2100. This will lead to heat waves that are very likely to increase in frequency and intensity over most land areas (especially in the Northern hemisphere).

Closely related to temperature change, *changes in precipitation* are also very likely to increase. Consequences are increased rainfall and heavy thunderstorms on the one hand, and water scarcity in semiarid and arid zones, with an increase in extreme droughts, on the other hand. According to the IPCC, the precipitation has already increased over land north of 30°N accompanied by a general decline in the tropics since 1970. This trend will continue, with an uneven regional distribution of regions with heavy rainfall and others with water shortages. The city of Lima (Peru), for example, which has already an extremely low precipitation rate of 9 mm per year, will suffer from the melting of the glaciers in the Andes, leading to water shortage in the future.

Temperature change and changing precipitation involves warming of sea waters and melting of glaciers, snow and ice, all leading to *sea-level rise*. During the periods 1961–2003, the sea levels rose by 8 cm as a direct consequence of the expansion of sea water (seas getting warmer) and glaciers melting. A further increase in sea levels is expected, leading, for example, to an estimated rise in the Netherlands of between 70 and 130 cm by 2100. Many low-elevation coastal areas may even disappear from the landscape by the end of the century. As many megacities are located close to the sea, with very low elevation levels, they are put at an extreme risk.

Many impacts of global climate change are closely connected with an increase of *extreme events*, such as floods, storms, long-term extreme droughts and hurricanes. Figures and calculations of the Munich Re's Geo Risks Research reveal that an increase in natural catastrophes can already be observed (Höppe and Grimm 2008). Although the strongest effects of climate change are expected in Africa and parts of Asia, Europe and the USA will also be victims of these changes. Many authors see major recent extreme events such as hurricane Katrina in the USA or the winter storm Kyrill in Europe as forerunners of increasing extreme weather changes.

In general, it is the exposure of regions and urban areas to global climate change which makes them vulnerable. However, determining urban vulnerability by exposure is only one aspect of the whole story. The vulnerability of cities is also influenced by the fact that they – and in particular megacities – comprise a great number of *elements at risk*. As we argue elsewhere (Heinrichs et al. 2009), megacities are characterised by more than population alone. Their size must also be seen in terms of the scale of exposed values, economic functions and importance for their nations' economic activities. Megacities are likewise places of dynamic changes; thus, vulnerability cannot be taken as a static or constant term. What is more, they are extremely complex, which also makes the interplay between climate change effects and vulnerability a complicated matter. We will take up these arguments in more depth in Sect. 2.4 below.

Moreover, vulnerability is characterised by the ability of cities and their residents to cope with adverse effects of global climate change. To illustrate this point, the working definition of vulnerability developed by Wisner et al. (2004, p. 9) is helpful. In their definition, vulnerability is 'the characteristics of a person or a group in terms of their capability to anticipate, cope with, resist, and recover from the impact of a natural hazard'. Instead of emphasising exclusively the characteristics of the natural or technological hazard itself or the exposure (structure, building, etc.) to the hazard, this definition focuses on how cities, communities, social groups or residents are able to deal with the impacts of a natural hazard.

Considering the coping capacity of cities, the situation is completely different between the cities of the global North and those of the global South. In countries of the South, urbanisation hotspots lack functions such as durable housing, access to sufficient and clean water, key resources and sanitation or waste management. Institutional settings in these areas are extremely weak, lacking the capacities to enforce laws and regulations, and residents are faced with rampant corruption. Criminal activities are high, and this again affects, first of all, vulnerable groups. All these factors, operating in conjunction with climate change, create *stress bundles* that increase the impacts of climate change in megacities (Fragkias 2009) and that decrease the coping capacity of their residents, in particular, that of vulnerable groups or persons.

In summary, the vulnerability of megacities is not only determined by the exposure to climate change risks but also by the ability to cope with negative climate change impacts. This insight highlights the crucial role played by economic and social characteristics and capacities of cities, in particular, with respect to adaptation. This aspect will be taken up in the subsequent section.

2.3 Adapting Cities to Climate Change Is Becoming an Unavoidable Task

In contrast to mitigation, the possibility for adaptation to climate change has for a long time been strongly rejected by both climate change researchers and policymakers. There was considerable concern that a far-reaching discussion on adaptation would reduce the pressure for mitigation – and thus foster (or at least increase acceptance of) the continuing anthropogenic emission of greenhouse gases into the atmosphere.

However, this perspective has changed. Today, in addition to strategies and measures referring to mitigation, adaptation is seen as an essential and integral part of climate policy. Hence, the above-mentioned Fourth IPCC Report deals with both mitigation and adaptation. In the current discussion on climate change, there are (at least) three arguments which are put forward in favour of adaptation (Pielke et al. 2007).

First, the intended positive effects of mitigation and the necessity to deal with climate change impacts occur at *different timescales*. It is quite obvious that reducing greenhouse gas emissions will take a long time – even if an international agreement is reached in the near future. Historic emissions of the industrialised countries are responsible for the fact that climate change is already occurring. It will take a long time to change this path. In the words of IPCC (2007a, p. 6), 'there is high agreement and much evidence that with current climate change mitigation policies and related sustainable development practices, global GHG emissions will continue to grow over the next few decades'. In the meantime, cities and their residents have no choice other than to adapt to the impacts of climate change.

Second, the *vulnerability of many societies and locations has increased* during recent years. Responsible factors include, for example, demographic growth, access to water resources, social inequality, urban segregation and wars. Although these factors occur irrespective of climate change, they make societies more vulnerable to climate change. An increase in the coping capacities of societies could improve their resilience and thus protect them from negative climate change impacts. We will take up this argument below with respect to megacities.

Third, there is *stronger influence of those who are mainly affected by climate change*. At international summits, the participating states, in particular, states mainly affected by climate change impacts, increasingly demand strategies and measures for adaptation enabling them to cope with climate change impacts. This demand is put forward irrespective of the extent of mitigation which is still required in order to avoid the most severe impacts of global warming. This argument is particularly put forward by developing countries as affected states. These countries state that instead of spending money on potential threats to future generations, one should spend the money on the improvement of the living conditions of the present generation.

It becomes obvious that the close connection between climate change and vulnerability makes adaptation action indispensable. Although international protocols (e.g. the Kyoto Protocol) have not yet been extended down to the local level, cities themselves have started to take up the challenge in recent years. Their examples show that adaptation planning on the regional scale comprises a wide range of tasks. The impacts of global climate change have to be known on the regional and local scale; thus, a downscaling of information to the regional and local scale is required. This does not only affect changing climate conditions, for example, temperature changes, but also changing regional water cycles and the emergence of flood risks, together with the identification of *elements at risk* (affected population, economic assets), of residents' and social groups' vulnerability and of their coping capacity. Furthermore, the development of a set of adaptation action to avoid risks, or emergency measures in case of hazards, is also required. What makes the task more challenging is the fact that adaptation action does not only cover parts of local policies (e.g. environmental policy), but a broad range of fields, such as housing, infrastructure planning, access to sufficient and clean water, sanitation, traffic, resource management and social policy. This cross-cutting nature requires new governance arrangements to integrate and mainstream adaptation strategies into and across existing sector policies, together with cooperation as well as coordination mechanisms. It further requires the development of regulatory instruments, compliance mechanisms, monitoring capacities and communication strategies to build general awareness on the subject.

The necessity of scaling down climate change impacts to the local level is supported by convincing arguments: Firstly, climate impacts are now being felt in cities, thus, there are pressures to act on the city level. Secondly, as effects are location-specific, they require local adaptation efforts that national adaptation plans can hardly provide. While climate change mitigation can be conducted at the national and international level, adaptation measures must primarily be localised (IIED 2007). Thirdly, in contrast to mitigation where the reduction of greenhouse gas emissions has no immediate and tangible local effects, adaptation measures create direct and visible benefits in the location which are an important prerequisite for local political support.

Surprisingly, most cities have so far opted to place an emphasis on mitigation strategies. For example, very few cities in the C40 climate group of the world's largest cities, originally sponsored by the Mayor of London in 2005, have explicitly adopted adaptation as a complementary action to mitigation. Furthermore, where this is done, it is mostly confined to cities of the global North. On the other hand, while it can be concluded quite safely that adaptation is far from the mainstream development agenda, some cities in developing and transition countries have taken the lead in adaptation issues. While there are some examples where cities have pressed forward in developing adaptation plans, e.g. Cape Town in South Africa (City of Cape Town 2006, see Sect. 2.5), we are still far away from a situation of mainstreaming adaptation into local development efforts in cities.

Greater risk of floods, landslides, erosion and the associated impacts on settlements, infrastructure and people pose a huge challenge for cities to adopt climate adaptation action properly. In order to cope with these challenges, risk management in all its broadness is required. While this task is already challenging for small- and medium-sized cities, it is even more so for megacities. This originates in the specific conditions of megacities. These conditions will be the focus of the subsequent section.

2.4 Megacities Face Unique Challenges for Adaptation

Megacities are most commonly defined by their population size, whereby thresholds vary from five to ten million people. In a recent report on the *State of the Worlds' Cities*, the UN system adds the category of metacities with 20 million inhabitants (UN 2007).

What justifies special attention on megacities with respect to climate change adaptation, and what particular challenges do they face? There are at least three arguments for this. Firstly, most of these large urban agglomerations are situated in locations that are and will be particularly affected by climate change effects. Secondly, climate change in megacities adds to already existing conditions of extreme vulnerability to non-climate-related factors, thus adding pressure on already huge potential losses (Pielke et al. 2007). Thirdly, the coping capacities of megacities in the global South (where most megacities are located) to confront these challenges are limited. It is worth exploring these three arguments and their relationships more closely.

1. The first argument is that many megacities are situated in locations that are and will be *particularly affected by climate change effects*. The most prominent examples are settlements in the so-called low-elevation coastal zones (LECZ), defined as a contiguous area along the coast that is situated less than 10 m above sea level. The IPCC has emphasised how these zones and in particular the coastal river deltas are among the world's most valuable, heavily populated coastal systems. A recent study by McGranahan et al. (2007) shows that although this zone represents only about 2% of the world's land area, it contains about 10% of its total population and 13% of its urban population. Almost two thirds of the world's large cities with more than five million inhabitants fall at least partly within this zone. Particularly, the megacities of low-income and middle-income nations are situated here. What is more, the population in these coastal zones continues to increase rapidly in most countries. China provides the most dramatic example. The coastal provinces of China experienced a net immigration of about 17 million people between 1995 and 2000 (Huq et al. 2007), creating pressures in already crowded coastal areas including the megacities of Shanghai and Guangzhou in the Pearl River Delta. Similar trends occur in Vietnam, Japan, the Philippines, India and Bangladesh. The city of Dhaka, for example, has over ten million

inhabitants and is central to Bangladesh's economy. It is very vulnerable to flooding – as shown by the devastation from floods in 1988, 1998 and 2004. The combination of sea-level rise and increased frequency and intensity of storms will increase vulnerability in future (Alam and Rabbani 2007). Mumbai is another example. As in Dhaka, large sections of the city are only 1–5 m above sea level. Much of central Mumbai is built on landfill (Huq et al. 2007). The probable long-term trend of sea-level rise is likely to prove very damaging for the city as this, combined with storm surges, may make large urban areas uninhabitable.

Many of the world's largest cities/metropolitan areas, in particular, in Asia, are in the floodplains of major rivers (e.g. the Ganges-Brahmaputra, the Mekong and the Yangtze). In India, three mega-urban regions – Delhi, Kolkata and Karachi – are situated on the banks of the Brahmaputra. The most serious regional impact of climate change is expected to be a change in the river hydrology due to glacier melt and regression of the Himalayan glaciers (Tangri 2003; Revi 2008). This will affect the availability of irrigation and drinking water for over 30 million people.

The likely link between change in temperature and precipitation, recession of glaciers and the effects on water availability is also a concern, for example, for megacities in Latin America. As pointed out above, Lima is likely to be critically affected by the melting of the Andean glaciers. In Santiago de Chile, the effects are expected to concern not only water supply for drinking, agriculture and industry but also the likelihood of periodic flooding.

Furthermore, most large cities will experience more heat waves and more problems with air pollutants. In larger and denser cities, the temperatures in central heat islands can be several degrees higher than in surrounding areas. Related to this, warmer average temperatures will allow the expansion of the area where many tropical diseases can occur – for instance, where mosquitoes that spread malaria and dengue fever can survive and breed. As a consequence, heat waves will put additional stress on cities and their residents. Health conditions will deteriorate, while the number of heat deaths will increase. These impacts will in particular influence the living conditions of temperature-sensitive vulnerable groups such as old-aged people or children.

2. Climate change in megacities, according to the second argument, adds to *already existing conditions of extreme vulnerability* to non-climate-related factors. This can be observed in the case of the supply with resources that are particularly climate-sensitive (water, agricultural production, soils).To return to the example of the National Capital Region of Delhi, there already exist water shortages today. Drinking water is being transported over 300 km to meet the demands of about 15 million people (Revi 2008). In consequence, a further climate-change-related reduction in water availability will heavily aggravate this problem. Another example relates to the sea-level rise. In Metro Manila in the Philippines, policymakers have started to acknowledge the threats associated with the gradual sea-level rise of 1–3 mm per year, projected to occur with climate change. While the extra impact on flood risk associated with this marginal increase might itself appear negligible, it adds to the ongoing lowering of land surface by several centimetres to more than a decimetre per year as a consequence of excessive groundwater extraction (Rodolfo and Siringan 2006). What is more, not only the vulnerability of megacities as a whole is extreme. It is also unevenly distributed across locations and population. In no other place is the contrast between rich and poor more obvious than in the megacities of the low-income and lower-middle-income countries. UN Habitat concluded in a recent report: 'Large cities tend to display wider inequalities than smaller cities, even in countries with low levels of socioeconomic inequality' (UN Habitat 2006, p. 110). This perspective can be verified by looking at child mortality rates. In Rio de Janeiro, the probability that a child in one of the many slums will reach the age of five is three times less than in other areas of the city,

where the conditions (water and sanitation, housing, etc.) are better (UN Habitat 2006). It must be recognised that in megacities, a large proportion of the urban population has extremely limited resources and options to cope with already ongoing and day-to-day stresses. Climate change effects will add critically to these problems.

3. This leads to the third argument that the *capacity* in the global South (where most megacities are located) to confront these combined challenges is limited. This refers first of all to the capacity of city governments to regulate development in a form that minimises potential vulnerabilities in the first place. For example, in Dhaka, flood risk has risen tremendously because buildings have encroached on or filled in drains, and many natural drains have been filled up to construct roads. Heavy and/or prolonged rainfall produces large volumes of surface water, which can easily overwhelm existing drainage systems. However, it would be too simple to blame city or local governments. Megacities present an aggregation of an extreme amount of functions and their interaction (land use, traffic, energy and water supply) that are difficult to predict, to plan for and to control. In addition, they are places of extreme complexity in governance structures. Mexico City is an instructive example. The metropolitan area is divided into four larger political/administrative units: the federal district (further divided into 16 subunits), the administrations of the states of Mexico and Hidalgo (with together 59 municipalities) and the national government administration. There are numerous coordination units between these entities, but because they are not linked to political and financial authority, their influence remains limited (MRC Mc Lean Hazel and GlobeScan 2007).

As demonstrated, climate change adaptation policy cannot concentrate on coping with the marginal additional effects of anthropogenic climate change alone. Instead, it needs to be integrated with broader environmental, social and developmental policy goals. This task seems to be particularly challenging for the capacity of governments of many megacities in several respects: (a) political, in terms of aggregating legitimacy and generating support for adaptation action and coordination, (b) financial/fiscal, in terms of capacities to equip adaptation action with resources, and (c) administrative, in terms of aggregating the necessary information on the effects of climate change, putting the institutions and mechanisms in place, integrating policies across sectors and developing compliance, evaluation and review as well as monitoring instruments. However, there are examples where cities are successfully taking up the challenge. The next section examines these cases of early adapters and their strategies to utilise and enhance their capacities.

2.5 Experience of Early Adapters for Utilising and Building Adaptation Capacities

Climate change adaptation has recently emerged as a field of policy at city level. Some examples for adaptation action in cities exist, among them a few megacities. First attempts are being made to derive lessons on several questions relating to early climate change adaptation (Carmin et al. 2009; Barton 2009; Heinrichs et al. 2011). What is driving cities to initiate climate adaptation action? What strategies do cities employ, and what opportunities do they seize for enhancing adaptation capacities?

Generally, several cities have embarked on climate action planning and have, for this purpose, started to engage in cooperation networks. The C40 climate group has expanded to more than 40 of the largest cities in the world. The International Council for Local Environmental Initiatives (ICLEI) Cities for Climate Protection Initiative includes more than 650 municipalities worldwide. However, most action so far has an emphasis on mitigation (Betsill and Bulkeley 2004) and is located in cities in the global North, for example, the city of London. It is interesting to note that European cities, in contrast to US cities, emphasise a joint approach focusing on mitigation as well as adaptation (Fragkias 2009). In contrast, there are only a handful of examples where large cities in the global South have developed dedicated plans for adaptation (e.g. Cape Town, Quito). Nonetheless, their action provides some tentative insights on the drivers, strategies and opportunities associated with adaptation.

First, one of the main drivers of adaptation action appears to be a clear *awareness by local stakeholders of local vulnerability to climate change* and the perceived and actual risk. Safety of population and minimising the impacts of natural disasters are major objectives found in adaptation plans (Carmin et al. 2009). Initiatives are often linked to concrete experience of disasters, which reinforces predictions about climate impacts and builds awareness of the need for adaptation. The creation of awareness and local knowledge is driven by locally relevant scientific information. The identification of risks by downscaling climate models and the analysis of vulnerability generate political interest in understanding how the local climate is likely to change, how the city will be affected and what local response options seem appropriate to confront predicted impacts. In the attempt to address existing uncertainties about climate change impacts, there is significant reliance on university scholars, centres and programmes.

Second, *downscaling of climate models and vulnerability analysis* does not only help to identify threats but also encourages cities to establish priorities. Adaptation plans often pursue a twofold approach: They develop a general strategy that is then translated into sector-specific goals and plans. The examples of Cape Town and Quito show that these strategies are not developed separately from other local development objectives but are tied closely to existing strategies. On the one hand, this ensures the integration and mainstreaming of adaptation action. On the other hand, it serves as an opportunity to advance existing (local) development goals and thus ensures continuity instead of radical change in local priorities. The focus, however, seems to vary significantly between cases. In some cases, adaptation is connected strongly to existing environmental programmes. In other locations, it supports the intention to build up reputation in order to create competitive advantages, for example, in tourism (Barton 2009; Carmin et al. 2009).

Third, early adaptation action involves strong *local leadership* that is motivated by the opportunity to become recognised as innovative and future-oriented. Such local championship relies on scientific knowledge and seizes opportunities for visibility at regional, national and international arenas (Barton 2009; Mukheibir and Ziervogel 2007). Local championship is often realised by persons at the interface of science and policy, thus translating scientific knowledge to regional politicians and stakeholders.

Fourth, local adaptation action strongly builds on interpersonal interaction to establish confidence in priorities. The transfer of ideas, knowledge and insight through *external networks*, that is, international or cross-country cooperation (ICLEI, C40, UCLG, etc.), is strong across early adaptors. Memberships in networks and attendance at conferences go beyond enhancing reputation, as these relationships and events are important sources of ideas and information for cities (Carmin et al. 2009). Furthermore, early adapters utilise diverse types of climate-related structures, including 'internal' networks in cities, making it possible for information to be exchanged between politicians and departments as well as fostering participation in events at regional, national and international level. This involves a strong presence and engagement of NGOs and CBOs.

Fifth, a common practice in the implementation of adaptation plans is the creation of *dedicated climate teams* working within a centralised office and not attached to one specific sector. This appears to be an adequate treatment of the cross-cutting nature of adaptation and avoids confining adaptation to the responsibilities of one sector (most likely the environmental department) alone. An alternative is the creation of a climate protection department within the office of the city mayor. This reinforces the interdepartmental character of climate impacts.

Table 2.1 Determining factors that motivate and drive early adaptation strategies

Determining factors	Description
Existence of climate-relevant information and critical levels of awareness	Relevant (local) scientific information exists
	Concrete recent (local) experience with disasters exists that connects the awareness/ attention of politicians and other stakeholders to climate-induced hazards
Integration of climate-hazard-related priorities in the local development agenda	A dedicated adaptation strategy (including goals, instruments, measures) exists
	The strategy has been translated into sector-specific actions and priorities
	Integration and mainstreaming adaptation action as a cross-cutting task
	Adaptation strategy/priorities converge with and reinforce (local) development strategies (win-win situation)
Political leadership (local champions)	(Charismatic) persons with clear understanding and with high acceptance push the adaptation agenda
	Ability of leader(s) to communicate scientific results to the public (thus working at the interface of science and policy)
	Embedded in external and internal networks (see below)
Involvement of external and internal networks	Linkages to outstanding external networks (internationally, partners from other countries and/or cities, business, NGOs, etc.)
	Excellent internal networks (connections to important local stakeholders, e.g. leading politicians, administration)
Coordination mechanisms to develop and implement adaptation strategy	Creation of a separate and dedicated office to acknowledge the interdepartmental character of climate action
	Provision of profile and political weight to this unit by associating it with city mayor's office
Access to additional external financial capacities	Access to external funds (accessible with low bureaucratic burden and not competing with other sector budgets)
	However, financial resources not decisive

Finally, enhancing *financial capacities* seems to play a role in driving adaptation responses, but to a lesser degree than expected (Carmin et al. 2009). Early adapters do have the opportunity to tap the Least Developed Country Fund, which requires a national adaptation plan as a prerequisite. More important seem to be the funds where access of municipalities/cities to unrestricted international funding allows setting and taking on of immediate priorities without the extra burden of ongoing bureaucratic negotiations (Table 2.1).

As the examples show, adaptation is an emerging field of policy in cities, mostly complementing mitigation strategies. In some cities, dedicated adaptation plans have come into existence. These cities confirm with their action that climate change is a mainstream planning task and that it presents the opportunity to enhance local capacities. As demonstrated, such opportunities may include the political recognition as local *champions*, innovation of local administration (e.g. by creating cross-departmental working groups and units), the development of new networks, the creation of new knowledge and the formation of a strong and broad political consensus on development priorities that support and confirm existing strategies.

However, does the experience from early adaptation also apply to potential later adapters and in particular to megacities? Will they be able to follow the same model, realise the same opportunities and pursue similar strategies? These questions will be addressed in the subsequent section.

2.6 What Can Be Learnt from Early Experience for Mainstream Adaptation in Megacities?

It is quite evident that the key factors that inspire the action of early adapters are to some extent case-specific, for example, the availability of access to particular networks and the opportunity to align the adaptation strategy to existing development goals and plans. Other driving factors are exclusively available for early movers and will no longer apply to mainstream adapters. The opportunity for leading actors to gain recognition as local champions and political innovators is one example. From this perspective, the incentives and opportunities for taking up emerging questions of climate change risks, vulnerability and adaptation early and for developing new and innovative city strategies to cope with adverse effects of global climate change may be of limited interest to mainstream adapters. Some factors are relevant, for example, those relating to accommodation of the integrative cross-cutting nature of climate change in the implementation arrangements or the need to localise and clarify the effects of climate change as clearly and exactly as possible. Overall, however, the experience of early adapters cannot be transferred as a model to their mainstream counterparts, in general, and to megacities, in particular.

But what are the factors that support the extension of climate adaptation issues to mainstream adapters? What prerequisites need to be in place for the diffusion of adaptation strategies – beyond the few examples of innovative early adapters? Where these factors and prerequisites are not present, what obstacles do cities face when they want to take up adaptation and how can these obstacles be overcome? We suggest a closer look at the portfolio of mainstream capacities and competences in megacities that enable the diffusion of climate adaptation issues. By the term *capacity* or *competence*, we refer to (1) the political and legal powers of cities to develop and implement adaptation strategies, (2) their financial capacities to do so and (3) their administrative ability to implement and sustain concrete action.

1. Political and legal capacities refer, first of all, to the *competences of political actors in megacities*. An adequate definition and distribution of competences are a key prerequisite for adaptation action. The city manager can only act if he is empowered to do so. In order to fulfil functions that arise from the necessity to adapt to climate change impacts, city managers must have the legislative powers for climate policy measures.

However, in many of the countries of the global South, the political power is solely allocated to the central level, and city managers are lacking discretionary powers to take responsibility for climate adaptation functions. As discussed above, climate mitigation issues are primarily related to the national policy level. Questions of defining national reduction targets, distributing the reduction targets among sectors (e.g. private households, industry, transport, public sector) or introducing measures are a national task. In contrast, adaptation issues are in most cases regional or local by nature. This means that risk management in its entirety is primarily a task of city managers. This includes, for example, infrastructure planning, sanitation, resource management and measures for residents' health. Equipping city managers with the political authority to deal with adaptation issues is a vital prerequisite. This requires a new division of functions between the national and the city level, whereby the (vertical) allocation of adaptation functions to the city level should be the rule. Shared functions between the national and the city level should be the exception and restricted to cases in which regional externalities are predominant.

The vertical distribution of functions between the national and the city level is not the only determining factor: An effective distribution of functions and responsibilities between the various subnational entities is also required. Returning to the case of Mexico City, the vertical distribution of functions between the national government administration, the federal district and the administrations of the states of Mexico and Hidalgo has produced overlapping competences resulting in weak political competences. Although numerous coordination units between these entities exist, they are not defined by a clear division of competences that empower the responsible level. Mexico City is a typical example for megacities around the world. Adaptation to climate change is harder to achieve in such a setting because the interests of the different entities tend to conflict and do not allow an overall planning process.

The vertical distribution of competences between political entities is one prerequisite for a broader diffusion of adaptation action in cities. The horizontal division of powers is another factor. Climate adaptation strategies are cross-cutting by nature; thus, the implementation of these strategies can only be successfully achieved if political decision-makers have competences that override those of included sectors. Providing a climate office with planning competences for developing adaptation targets and measures is worthless if the affected infrastructure, health or social branches are not included and do not align their action.

2. If city managers are empowered to develop adaptation strategies, they also need the financial resources to do so. Political and legal competences are of no value if not accompanied by *competences for managing financial resources*. Thus, political capacities have to be matched by financial capacities.

   One could argue that it is sufficient for national governments to allocate funds to city managers for financing public expenditure and meeting the cities' adaptation requirements. This, however, is not adequate. First, national governments also suffer from scarcity – they do not have sufficient revenues at hand to pass them to their sublevels. Instead, they rely on these revenues to fulfil their own – national – tasks. This applies in particular to the poor countries of the global South. Second, and more important, city managers need some degree of autonomy to decide on adaptation action according to priorities. The examples of early adaptation discussed above highlight that setting priorities is a decisive element of adaptation strategies. Setting priorities is primarily a political process, and it is more difficult to define priorities if the resources come from an external source, that is, the national government. Experience teaches us that donors try to influence priorities according to their own preferences. These do not necessarily reflect the preferences of citizens living in the affected cities. In contrast, it is easier to choose citizens' priorities if the benefits of certain measures are balanced against the respective costs. To achieve this, it is decisive that city managers consider the trade-offs between adaptation measures and the financial burdens. This can be achieved best when they have responsibility to raise revenues from their own sources. This implies local competence in raising taxes, fees and charges. Grants from national governments should only be given if the city managers' revenues are not sufficient, or in the case of external effects.

   The proposal of giving city managers – as decision-makers – the competences over both expenditures and revenues from their own sources – and the right to decide upon both – is based on the principle of *fiscal equivalence* (Olson 1969; Hansjürgens 2003). If properly applied, this principle guarantees more efficiency in the public sector. The weighing up of advantages (expenditures) and disadvantages (revenues) eases the process of identifying political priorities. However, this condition is violated in many countries, even in the federal states of the global North. It becomes clear that this prerequisite is very challenging and can only be fulfilled within a mid- or long-term process.

3. A third capacity which is decisive for developing and implementing adaptation strategies to cope with climate change risks is related to *administrative competences*. Developing adaptation strategies and implementing them is primarily a task of the public sector that includes the provision of public goods, such as infrastructure, sanitation, public transportation, housing and social policy. Although individual preferences for these goods exist, the private sector is not able to provide them due to free-ride behaviour. A second task for the public sector consists of setting the framework for private behaviour and enforcing compliance. This implies the development and enforcement of rules for, for example, access to water, building standards, urban spatial planning and so on.

Table 2.2 Capacity prerequisites for mainstream adaptation

Major prerequisites	Description
Political capacities (competences of city managers)	Sufficient political power and responsibility allocated at the city level
	Adequate vertical division of functions and responsibilities between governmental levels
	Adequate horizontal distribution of competences (across different sectors)
Financial capacities (access to own resources)	Ensuring the principle of fiscal equivalence with local competences over revenues and expenditures
	Limiting transfers (grants) to improving revenue situation (general grants) or for regional external effects (special purpose grants)
Administrative capacities (capacity building, good governance, organisational issues)	Capacity building with respect to (local) information about climate change impacts and adaptation; awareness raising
	Building linkages with research organisations, both local and external
	Following principles of good governance: participatory governance, including stakeholders in decision-making, while reducing special interests
	Organising climate adaptation office across various sectors, close to city manager's office

With respect to adaptation issues, it is not only necessary that a functioning administration exists. A central prerequisite for fulfiling these tasks is adequate knowledge and information about (local) climate change impacts. The administration must possess the knowledge about climate change impacts in their concrete field (e.g. infrastructure planning, water provision) in order to properly address the problem of adaptation. In this context, the factors mentioned for early adapters can be applied: Climate adaptation strategies will only be developed if sufficient (local) knowledge is available and if the local administration is aware of the adaptation necessity. Capacity building within the public administration, at a scale that exceeds the few cases of early adapters, is therefore a key for the diffusion of climate adaptation action. The experiences of early adapters show that strengthening the linkages with local and external research organisations and initiatives can make a vital contribution.

Given the situation in many megacities of the South, with the interests of special groups or individuals with great influence dominating the interests of the general city population, a decisive factor for developing and implementing adaptation strategies is good governance. Good governance represents the ideal that both politicians and the administration are only responsible to and serve public, not private interests (for details, see Bulkeley et al. 2009). It is therefore important that the adverse effects of special interest groups or individuals can be reduced and the interests of citizens considered. An important step in this direction could be the strengthening of participatory government. By including stakeholders in decision-making processes, at least three advantages can be realised. First, stakeholder knowledge can be included in decision-making. In many cases, local stakeholders are more knowledgeable than the local administration. Second, proposals for resource use can be developed jointly. Although this may imply higher transaction costs of decision-making, it has the great advantage that the solutions found have higher acceptance. Third, this may lead to a better monitoring and enforcement of certain measures.

A final prerequisite for strengthening administrative capacities is related to how adaptation issues are embedded in public administration. As the experiences of early adapters show, the chances for successful adaptation are greater if the climate adaptation office is an independent office and if it is closely related to the city mayors' office. In as much as such an arrangement reflects the cross-cutting nature of adaptation strategies and measures, it is suited for the diffusion of adaptation issues. A summary of factors relevant for mainstream adapters in cities of developing countries is given in Table 2.2.

Mainstream adaptation requires quite significant competences and capacities: political, financial and administrative. Most of these prerequisites are not entirely new. Rather, they are closely connected to the ongoing debate on creating the conditions for local development as such. This is particularly the case for the example of the adequate division of competences between the national level and the city level or for the distribution of authority to generate financial resources. However, for adaptation to climate change, these concerns reappear with a new urgency. As has been demonstrated, some additional insights can be gained from examples of early adapters. This suggests that keeping track of their experiences in the future may yield further insights.

2.7 Conclusion: Can Capacity in Megacities Meet the Adaptation Challenges?

In this chapter, we have analysed the relationship between the necessity of making urban adaptation responses and the competences and capacities in megacities. Three questions were of particular interest. What climate change adaptation challenges do megacities face? What opportunities but also obstacles exist for meeting these challenges? What are the key competences and capacities that adaptation strategies and action can and should build on?

Megacities play a particular role both as contributors to climate change and likewise as areas that are affected by its consequences in particular forms. With respect to adaptation, they are characterised by the combination of climate change impacts, extreme levels of already existing vulnerability and quite distinct challenges for coping capacities. Megacities owe this to their unique attributes including size, the high speed of change and their complexity. Due to these characteristics, capacity for adaptation might even decrease over time. However, adapting to climate change is becoming an unavoidable task for cities – and in particular for megacities. Furthermore, there are convincing general reasons why adaptation action should be given greater attention as a complement to mitigation.

The experience of early adapters (cities that have already initiated local adaptation strategies and plans) shows that cities are beginning to take up the challenge of adapting to climate change. The review in this chapter confirms that adaptation is an emerging field of urban policy, mostly as a complement to mitigation strategies. In some cities, dedicated adaptation plans have come into existence. The experience likewise reveals that early adapters are able to seize several opportunities. As demonstrated, such opportunities may include the political recognition as local champions, innovation of local administration (e.g. by creating cross-departmental working groups and units), the development of new networks, the creation of new knowledge and the formation of strong and broad political consensus on development priorities that support and confirm existing strategies.

However, these determining factors that motivate and drive action of early adapters do not necessarily work for potential later adapters and in particular for megacities. They are either case-specific, for example, the availability of access to particular networks and the opportunity to align the adaptation strategy to existing local development goals and plans, or they are exclusively available for early movers only and will no longer apply to mainstream adapters, such as, for example, the opportunity for leading actors to gain recognition as local champions and political innovators. Some other factors are of instructive value to mainstream adapters as good practice, for example, those relating to accommodating the integrative cross-cutting nature of climate change in the implementation arrangements or the need to localise and clarify the effects of climate change as clearly and exactly as possible. Overall, however, the experience of early adapters cannot be simply transferred as a model to their mainstream counterparts, in general, and to megacities, in particular.

From this perspective, the diffusion of adaptation action to megacities will rely to a great extent on mobilising potentials and overcoming obstacles with respect to a range of political, financial and administrative competences and capacities. As adaptation is an explicitly local affair, emphasis needs to be placed on allocating sufficient competences to address adaptation at the (local) city level. This refers not only to the political power as well as vertical and horizontal distribution of functions but also to financial capacities over both revenues and expenditures. With respect to administrative capacity, it refers, for example, to principles of good governance and the development of networks for the generation of locally relevant knowledge.

It can be concluded that many megacities around the world suffer from significant deficits with respect to these capacities and competences. These shortcomings are probably among the causes why initiatives towards strengthening adaptation action are still the exception rather than the rule. What is more, the obstacles cannot easily be overcome in the short term. Therefore, the adaptation challenges are very likely to remain unmet in many places, at least in the near future. However, the examples of the early adapters show that while a local vision is a primary requisite for adaptation action, successful cases also rely on external competences and capacities. Thus, there are a range of other arenas with potential to make adaptation more prominent in order to support city initiatives. Giving adaptation in cities a higher profile in the international debate is one of them.

References

Alam M, Rabbani MDG (2007) Vulnerabilities and responses to climate change for Dhaka. Environ Urban 19(1):81–97

Bartlett S, Dodman D, Hardoy J, Satterthwaite D, Tacoli C (2009) Social aspects of climate change in urban areas in low- and middle-income nations. Contribution to the fifth world urban research symposium. The World Bank, Marseille, June 28–30, 2012 http://siteresources.worldbank.org/INTURBANDEVELOPMENT/Resources/336387-1256566800920/6505269-1268260567624/Satterthwaite.pdf

Barton J (2009) Adaptación al cambio climático y la planificación de ciudades-regiones, quoted from manuscript

Betsill M, Bulkeley H (2004) Transnational networks and global environmental governance: the cities for climate change program. Int Stud Q 48:471–493

Bulkeley H, Schroeder H, Janda K, Zhao J, Armstrong A, Chu SY, Ghosh S (2009) Cities and climate change. The role of institutions, governance and urban planning. Contribution to the fifth world urban research symposium. The World Bank, Marseille, June 28–30, http://siteresources.worldbank.org/INTURBANDEVELOPMENT/Resources/336387-1256566800920/6505269-1268260567624/Bulkeley.pdf

Carmin J, Roberts D, Anguelovski I (2009) Planning climate resilient cities. Early lessons from early adapters. In: Contributions to the fifth world urban research symposium. The World Bank, Marseille. June 28–30

City of Cape Town (2006) Framework for adaptation to climate change in the City of Cape Town. City of Cape Town, Cape Town

Fragkias M (2009) The critical role of good urban governance for cities and climate change, quoted from manuscript

Hansjürgens B (2003) Äquivalenzprinzip und Staatsfinanzierung. Berlin, Duncker & Humblot

Heinrichs D, Aggarwal R, Barton J, Bharucha E, Butsch C, Fragkias M, Johnston P, Kraas F, Krellenberg K, Lampis A, Ling OG, Vogel J (2011) Adapting cities to climate change: opportunities and constraints. In: Cities and climate change urban development series. The World Bank Publications, pp 193–224

Heinrichs D, Kuhlicke C, Meyer V, Hansjürgens B (2009) Mehr als nur Bevölkerung: Größe, Geschwindigkeit und Komplexität als Herausforderung für die Steuerung in Megastädten. In: Altrock U, Kunze R, Pahl-Weber E, Schubert D (eds) Jahrbuch Stadterneuerung. Arbeitskreis Stadterneuerung an deutschsprachigen Hochschulen und dem Institut für Stadt- und Regionalplanung der Technischen Universität Berlin (ed.) Universitätsverlag TU Berlin, Berlin, pp 47–57

Höppe P, Grimm T (2008) Rising natural catastrophe losses – what is the role of climate change? In: Hansjürgens B, Antes R (eds) Economics and management of climate change: risks, mitigation, and adaptation. Springer, Berlin, pp 13–22

Huq S, Kovatts S, Reid H, Satterthwaite D (2007) Reducing risks to cities from disaster and climate change, Editorial. Environ Urban 19(1):3–15

International Institute for Environment and Development (IIED) (2007) Reducing risks to cities from climate change: an environmental or a development agenda, Environment and urbanization brief 15. IIED, London

IPCC (2007a) International Panel on Climate Change. The physical science base. In: Contribution of working group I to the fourth assessment report. Cambridge University Press, Cambridge

IPCC (2007b) International Panel on Climate Change. Impacts, adaptation and vulnerability. In: Contribution of working group II to the fourth assessment report. Cambridge University Press, Cambridge

McCarney P, Alfani M, Moreno E (2009) City indicators on climate change: implications for policy leverage and governance. Contribution to the fifth world urban research symposium. The World Bank, Marseille, 28–30 June, http://siteresources.worldbank.org/INTURBANDEVELOPMENT/Resources/336387-1256566800920/6505269-1268260567624/McCarney.pdf

McGranahan G, Balk D, Anderson B (2007) The rising tide: assessing risks of climate change and human settlements in low elevation coastal zones. Environ Urban 19(1):17–37

MRC Mc Lean Hazel and GlobeScan (2007) Megacities und ihre Herausforderungen. München, Siemens AG,

Mukheibir P, Ziervogel G (2007) Developing a Municipal Adaptation Plan (MAP) for climate change: the city of Cape Town. Environ Urban 19(1):143–158

Olson M (1969) The principle of fiscal equivalence: the division of responsibilities among different levels of government. Am Econ Rev 59:479–487

Pielke R, Prins G, Rayner S, Sarewitz D (2007) Lifting the taboo on adaptation. Nature 445:597–598

Revi A (2008) Climate change risk: an adaptation and mitigation agenda for Indian cities. Environ Urban 20(1):207–229

Rodolfo KS, Siringan FP (2006) Global sea-rise is recognised, but flooding from anthropogenic land subsidence is ignored around northern Manila Bay, Philippines. Disasters 30:118–139

Tangri AK (2003) Impact of climate change on Himalayan Glaciers. In: Proceedings of V workshop on water resources, coastal zones and human health. NATCOM, New Delhi

UN Habitat (2006) State of the world's cities 2006/2007. The millennium development goals and urban sustainability. Earthscan, London

United Nations (UN) (2007) World urbanization prospects, The 2006 revision. United Nations Publications, New York

Wisner B, Blaikie P, Cannon T, Davis I (2004) At risk: natural hazards, people's vulnerability and disasters. Routledge, New York

World Bank (2008) Climate resilient cities: a primer on reducing vulnerabilities to climate change impacts and strengthening disaster risk management in East Asian Cities. World Bank, Washington, DC

Underground Cities

3

Eduardo F.J. de Mulder, Jacques Besner, and Brian Marker

Abstract

Space is a rare commodity in the built environment. This drives urban developers to seriously explore the underground as an option for urban expansion, at least for some specific facilities. For many centuries underground and cave cities have been constructed at scattered locations around the globe, particularly in harsh climatic zones. Underground urban expansion was strong in Canada (Montreal and Toronto) in the 1960s–1990s, and today such development is booming in China. This goes in parallel with rapid improvement of tunnelling techniques which have become cheaper, safer, and faster. Together with economic growth and vision, these improvements have supported significant expansion of underground facilities particularly in Japan, Italy, and Norway and in China which is now the world's leading nation in tunnelling. One of the boundary conditions for safe underground development is proper knowledge of the geoscientific conditions of the subsurface. Since subsurface information can now be rapidly digitised, stored in databases and coupled with GIS in many countries and cities, geological and geotechnical information has become more accessible, and more detailed interpretation of subsurface characteristics has become possible. With regards to the future, this chapter describes four global developments and trends (population, urbanisation, quality of life/economic growth and environmental awareness) which, in combination, will probably contribute to increased urban land prices. A fifth trend (progress in science and technology) will make the underground more accessible as a feasible option for urban expansion and

E.F.J. de Mulder (✉)
Geoscientists, Executive Director, Earth Science Matters Foundation, Wageningen, The Netherlands
e-mail: director@earthsciencematters.org

J. Besner
Associated Research Centers for the Urban Underground Space, ACUUS Secretariat, 34 Seville, D.D.O. (Montreal) Quebec, Canada H9B 2S5
e-mail: info@acuus.qc.ca

B. Marker
Independent Consultant, 40 Kingsdown Avenue, London, W13 9PT, London, United Kingdom
e-mail: brian@amarker.freeserve.co.uk

F. Kraas et al. (eds.), *Megacities: Our Global Urban Future*, International Year of Planet Earth,
DOI 10.1007/978-90-481-3417-5_3, © Springer Science+Business Media B.V. 2014

reduce uncertainties in geoscientific modelling. However, psychological barriers and legislative restrictions may frustrate or delay further underground expansion. With time, these will be overcome in places where the need for such development and the benefits for society are greatest.

Keywords
Underground construction • Urbanisation • Tunnels • Urban planning • Sustainable development • Subsurface management • Indoor environments • Pedestrian networks

3.1 Introduction

As space became a scarce commodity in the built environment, urban developers and architects found room for urban expansion by increasing vertical development. Starting in the USA in the late nineteenth century, and especially since WWII, skyscrapers have increasingly dominated the urban centres of Europe and Asia. High land prices in the 'golden mile', the increased need for dense and rapid public transport systems and the growing reality of the law of the diminishing returns for ever higher buildings have begun to force alternatives. In recent years, the other vertical dimension, underground, has become an increasingly significant option for urban expansion. With greatly improved, safer, faster and cheaper tunnelling techniques, that option has become more and more economically viable. Helped by well-designed, spacious, comfortable and light underground spaces, a new phenomenon is beginning to enter our vocabulary: underground cities.

Underground cities can only be built safely and at reasonable cost if based on sound scientific understanding of the subsurface. The necessary knowledge is widely available in various scientific organisations. The International Tunnelling Association (ITA) is a strongly technically oriented body that has recently expanded its scope into underground space *sensu lato*. This field is particularly addressed by one of ITA's Working Groups: 'Urban problems, underground solutions' and more recently by the ITA's Committee on Underground Space – ITACUS. The Associated research Centres for the Urban Underground Space (ACUUS), created in 1997, has a strong focus on cities and the sustainable development of urban underground spaces. The International Working Group on Urban Geology, founded in 1992, included within its goals the promotion of studies on the use of the urban subsurface. This was a joint initiative of the International Union of Geological Science (IUGS), the International Association of Engineering Geology and the Environment (IAEG) and the International Association of Hydrogeologists (IAH) (de Mulder 1996). Also, the International Union of Geodesy and Geophysics (IUGG) has a working group that focuses on the urban subsurface. Finally, the Federation of International Geo-engineering Societies (FedIGS) established a joint technical committee on the sustainable use of underground space (SUUS) in 2006. The committee completed its work in 2010.

This chapter briefly describes the concept of underground cities against a background of scientific aspects linked to sustainable development. Exploring one of the last frontiers of humankind in a balanced way is one of the key issues dealt with in the megacity theme of the International Year of Planet Earth (Woodfork and de Mulder 2011). This chapter is a contribution to the legacy volume on this topic.

3.2 Underground Cities in the Past

Urban settlement began about 8,000 years ago (Mithen 2003). Initially, humans used the underground in their settlements to store water and food, for protection against extreme weather conditions, to bury their dead or to escape from hostile tribes. Al Najaf, a city in today's Iraq, is an interesting example. Early Arab settlers founded a town at the location of the modern city in the

Fig. 3.1 Cave houses in Dunhuang, China

desert. Rooms were dug out in three levels below the surface (de Mulder et al. 2012). Other examples of underground settlements serving as shelter against harsh climatic conditions can be found in other deserts, such as at Coober Pedy (Australia) where an underground town developed as a result of opal mining operations. Many other underground facilities developed as a result of mining activities, for example, in connection with salt mining in Poland. Underground cemeteries are well known from Rome (Italy) and Xi An (China). Underground towns built for the prime reason of defence include Turin (Italy) and Oppenheim (Germany). There are many examples of smaller settlements built as underground defence systems, for example, settlements built along the Maginot Line and the Atlantic wall (France), the Adlernest (Germany) and the Pentagon in the USA. The introduction of rail transport has strongly stimulated underground construction since the mid-1800s.

In some areas, cave buildings, partly or wholly underground, are frequently found in cliffs along river banks or lakes. The use of many cave shelters may date back several tens of thousands of years, well into the last glacial period, as indicated by human remains found in caves around the word. In 2009, a row of 5,500-year-old houses which were found in a cliff at the loess plateau near Xi An (China) was excavated. The 'Thousand-Buddha Caves' or Mogao Caves, near Dunhuang City, Gansu Province (West China), are 1,600 years old. Many of the caves in this UNESCO World Heritage site are carved into soft sandstone (Fig. 3.1) and beautifully decorated. Even today, a considerable number of people lives in caves dug into the loess deposits in China, into volcanic ashes in Turkey or into soft limestone along the Loire River (France). Cave dwellings are being promoted for ecologically friendly living purposes.

In the past, options for underground development were determined by geographical and geological conditions. Geography controlled the selection of the (partly) underground sites, while the composition of the rock determined its excavation potential, using simple tools, and the stability of the walls of the underground construction. Loess, soft limestone and sandstone, as well as volcanic tuff often proved the best rocks for such options.

3.3 Underground Cities Today

Since the mid-1970s, underground construction has boomed around the world for a variety of reasons: a rapid increase in population (from four billion in 1975 to seven billion in 2011) accelerated worldwide urbanisation (from 37.3% in 1975 to 51% in 2010); the GDP per capita more than doubled worldwide since 1950 and an exponential rise in urban land prices and technological improvements. Over the same period, public environmental concerns have increasingly influenced and been translated into policy making, in particular in Europe, Canada and the USA. Public voices and resistance against burning fossil fuels have grown, provoked in part by an increase in air pollution, as well as the pollution of soils and water resources, and the resulting alarming costs of cleaning up contaminated industrial sites. Together with increasing traffic congestion, the deterioration of environmental conditions has provided a fertile ground for some pioneer urban authorities to consider alternative solutions such as underground development.

Reasons for such development, however, seldom came from a lack of space (at the surface) but rather from a mix of pre-existing opportunities as, for example, the availability of abandoned underground space under a city (e.g. Kansas City), abandoned mines (e.g. Coober Pedy City) or the occurrence of dense underground infrastructure networks and particularly subways (e.g. Montreal). In general, decisions for such underground development were individual, taken by a developer, not the result of long-term urban planning. Innovative strategic planning is more likely to develop once scattered, and 'autonomous' underground structures are in place, encouraging municipalities to incorporate and extend these existing facilities through municipal planning practices.

3.3.1 Tunnelling

Although underground infrastructure has expanded significantly since the mid-1970s, it is still lagging far behind urban development at the surface. Most underground constructions occur in and around large cities, starting with tunnelling. In Europe, tunnelling began relatively early, especially for trains. Prior to 1975, extensive tunnel construction occurred in countries with mountainous terrain such as Italy, Switzerland, Austria, France and Norway. In Norway, four out of ten longer railway tunnels were built and completed before 1950. In the 1970s emphasis shifted from rail to road tunnels, and 16 of 20 longer road tunnels have been completed in Norway since 1985. Norwegian tunnelling was fuelled by a significant increase in oil and gas income and a policy of investing in long-term sustainable development goals. In contrast, Italy, the other leading European tunnelling nation, puts its emphasis on public transport and built mainly rail tunnels.

By 1975, some 50 long (>5 km) rail and road tunnels were being operated in Europe. In the following decade, that number grew much faster than in the previous 25-year period. This trend continues even now. Traditionally, there have been fewer tunnelling activities in North America than in Europe. They mainly date back to the period before 1975. Most of the longer (2–5 km) road tunnels in the USA were built between 1940 and 1960. The most recent long (>5 km) railway tunnel in the USA was built in 1970. In Canada, large-scale railway tunnelling continued over a longer period and peaked between 1975 and 1985 when 108 out of the 173 existing tunnels were completed (de Mulder et al. 2012).

In Asia, major tunnelling began in Japan but much later than in Europe and North America, with peaks in 1975 and 1982. Economic recession resulted in a sharp decline in the construction of tunnels from 1985 to 1995. Subsequently, the numbers increased again, making Japan the world's leading nation in longer traffic (mostly railway) tunnels. Since 2005, China took over that position with the completion of 25 long tunnels in that year alone. Also, in the Republic of Korea, underground transport infrastructure grew rapidly during the last decade of the twentieth century (Shin and Park 2007). In the case of India, tunnels are generally not very long; however, the total sum of the lengths of all tunnels is comparable to the cumulated length of tunnel systems in most European countries. The railway tunnels in the Himalayan (Himachal Pradesh and Kashmir) and Western Ghat regions (Konkan railway) provide good examples for major tunnels in modern India. Less significant Asian tunnelling operations are reported from Singapore and Vietnam. Very few longer traffic tunnels or other major underground structures other than mines have been built in Australia, New Zealand, Africa or South America.

3.3.2 Underground Space

Underground cities are expanding rapidly and on a global level, but China is definitely leading in this respect. In the largest Chinese cities, more than 1,200 km of new subways lines are in construction. In 2007, Beijing had 30 km^2 of underground space with an annual increase of approximately 10% (Qian and Chen 2007). Shanghai, the largest Chinese metropolis (>20 million people) had one single subway line in 1995, while in 2010 Shanghai had 11 lines with 267 stations and 410 km of tunnels. The overall plan calls for seven new subway lines, mostly underground, to be completed by 2025.

Construction of underground public and/or commercial facilities often spreads from subway stations, which act as nodal points in the urban fabric. This is clearly demonstrated in the Canadian cities of Toronto and Montreal. In Montreal, urban dwellers face a harsh climate with extremely cold winters and very hot, humid summers. Climate has been one of the primary drivers for the successful development of an underground city, a process which began in 1958 with the planning of Place Ville-Marie. Three visionary architects, including Leoh Ming Pei and the urban planner Vincent Ponte, started to design and plan the redevelopment of the Montreal city centre on top of a very large railway marshalling yard. The new area was planned to accommodate shopping centres and offices, as well as separate walkways for pedestrians. When the area around Place Ville-Marie was redeveloped in 1962, an underground pedestrian network of 500,000 m^2 was created. Next, two parallel subway lines, with future stations at a walking distance from the plaza, were constructed.

The stations were planned to be accessible from the street through the lobbies of buildings connected to spacious mezzanines over the subway stations and to shopping galleries. The stations were decorated with works of art, creating a pleasant indoor climate in the winter. When the subway started running in 1966, ten major buildings were connected underground with train stations. By then, the underground city, known in Montreal as the indoor city, had 240 boutiques and department stores, 4,000 indoor parking places, 2,200 hotel rooms, 36 restaurants and 4 cinemas. From then on, the underground city grew continuously. In 2011, the indoor pedestrian network, with its indoor squares, cinemas and shopping galleries, had extended to 32 km. Ten subway stations, two train stations and two bus stations have been connected, and access has been secured to 63 real estate complexes representing 80% of total office space in the city centre. There is also access to 1,615 apartments, a convention centre, many museums, 3 universities and a college campus and to 14,500 public car parking lots. The network has 178 entry points, giving 500,000 pedestrians daily access to the underground city. Further development of this indoor city was enhanced by excellent communication between the municipality and urban developers.

For example, the private sector was involved in connecting the subway stations with nearby high-rise buildings, allowing public access to private property. This was negotiated through zoning easement and made possible by a reduced number of indoor car parking units. However, the most successful aspect of the Montreal underground city has been the exclusive financial participation of the private sector which agreed to pay for the connecting tunnels, maintenance and surveillance, the City merely having a supervising role, a kind of PPP realised long time before the concept was invented. Another success factor is that the indoor pedestrian network is opened to the public during almost the same operating hours as the subway. Such public private partnership arrangements have been indispensable for successful underground development (after Galipeau and Besner 2003). In contrast to North America, where underground construction normally starts as a private investment, public initiatives in Europe have taken considerably longer to prepare thus extending pay-back times.

3.4 Underground Cities in the Future

To explore what may be expected in underground cities for the next 50 years, trends in land pressure should be further analysed. Other factors, such as public and political perceptions of the underground, may impact such developments considerably. But even if these analyses point to the need for underground development and for society to accept such underground solutions, the subsurface itself, in most areas, should physically be able to provide and accommodate such solutions in a sustainable and economically viable manner. Geological properties of the underground should fulfil the requirements for the desired applications.

Five relevant developments and trends concerning the need for physical space for future generations are reviewed here: population pressure, growing urbanisation, enhanced quality of life, environmental awareness and progress in science and technology. The UN (2011) predicts for its medium fertility variant, 9.3 (high 10.6 and low 8.1) billion people living on this planet by 2050. That trend will probably be followed by a slow further rise to 10.1 billion in 2100 (medium fertility variant) and may enter a long period of relative stability in population until approximately 2300 (UN 2004). By November, 2011, the world population reached 7.0 billion, and according to the latest UN estimates, almost all of the world's population growth between 2000 and 2030 will be in urban areas in developing countries. If present trends continue, urban population will equal rural population by around 2017 in the Third World, and almost 70% of the world population will be living in cities by 2050 (United Nations Population Division 2008). Urban space will be under much higher pressure than today as horizontal urban expansion is limited by geographical factors. In combination, both trends will, in the long term, cause urban land prices to rise continuously, particularly in the city centres. The number of megacities (>10 million inhabitants) spread over the globe will rise from 19 in 2000 to 27 in 2025. There is no reason to expect the scarcity of urban land to significantly diminish before the projected stabilisation of the world population by 2050.

From 1950 to 1995, average incomes in the industrial nations rose 218% and in the developing countries 201% (Lomborg 2001). At the same time, we also notice a significant increase in life expectancy, caloric consumption and a reduction in global poverty. Wealthier people demand higher quality housing, while fast economic growth leads to larger industrial areas and economic zones. During the past half- century, increased prosperity generated very large volumes of domestic and industrial waste. Cities increasingly became surrounded by waste disposal sites, which were often used as building land in the context of successive urban expansion. Industrial expansion, waste disposal and demand for increased food and natural resources also lead to major pressures on physical space. Although commodity prices have fluctuated strongly since 2002, these trends are not likely to change dramatically in the near future. This assumption is supported by data that the world's average GDP growth per capita rose from 2.2% since 1970 to slightly above 3% per year in 2008 (IMF 2001). Despite the current (2011) global economic crisis, there are no reliable indications that this long-term trend will be reversed in the next 50 years.

Environmental concerns have driven political action to protect peripheral land from further urban development. That contributed to scarcity of available land for city expansion. Also, many hectares of degraded land in and near cities remain unsuitable for any development without expensive remediation. Both factors put severe limitations on urban expansion, at least in the short term.

Technological innovations, including the green industry, reuse of natural resources, land reclamation, engineered waste disposal sites and better incineration techniques, have helped to make more land available and have resulted in more efficient land use patterns. New generations of tunnel boring machines have made underground construction safer, cheaper and faster since the early 1990s. The authors expect this trend to increase in the course of the twenty-first century. Simultaneously, geoscientific knowledge increased rapidly, in particular in terms of data accessibility, modelling and understanding of geological and geotechnical processes (de Mulder and Jackson 2007).

The first four trends, either alone or in combination, will probably make urban land a more valuable and more expensive commodity. Science and technology may counteract this trend and may make degraded, unsuitable, mountainous land and even water bodies available for urban expansion. These factors point to a need for a more efficient use of urban land, including options for vertical and underground expansion and the use of land in multiple ways. High-rise buildings may continue to be the dominant construction type in the city centres for the next few decades but as these have their own intrinsic limitations in terms of space and energy efficiency, the authors expect the underground space to become an increasingly important domain for solving land scarcity problems in the coming years. Improved tunnelling and other excavation techniques will further contribute to such development, including that of underground cities. The rapid expansion of underground construction we observe today particularly in China underpins this statement.

However, psychological barriers may frustrate the expansion of underground cities. This is not the preferred domain for most people, perhaps partly because of unfamiliarity with its benefits and inherent cultural mindsets. Reluctance to being underground may depend on cultural differences as described by Nishida et al. (2007). Legislative factors may also hamper the expansion of underground cities (de Mulder et al. 2012). Vision and leadership of urban authorities may also strongly influence underground development. Some authors found that underground development flourished best where strong, local governments took the lead, backed by public and political interest for a better quality of (urban) space.

3.4.1 Urban Underground Planning

Our current limited knowledge of the composition and structure of the underground may be another hampering factor for successful underground urban development. Geological conditions such as rock quality and physical and hydrogeological conditions might either frustrate or facilitate the realisation of visions. Also, bearing capacity and thermo-isolation capacity of rocks for underground storage are determined by geological factors. Thick beds of impermeable rocks under waste disposal sites may favour their construction. Geological factors also control processes affecting the environment (e.g. groundwater flow). They also may have very significant impact on construction costs, in particular for underground operations. Construction in and on solid ground is much cheaper than building in swamps.

Proper information about the subsurface is essential for planning underground development. More and more cities have begun generating and storing subsurface data for future urban development and planning. Coupled with geographical information systems and dealt with by urban geoscientists, these will provide urban planners with a solid information base for effective and transparent decision-making at least for underground parking garages, pipelines and tunnels. Apart from these classical examples, such information may also be used for more innovative ways of underground development such as 'Geo-domes' (Taselaar 2002). This prize-winning concept has promising potentials for underground cities and industrial sites. Initially, construction of such major underground objects will disturb the groundwater flows, even in a wider area. Within years to decades, new balances in flow directions and velocities will be established around such structures. Soon, our growing understanding of the complex and interacting underground processes, together with improved modelling capabilities, will allow precise prediction of the impact of such underground operations on these processes. Hydrological engineering may modify and even replace current balances with more environmentally favourable conditions.

Although options for future use of the subsurface seem almost endless, there are still constraints that hamper rapid expansion of underground cities today. It is reasonable to suggest, however, that at least several of these constraints will be overcome in the future with accessible technologies, growing cultural acceptance and new attitudes in urban planning. This may happen first in cities where the pressures on urban land are highest. By creating comfortable and light environments in underground spaces with sufficiently reassuring safety and security measures, psychological barriers may be overcome as already demonstrated in several underground sites (Duffaut 2008). Trends in technological development and spatial planning as well as psychological barriers make it more likely that the topmost 100 m below surface will be developed first. Deeper levels (100–1,000 m) may offer comfortable working environments, good storage conditions and save energy and will probably be the next challenge that will have to be considered.

Fig. 3.2 Artist's impression of a partly underground city by the end of the twenty-first century

Climatic factors and investors may also drive underground development, as demonstrated in the City of Montreal, Canada. Climate change may also make the use of the subsurface more imperative in some areas. Together with positive geological factors and energy saving aspects, the best conditions and highest expectations for future underground urban development (Fig. 3.2) may be found in big cities in continental climatic zones in countries where sufficient investment capital and municipal leadership are available. Currently, this is occurring along the east coasts of the USA and Canada, the central part of the USA, major cities in Northern and Eastern Europe, Japan, Korea and China.

Acknowledgments The authors acknowledge the contributions by Robert Hack (ITC) and Derk van Ree (Deltares) to this chapter.

Literature References

de Mulder EFJ (1996) Urban geoscience. In: McCall GIH, de Mulder EFJ, Marker BR (eds) Urban geosciences. Balkema, Rotterdam, pp 1–11

de Mulder EFJ, Jackson I (2007) Data and information in the international year of planet earth (2007/2009). EOS Trans. AGU, 88 (23) Jt. Assembly Suppl. Abstract IN31A-02

de Mulder EFJ, Hack R, van Ree CCDF (2012) Sustainable development and management of the shallow subsurface. Geological Society, London, pp 211

Duffaut P (2008) L'espace souterrain, un patremoine à valoriser. In: Géosciences, Spec. Publ. 7 & 8, pp 224–235

Galipeau G, Besner J (2003) The underground City of Montreal – a win-win approach in the development of a city. Presentation at the 1st international conference on sustainable development & management of the subsurface, Utrecht, The Netherlands

International Monetary Fund (2001) International statistical yearbook: 2001 update. IMF publication

Lomborg B (2001) The skeptical environmentalist: measuring the real state of the world. Cambridge University Press, Cambridge

Mithen S (2003) After the ice, a global human history 20,000–5,000 BC. Harvard University Press, Cambridge.

Nishida Y, Fabillah H, Ichihara S, Nishi J, Cho KD (2007) The underground images in Japan, Korea and Indonesia. In: Kaliampakos D, Benardos A (eds) Underground space: expanding the frontiers. 11th ACUUS conference, Athens, pp 169–174

Qian Q, Chen X (2007) Evaluation of the status quo and outlook of the urban underground space development and utilization in China. In: Kaliampakos D, Benardos A (eds) Underground space: expanding the frontiers. 11th ACUUS conference, Athens, pp 15–21

Shin H-S, Park E-S (2007) The current status of underground space utilization in Korea. In: Kaliampakos D, Benardos A (eds) Underground space: expanding the frontiers. 11th ACUUS conference, Athens, pp 169–174

Taselaar (2002) Geokoepels, de ruimte in nieuw daglicht. Ingenieursbureau Amsterdam, Amsterdam, The Netherlands (in Dutch)

UN Department of Economic and Social Affairs, Population Division (2004) World population in 2300. United Nations

UN Department of Economic and Social Affairs, Population Division (2011) World population prospects: the 2010 revision. United Nations

United Nations Population Division (2008) World urbanization prospects: the 2007 revision. United Nations

Woodfork LD, de Mulder EFJ (2011) The international year of planet earth, Final report. Earth Science Matters Foundation, Wageningen

Web References

China caves: http://news.xinhuanet.com/english/2009-01/25/content_10717077.htm

Dunhuang: http://www.nytimes.com/slideshow/2008/07/06/arts/0706-COTT_2.html

Ecocaves: http://www.living-abroad-consultants.com/id21.html

Water Quality and Socio-Ecological Vulnerability Regarding Urban Development in Selected Case Studies of Megacity Guangzhou, China

4

Rafig Azzam, Ramona Strohschön, Klaus Baier, Lin Lu, Katharina Wiethoff, Anna Lena Bercht, and Rainer Wehrhahn

Abstract

The megacity Guangzhou in the South-Chinese Pearl River Delta is one of the most economically dynamic and rapidly urbanizing areas in the world and meanwhile home to some 15 million people. The urban growth, which also includes various spatial structural changes in the city center as well as in the peri-urban area, has created severe deterioration of water quality within the last 30 years.

This chapter identifies on the basis of various morphological city patterns (urban units) along the new city axis which traverses the modern center of Guangzhou from north to south, different stages of mega-urban development, as well as factors influencing the local water resources and examines spatial variations of water quality. Surface water and groundwater within the investigation areas of Liede, Xincun, Yuangangcun, and Shibi was sampled in the rainy as well as in the dry seasons between 2007 and 2009, leading to the result that the main factors affecting water contamination in the urban and peri-urban creeks are still domestic sewage, resulting in very high concentrations of total coliform bacteria up to 1.3×10^7 MPN/100 ml regarding surface water and just as high concentrations up to 9.2×10^5 MPN/100 ml in groundwater. As the measuring results of ammonium, nitrate, and heavy metals mostly comply with the Chinese and international WHO standard, this chapter only focuses on total coliform bacteria.

R. Azzam • R. Strohschön (✉) • K. Baier • L. Lu
Department of Engineering Geology and Hydrogeology,
RWTH Aachen University, Lochnerstr. 4-20,
52064 Aachen, Germany
e-mail: azzam@lih.rwth-aachen.de;
strohschoen@lih.rwth-aachen.de;
baier@lih.rwth-aachen.de; lu@lih.rwth-aachen.de

K. Wiethoff
Institute of Landscape Architecture,
RWTH Aachen University, Jakobstr. 2,
52064 Aachen, Germany
e-mail: katharina_wiethoff@web.de

A.L. Bercht • R. Wehrhahn
Department of Geography, University of Kiel,
Ludewig-Meyn-Str. 14, 24098 Kiel, Germany
e-mail: bercht@geographie.uni-kiel.de;
wehrhahn@geographie.uni-kiel.de

F. Kraas et al. (eds.), *Megacities: Our Global Urban Future*, International Year of Planet Earth,
DOI 10.1007/978-90-481-3417-5_4,

By means of a transdisciplinary research design developed by engineering geologists and hydrogeologists, landscape architects, and urban geographers, structural deficits in water infrastructure and city planning, as well as ecological and social vulnerability become obvious.

Keywords
Water quality • Water infrastructure • Land use (change) • Structural city pattern • Social vulnerability • Urban units • Guangzhou

4.1 Introduction

At least since the United Nations Conference on Environment and Development in 1992 in Rio de Janeiro, the sustainable use and development of land and water resources should be of central importance (cp. Weng and Yang 2003). However, the relationship and interactions between land use and water resources had been underestimated in interdisciplinary studies for a long time. The current ecological discussion increasingly focuses on these relationships and the building of integrative approaches (e.g., Hiwasaki and Arico 2007; Tong and Chen 2002). In industrial societies, as well as in newly industrialized and developing countries, water has very diverse functions; it serves, for example, for the production of drinking water, as traffic routes and is also used for wastewater disposal. These functions are frequently not compatible with each other and can therefore substantially affect water quality. Since both water quantity and quality can be greatly influenced by various parameters such as soil sealing or infiltration of pollutants, the specific type of land use is of special relevance for water resources. The processes, interactions, effects, and phenomena of the problem area "resource flows," "land use," and "land use modification," respectively, are most complex and dynamic especially in megacities. Since these factors overlap with one another and (can) lead to a loss of governability and control and are therefore of special ecological, social, and economic relevance, the German Research Foundation has supported the Priority Program 1233 "Megacities – Megachallenge: Informal Dynamics of Global Change" since 2007. In its subproject, "Analysis of Informal Dynamics in Mega-Urban Areas – Based on Spatial Structure and Steering Mechanisms Focused on Water in the Pearl River Delta," these very processes and reciprocal effects of land use and water resources typical in the southern Chinese megacity of Guangzhou are examined in cooperation with engineering geologists and hydrogeologists, landscape architects, and urban geographers.

Against this background, the hypothesis of the chapter states that ecological vulnerability, vulnerability with respect to insufficient water supply and of men, is increased by unorganized spatially structured changes. Various stages of development intensity are recognizable and representable with regard to hydrological, morphological, and socioeconomic changes.

In response to the hypothesis, the following research questions are examined:

1. Which kinds of land use changes are present in the area of the Guangzhou new city axis, taken as an example in this chapter?
2. How do these changes affect the quality of the surface water and groundwater?
3. To what extent are infrastructural differences present in the transect of the city axis relating to water supply and wastewater disposal?

To analyze this complexity, engineering and hydrogeologists, landscape architects, and urban geographers conceptualized a trinomial research approach considering macro-, meso-, and microscale relating to the impacts of urbanization on the urban water household and the reciprocal effects between settlements, humans, and the environment (cp. Fig. 4.1).

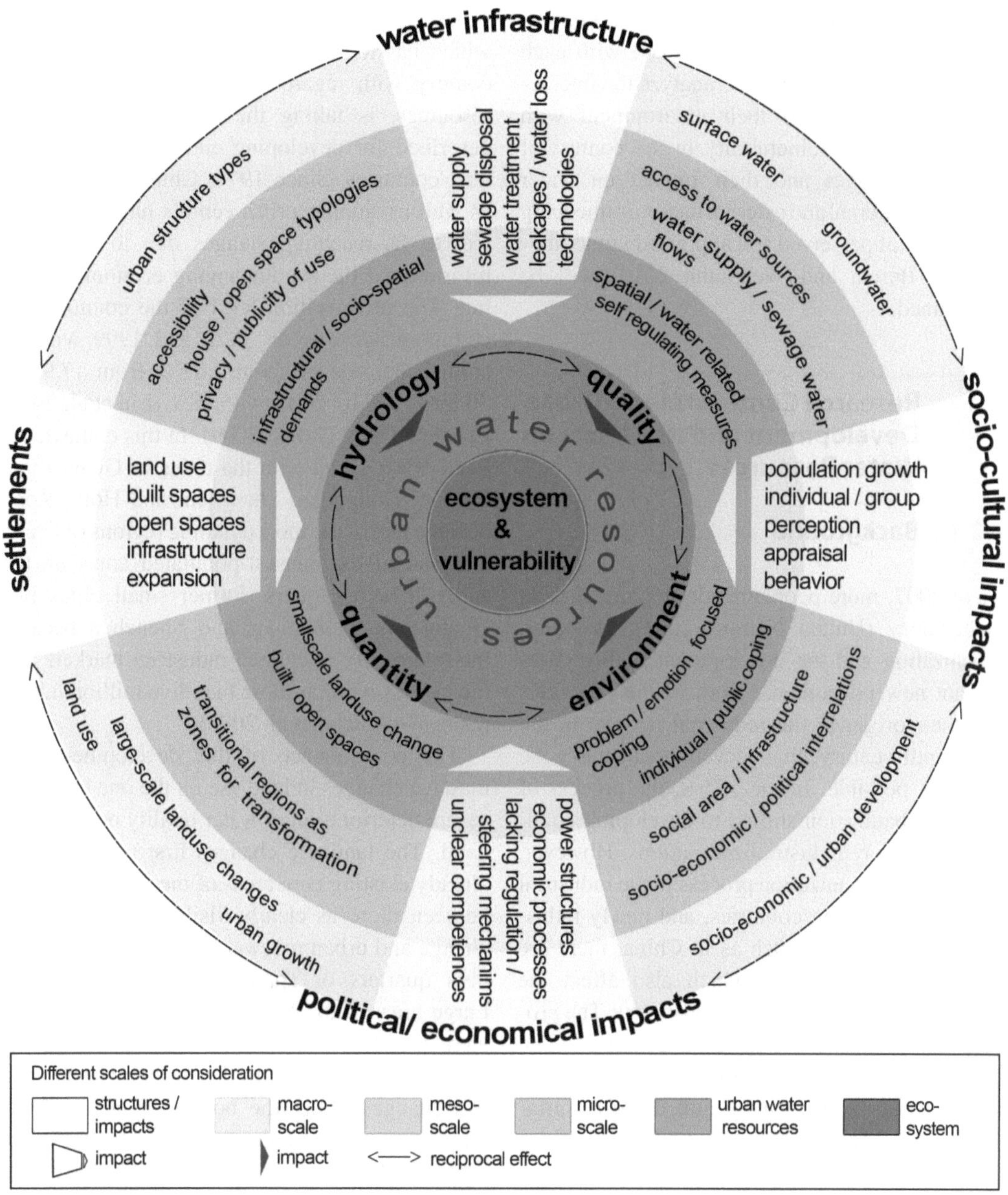

Fig. 4.1 Conceptual research approach: interrelations between urbanization and water resources (Strohschön et al. 2013)

With regard to the hypothesis, it is first relevant to determine endogenous and exogenous transformation variables (i.e., reasons, influences, and reciprocal effects), especially emerging in China since 1978. Based on these variables, it is analyzed whether the identified changes and transformations follow determined samples, whether "standards" can be deduced and which kind of regional structural urban patterns can be differentiated. Regarding the structural changes in Guangzhou, to what extent these macroscale transformation processes recur on the microscale (i.e., in the investigation areas) was examined and how high and how direct the efficiency range is on the smallest units was evaluated. As humans and environment are strongly linked with each other, it is after all relevant to analyze the interactions of humans and their environment with respect to social vulnerability in the context of land use changes and their impact on water resources: Internal and external factors influencing the quality of perceived and appraised vulnerability (cp. Bercht and Wehrhahn 2010) will be determined.

4.2 Research Context: Mega-Urban Development and Its Effects on Water Resources

4.2.1 Background

As of 2007, more people reside in cities than in the country (United Nations 2006). However, urbanization and the development of big cities are not new phenomena: Both of the processes had their origin in the industrial nations in the nineteenth century; their development has gone as far as possible. In the 1950s, the process of (mega-)urbanization shifted to developing countries and newly industrialized nations. However, between the urbanization process in the industrial nations, developing countries, and newly industrialized nations as much as in China, there are two serious differences which also affect the water resources and water management: The process ran gradually for about a century in the industrial nations. This enabled the cities to build the necessary water supply and disposal infrastructure as well as management capacities (Biswas and Tortajada 2004).

In comparison, the growth of cities exploded in the developing and newly industrialized countries. Due to this growth, the administrations there were not able to keep the rising need of water supply and disposal on pace with the construction of infrastructure and management capacities. While gradual growth enabled targeted planning of land use, uncontrollable growth over short periods causes massive land use changes (Lin et al. 2009; Kraas and Sterly 2009) which represent another major challenge for water management there.

The development of urbanization in China, which has to be seen as a newly industrialized country with regard to the status of its water resources, is taking the same course as that described for developing and newly industrialized countries. Since 1978, Chinese megacities as well as smaller urban centers have been subject to far-reaching changes due to economic reforms and the accompanying economic opening. The urbanization level of the country rose continuously between 1978 and 1999 with an annual increase of about 0.61%, from 17.92 to 30.89%, and by 2008 to 45.68% (Liu et al. 2003; People's Daily Online 2009). In this context, the Pearl River Delta with the cities of Guangzhou, Foshan, Dongguan, Shenzhen, and Hong Kong became one of the most dynamic regions of China and one of the densest-populated areas on the earth. In a few years, former small cities like Guangzhou, Dongguan, and Shenzhen became internationally integrated industrial markets and megacities with far more than five million inhabitants (Wehrhahn et al. 2008).

The consequence of this development was massive changes in land use on the one hand and severe deterioration of water quality on the other hand. The land use changes first took place in already existing core areas of megacities and can be seen there as clearly visible signs of urban change and urban renewal, such as modern business quarters or apartment tower complexes. Large areas with older and traditional buildings were mostly replaced with high-rises used for apartments or offices. In the further course, land use changes affect the bordering areas of the megacities; these are the most important zones of urban land use change within the framework of progressive urbanization, since potential building and development areas often still exist there. In most cases, the consequences resulting from urban development are drastic interferences into the ecosystem and the hydrological cycle. Especially large-scale soil sealing, increased water consumption, and water pollution from sewage play a major role. Research shows that in China as early as 1997, urban sewage and wastewater treatment systems were no longer able to cope with wastewater of the urban population (Ito 2005).

According to the authors' conceptual research approach as well as to Kraas and Sterly (2009), five main causes can be demonstrated for land use change in mega-urban regions:

- Strong population growth, especially due to migration
- Integration of the region into global economic processes
- The development of an urban middle class
- Excessive spatial expansion of mega-urban regions as effects of frequently lacking city planning and regulation, different local rules, and political decision makers
- Power structures

The change of land use by urbanization has four immediate effects on the hydrological cycle and the water quality of an area. These include flooding, water shortage, changes in the river/groundwater regime, and water pollution (Rogers 1994). In addition, indirect effects such as variations in temperature can be identified (Wenig 2001). The process of urbanization therefore has substantial effects on the hydrology of an area in the form of changes in natural drainage, the natural recharging of groundwater as well as by the input of pollutants into groundwater and surface water. Surface runoff is increased by extensive soil sealing, reducing the natural recharging of groundwater (Goudie 1990). The entry of pollutants into groundwater and surface water is aided by the interaction of ground, surface, and wastewater systems in urban areas (Strauch et al. 2009). This is particularly the case in cities having no adequate wastewater systems (Putra 2007; Putra and Baier 2009). However, the urban water cycle is still much more complex, and other components such as urban microclimates like urban heat islands, health aspects, and responsible institutions must also be taken into consideration. According to Breil et al. (2008), the urban water system consists of the following main components:

- Urban climate
- Water supply and disposal systems
- Rainwater, groundwater, and surface water
- Health aspects
- Institutions

These facts indicate high vulnerability of water resources through mega-urbanization. To be able to make statements about the vulnerability of water resources and to develop concepts for sustainable water management in megacities, it is therefore necessary to understand the processes of mega-urbanization and land use changes.

4.2.2 (Hydro-) Geographical Overview of the Study Area

The megacity of Guangzhou is the capital of South China's Guangdong Province, the economic, political, scientific, and cultural center of the province with a GDP of 821.6 billion in 2008. Next to Shanghai and Beijing, Guangzhou is one of China's boom regions (cp. Hong Kong Trade Development Council 2009). By the end of 2006, its population was approximately 11.52 million – including about 3.91 million nonofficially registered migrants (People's Government of Guangzhou Municipality 2007). By 2010 the total population had already increased to some 15 million inhabitants (Huang and Keyton 2010). The population density varied from 1,903 people in the peri-urban Panyu District to 30,364 people per square kilometer within the city core of Yuexiu District in 2006 (People's Government of Guangzhou Municipality 2007). However, only the density of "permanent population" is statistically acquired, so it can be assumed that the actual density and thus a higher water consumption and sewage production are even higher within some areas.

Geographically, Guangzhou is located between 22°26′–23°56′N and 112°57′–114°3′E. Due to the Asian monsoon, average annual precipitation amounts to 1,689–1,876 mm with maximum rainfall occurring in the months from April to September and an annual average temperature of 22.8°C (295.95 K). Guangzhou is located at the confluence of the Xijiang, Dongjiang, Beijiang, and Liuxi River systems, which form one of the most complex deltas in the world. Moreover, in the urban districts of Guangzhou alone, the channel network is made up of 231 surface and subsurface river channels together leading to a *quantitatively* sufficient supply of natural water resources.

The water supply is predominantly met by surface water since the groundwater is increasingly burdened by the intrusion of seawater, specifically

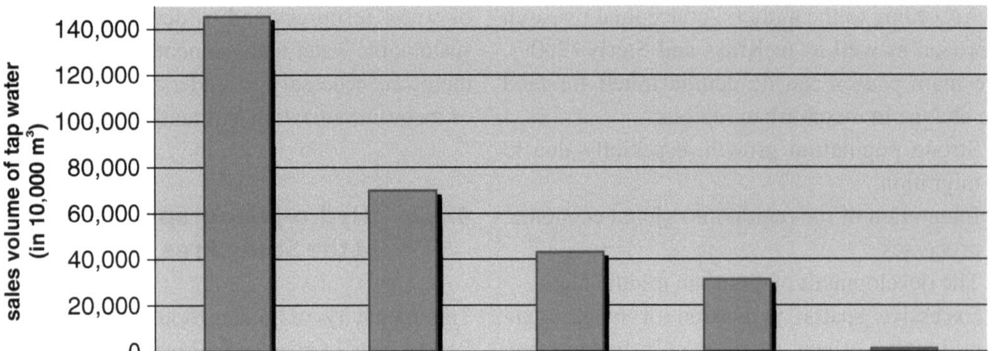

sales volume	private use	production use	public use	other use
145,381	69,850	43,007	31,240	1,284

Fig. 4.2 Tap water supply and consumer groups and consumption in Guangzhou, 2006 (based upon Guangzhou Municipal Statistics Bureau 2007; cf. Strohschön et al. 2013)

in the dry season (Liu et al. 2010). Raw water is obtained in a tributary of the Xijiang and the Shunde River at Beijiao and piped to Nanzhou waterworks in Lijiao about 27 km away. According to the Guangzhou Municipal Statistics Bureau (2007), the production capacity of water in 2006 was more than 6.4 m^3 per day, only 1,000 m^3 of which were obtained from groundwater resources. As shown in Fig. 4.2, private use accounts for the largest proportion of total sales volume of tap water, followed by industrial, public, and other use.

In the context of Guangzhou's urbanization and its population growth, the need of quantitatively rising water supply, as much as more supply and sewerage capacities, becomes relevant. To meet growing demands, the city's municipality is planning a capacity increase of about 2.3 billion m^3 per year of additional natural water withdrawals from the Beijiang (Xinhua News Agency 2006). In addition, four artificial lakes are to be created within the next years, including "Haizhu Artificial Lake" and "Baiyun Lake," which are to serve not only for the storage of and protection against floods but also as drinking water reservoir (Zhu 2009; Chen 2009). The expansion of the city's water infrastructure is indicated in Table 4.1.

Inadequate treatment capacity and therefore very severe deterioration of surface water and groundwater *quality* are the main problems facing Guangzhou's water supply. According to details from Zhu et al. (2002) or He (2005), only 10–25% of domestic wastewater in the Pearl River Delta is treated. This led to all monitored urban sections of the Pearl River being evaluated with the surface water quality class V (Shao et al. 2006), which was formulated by the Ministry of Environmental Protection of the People's Republic of China in 2002 (cf. Sect. 4.3.2), indicating the highest possible degree of pollution of surface water. The water assigned to this class is no longer suitable as drinking water. According to Hwang (2006), however, it has been possible to improve the water quality through improved treatment capacities to class IV which is suitable for industrial and agricultural purposes as well as for drinking water after treatment processes.

4.3 Analytical Framework and Interdisciplinary Research Design

4.3.1 Concept of Urban Units

In the complex aggregations of mega-urban areas, the different scales of consideration represent a major challenge. On the one hand, the very costly examination of the entire space or the mostly administrative units does not provide a detailed

Table 4.1 Quantitative rise of water supply[a] and expansion of supply and sewer systems from 1990 to 2006

Item	1990	2000	2005	2006	2007
Total volume of water supply (in 10,000 m^3)	86,136	124,958	170,969	172,997	178,809
Length of water supply pipelines (in 1,000 m)	2,947	5,169	11,532	12,495	13,750
Length of sewer pipelines (in 1,000 m)	1,217	1,952	4,827	5,157	5,548

[a]Refers to the total volume of water supplied by waterworks during the reference period, including both the effective water supply and loss during the water supply. Based upon Guangzhou Municipal Statistics Bureau 2007; www.stats.gov.cn

picture, while on the other hand, complexity is no longer manageable when examining at the smallest level, such as at the level of individual buildings. To understand mega-urban areas, it is therefore necessary, according to the authors, to divide large urban systems into smaller units.

Over the last 50 years, actual "city sample catalogues" have been created by many scholars, dealing with morphological and morphographical descriptions of the city's transformation (Larkham and Jones 1991; Moudon 1994; Whitehand 1981; Albeverio et al. 2008). It is common to all these methodologies that they use various common-known archetypes, such as street patterns (radial, network, linear, etc.) or existent spatial structures (polycentric, decentralized concentration, etc.) as works of Marshall (2005) and Rowe and Koetter (1978) can demonstrate. The urban form(ation) of the city cannot be described solely by those physical characteristics since there are no universal standards or a similarly well-defined understanding for urban spatial patterns such as "formal variables" based on numerical values as Stead and Marshall (2001) stated. However, they may help to provide a structural basis for further research by repetitive microstructures such as some types of buildings and their adjacent outdoor spaces are combined into groups of similar, homogeneous characteristics (urban units). Using these city blocks, it is possible to reduce the confusing system of the megacity into significant main elements which are characteristic and informative for developmental analysis.

In Guangzhou, such spatial units can be easily identified on the basis of "closed" settlement structures as relicts of the Mao and post-Mao era as well as actual political planning for urban expansion. In the years 1950–1978, the number of settlements increased on sparsely populated agricultural areas. The individual units were strictly separated from each other, composed as self-contained living and working quarters, so-called danwei, usually arranged as courtyard structures with clearly defined structural boundaries that meant not only adequate supply for the inhabitants but also strong social control. The onset of strong urbanization after the economical opening created a vast urban network, made up of urban and rural municipal and industrial centers. Based on today's urban landscape and seen from a city-morphological and morphographical perspective, various development stages are apparent, making it possible to speak of additive urban growth: single clusters – morphologically similar subunits – whose composition gradually forms the urban area of Guangzhou. The units are initially self-contained, but are expanded or merged into larger units within the course of transformation processes. Meanwhile, instead of housing developments, which are designed as a multifunctional living space for work, home, and life, more and more monofunctional designs are being implemented and implanted in the urban landscape.

It seems to be evident that structural rules of urban forms are closely related to local topography, ecology, technology, building resources, lifestyle, and aesthetic preferences. The existing complexity of the mega-urban system calls for a multidisciplinary research methodology, involving the study of "formal variables" and their impact on resource flows taking into consideration ecological determinants as well as the socioeconomic dynamics (Banister et al. 1997). The project mentioned in Sect. 4.1 is under way to develop a research design for small-scale analysis of mega-urban land use and its changes. In terms of the establishment of a transdisciplinary research approach on microscale, there are no adequate references in literature yet. Most of the scholars focused on the macroscale, taking the macroscaled land use change and its analysis by, for example, satellite images, as a basis for further research.

The research approach used in this chapter intends to get a deeper insight in the interdependencies between water cycle and urbanization focusing on the impacts of both on the population. Therefore, a common research basis has been set up, following the city's morphology. The clustered and additive city shape, as it is found in Guangzhou, offers an ideal basis for investigations on microscale, since the individual subunits are almost homogeneous itself. Thus, they can form the basis for a morphologic typology and in the further process serve as exemplarily investigation units for the reciprocal effects between urban structure and water regime on microscale.

The method is examined using the reference medium of water with a view to manifest impact of land use on the quantity and quality of water resources and thus helping to ensuring effective protection of water resources in mega-urban areas. Several different types of units could be identified during the main first research period between 2007 and 2009. The basis for further research will be the heterogeneous structured unit types as those areas promise a noticeable higher transformation rate than the homogeneous unit types as, for example, gated communities.

4.3.2 Research Design for Environment-Settlements-Humans Interactions

To link land use and its effects on water resources and to gain knowledge on microscale basis embedded in the mega-urban context on the one hand as well as to identify possible sources for surface water and groundwater contamination on the other hand, water-related land use structures were surveyed on-site. In addition, a total of 68 samples of surface water and 17 samples of groundwater were taken in the dry and rainy seasons between 2007 and 2009. From these, a total of 22 samples (13 surface water samples, 9 groundwater samples) taken in autumn 2007 and 2008 is relevant for this chapter. The water was collected in rinsed 555-ml and 1.5-L plastic bottles and brought to the laboratory as soon as possible after collection where the samples were kept refrigerated. The samples were not filtered and were examined in the laboratory for concentrations of total coliform bacteria in accordance with the standard examination method GB/ T 5750.12-2006, which has been implemented by the Ministry of Environmental Protection of the People's Republic of China. Immediately after collection of water samples, measurements were made in a mobile laboratory of the physicochemical parameter pH, electrical conductivity [μS/ cm], oxygen content [mg/ L], oxygen saturation [%], and redox potential [mV].

As environment, settlements, and humans are linked closely, the research of engineers, hydrologists, and landscape architects is supplemented by urban geographers who retain a broad and human-centered focus in the analysis of perception and appraisal and examine the effects of environmental conditions on human performance and the mediating social-psychological processes (cp. Bercht and Wehrhahn 2010) that explain why appraisals and behavior are interfaced with the environment in the way they are. Extensive site inspections, 51 in-depth interviews with Shibi's inhabitants, averaging 60–90 min in length were carried out from 2007 to 2009. Additionally, the method of auto-photography was applied in Shibi in order to better capture people's emotional states and personal appraisals of their living conditions. Five interviewees were given single-use cameras to photograph anything in Shibi that they related to the transformation processes and associated with positive or negative outcomes or feelings. The pictures were discussed afterward in detail. A qualitative research design was thus applied to facilitate an in-depth understanding of social aspects such as people's cognitive and behavioral response to water-related vulnerability factors. The analysis of those complex and diverse processes made the local-level investigation of the household and the individual itself indispensable.

The underlying morphogenetic approach can help to reveal formal and informal characteristics and puts special emphasis on the development of built and open space on the transitional areas. Formal and informal, behavioral and structural patterns could be described by changes of the built and open space. The open space includes public, private, and semipublic areas as well as the interspaces between different urban units. Apart from population density and constellation, also the development and the quality of the spatial density can be seen as indicators for the progressive development process. In the same way, self-regulating processes such as micro-urbanization phenomena show on the one hand the degree of vulnerability and on the other hand the variety of already existent coping strategies.

For this reason, the landscape architects examine:

1. The density which is based on densification of built surfaces, open surfaces, and spatial proportions
2. The functionality of the existing area based on perception and utilization and analyze based on these findings
3. The coping strategies in terms of spatial demands (open space and living area) and infrastructure (traffic and accessibility)

A transdisciplinary in-depth analysis may predict to a certain degree the long-term effects of urbanization on Guangzhou's water balance and on the citizens' living conditions.

4.3.3 Selection of Urban Units

Considering the urban morphogenesis in the Pearl River Delta since 1978, at least four main types of structural patterns could be identified beneath the heterogeneously structured units due to their changing morphologic shape and (water) infrastructural and population status. These types are represented by the units Liede, Xincun, Yuangangcun, and Shibi (cp. Fig. 4.3):

- *Liede*: The former traditional 800-year-old village, which was located in Tianhe District, consists of small groceries and residential houses with four to six storeys providing homes to approximately 7,000 locals and 10,000 migrants in 2007 (Wehrhahn and Bercht 2008). In August 2009, the "Liede renovation project" was started which is Guangzhou's first urban village "renovation" project, giving way to an integrated CBD complex project with a five-star hotel, luxury service apartment, and grade A office buildings as well as a shopping mall (Bureau of Foreign Trade and Economic Cooperation of Guangzhou Municipality 2009). Being Guangzhou's first urban village to be removed completely, all houses are demolished by now. Owing to these processes, a cleanup of the banks of the nearby creek on the one hand but also rising water consumption and disposal patterns as a result of the planned building complex on the other hand can be assumed.
- *Xincun*: Urbanized village with settlement origins of about 2,100 years. It belongs to the Haizhu District, is close to the site of "Haizhu Artificial Lake," and thus situated at the core of the southern part of the city axis. Hence, it is one of the targets of the real estate market looking for new areas suitable for the construction of housing and commercial infrastructure (Wehrhahn et al. 2008). It has undergone multiple transformation phases since 1990 and developed from an original village to an area, which is affected more and more by the current urbanization processes in recent years. It also finds itself under enormous outside development pressure today. A creek, several ponds and an agricultural area have – in the meantime – been transformed into settlement area.
- *Yuangangcun*: In 1992, this former traditional 800-year-old village located in Panyu District became part of Nansha Development Zone. This led to large-scale land use changes from agricultural to residential, industrial, and trade areas such as for sewing, collecting points of plastic,

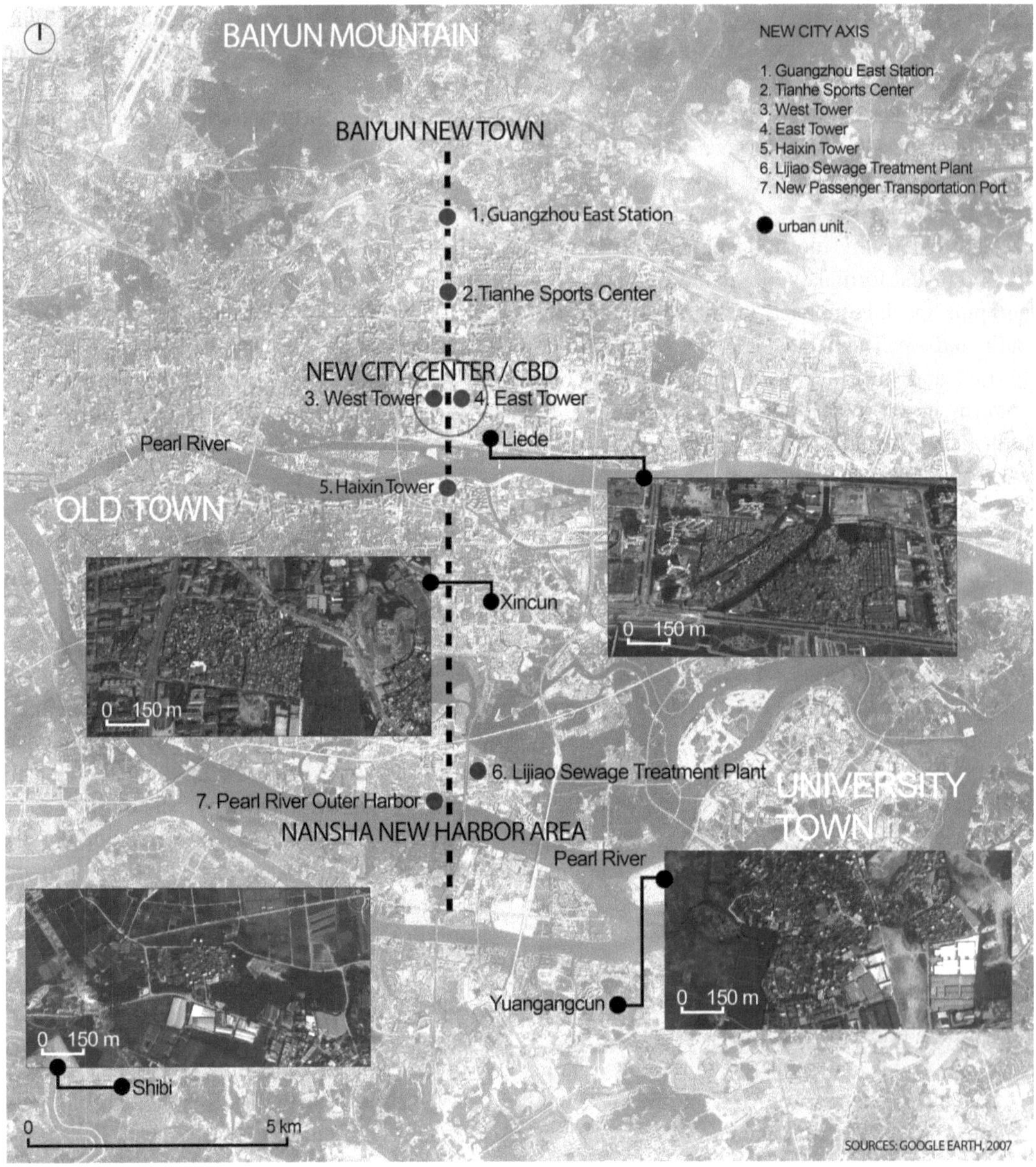

Fig. 4.3 Position of urban units and their proximity to Guangzhou's new city axis (Strohschön et al. 2013)

harbor engineering, and packaging. Direct domestic and industrial discharges are currently leading to serious water quality deterioration. Since 2008, more small trade buildings have been demolished; a modern high-rise complex and related traffic infrastructures are being built in the village's northern part, indicating the ongoing rapid urbanization progress in southern direction. Meanwhile, two out of every 6,000 inhabitants are migrants.

- *Shibi*: The village is located 17 km south of the city center in Panyu District and in 2008 was inhabited by about 10,000 permanent residents and about 10,000 migrants from other Chinese provinces working mainly in the secondary sector (esp. in garment manufacture). It is characterized by a comparatively rural way of life, an open structure, and dominated by one- to two-storey houses with saddle roofs, arranged mainly in courtyard structure and

Fig. 4.4 Conceptual design of the Guangzhou South Railway Station (**a**) and construction sites of the railway station project in Shibi Village in 2009 (**b** and **c**) (Strohschön et al. 2013)

adjacent to agricultural areas. The development of this still traditional village suggests that over time, Shibi will change to an urbanized village with predominantly large-scale urban land use structures. The Guangzhou government has adopted various pro-growth strategies since the late 1990s to strengthen its leading role as the regional center in southern China and to enhance the city's development to the south by creating an area for commerce, trade, and tourism (Guangdong News 2010; Xu and Yeh 2003, 2005). Though not directly located in the axis, but in its southern sphere, the construction of the Guangzhou South Railway Station in Shibi Village, which has started in 2004, is one project of particular importance (see Fig. 4.4). The station will be "the most modern and largest passenger railway station in Asia" (China Daily 2004) on a development area that spans 35 km^2, part of which is still used as farmland. It is estimated that the station will handle an estimated passenger capacity of over 80 million people per year (Pan et al. 2006), who also have to be provided with modern water infrastructure. The station was opened in 2010.

4.4 Results: Evaluation of the Effects of Urbanization on the Urban Water Household

4.4.1 Land Use Change in Guangzhou from 1988 to 2005

The conversion process of land use in Guangzhou is a reflection of the urbanization: In the last 20 years, Guangzhou's total built-up area has increased by about 1.8-fold and currently amounts to 7,434.4 km^2 (Zhao et al. 2009). One trend in the course of land use is apparent: water surface or arable land → minimally concentrated urban area → highly concentrated urban area; each classification type was determined by applying the maximum-likelihood method (for theoretical background, cp., e.g., Settle and Briggs 1987; Jia and Richards 1994).

As late as 1988, 80.8% of Guangzhou's area was dominated by intense agriculture, orchards, and forest areas (Yu and Ng 2007). Landsat 5-TM and Landsat 7-ETM images have been used to evaluate the land use changes in Guangzhou between 1990 and 2005:

In 1990, the total area of water within the city was 155.65 km^2, 7.46% of the total area. By 2000, the total area increased up to 162.34 km^2 and thus to 7.78% of total area. Thereafter, the proportion of water in Guangzhou decreased significantly and was as low as 131.24 km^2 in 2005 (6.29% of the urban area). Over the period 1990–2005, the city's water was reduced by an area of 24.42 km^2 due to the conversion of farmland for development and other urban functions such as enlargement of the transport network. With the beginning of economic development, the resulting sharp rise in population due to immigration of migrant workers coming mainly from interior provinces, and a resulting increasing demand for residential and settlement areas, large areas of agriculturally used land and minimally concentrated areas were converted into a highly concentrated urban area or reconcentrated: Highly concentrated urban areas made up an area of 522.98 km^2 in 1990, corresponding to 25.08% of the total area of Guangzhou.

As early as the year 2000, the area was 890.70 km^2, corresponding to 42.71% of the total area. In 2005, 42.75% of the urban area was classified as a highly concentrated urban area – this type of classification increased to a total of 891.50 km^2. Thus, the total densely concentrated urban area increased in the period from 1990 to 2005 by 368.52 km. The relatively small increase of only 0.04% within 5 years can be justified by the fact that particularly urban areas with a low density have risen sharply from 180.08 km^2 (8.63%) to 465.20 km^2 (22.31%) in the same period due to land use change from agricultural areas, water bodies, and green spaces into settlement areas, which at last illustrates the continuing process of urbanization.

4.4.2 Current Urban Development: The New City Axis

Guangzhou faces the same problems of overcrowding as other large cities in the developing world. In order to cope with the expected urban sprawl and diffusion of the already polycentral system of the city, the urban planning institutions of Guangzhou specified the directions of further urban expansion by city-internal development zones in 1999: "small changes in 1 year, medium sized changes in 3 years and a great change in 2010" as the urban government constitutes with the strategic master plan of 2005 (Jin 2007).
The goals are:

1. Optimal distribution of the function zones
2. Reduction of the population and building density
3. Traffic discharge in the historic town center
4. Ending up as an international eco-metropolis

The goals follow the five general principles: "exploration in the South, optimization in the North, extension in the East, adjustment in the middle and coordination in the West" (cp. Chen et al. 2006; Jin 2007). However, due to spatial limitations in the north (nature reserve of Baiyun Mountains) and east (Foshan), Guangzhou's main expansion directions will be toward the southern and eastern part of the urban agglomeration. The example of Guangzhou's new city axis, which is intended to regulate the future development between north and south and to create a new urban center by the strong planning emphasis of building activities in the mid part of the axis, gives an impression of the enormous transformation process and its impacts on settlements, environment, and humans. Based on this mesoscale consideration and description of the overarching impacts, this chapter focuses on the reciprocal influences between this development project and the case study areas on microscale.

The north-south axis, which is divided into two sub-axes by the northern branch of the Pearl River, is 12 km long and reaches from Baiyun New Town and Guangzhou East Station to the Pearl River outer harbor (cp. Wu et al. 2007). The entire planning surface covers a total of approximately five million m^2. Within the wide range of urban reconstruction measures, this megaproject shows the city's intention of maintaining to be the province's center and the market position next to Hong Kong. In order to realize the project by the Asian Games, which took place in Guangzhou in November 2010, and thus to reach the "great change in 2010," enormous areas must vacate for future development zones; radical demolition measures and expropriations were and will be inevitable (Fig. 4.5).

Due to the rising market value of the properties in the city center, many older adjacent settlement areas and urban agricultural areas risk becoming part of the big pool of realty venture (Nanfang Daily 2008). The realization of the axis is linked to the development of traffic infrastructure such as the opening of "Guangzhou Southern Railway Station" in shibi, as well as to a huge number of other prestige projects being constructed such as symbolic high-rises, the world's highest TV tower called "Haixin or Canton Tower," a designed greenbelt including the transformation of a former military area into a leisure park with large green spaces, and – as part of the city's water treatment program – the creation of Guangzhou's second largest artificial lake "Haizhu Artificial Lake" that formerly was an orchard covering 3,687 ha of which 1,120 ha were used for agricultural purposes (People's Government of Guangzhou Municipality 2008), partly also providing work and homes to about 50 families living in frugal dwellings without adequate water supply and sanitation (cp. Fig. 4.6). The lake is integrated in a wetland park consisting

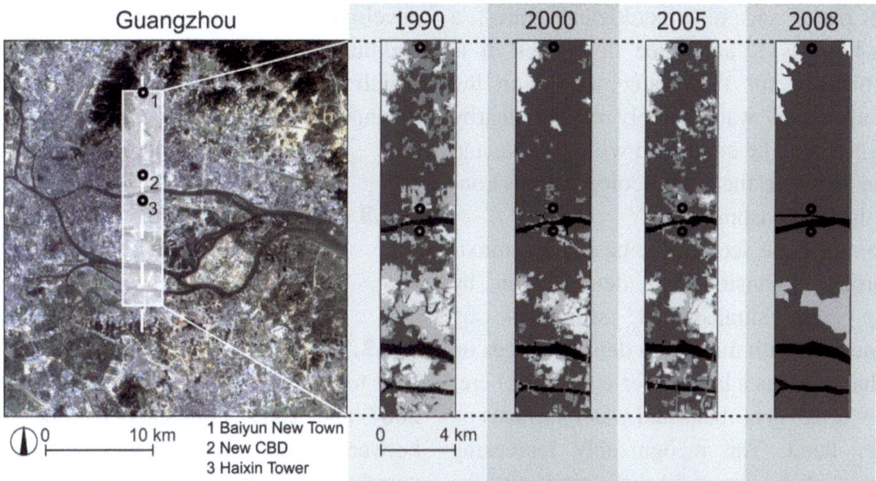

land use type	1990 area km²	1990 area %	2000 area km²	2000 area %	2005 area km²	2005 area %	2008 area km²	2008 area %
water	4.55	6.00	4.59	6.00	4.45	5.82	5.40	7.01
urban area, high building density	34.76	45.47	57.04	74.61	49.60	64.88	63.64	83.24
urban area, low building density	11.65	15.24	1.85	2.42	7.45	9.74	0	0
agriculture	17.39	22.75	7.25	9.48	7.84	10.26	4.41	5.77
forest, green space	8.10	10.54	5.72	7.49	7.11	9.30	3.00	3.98
total	76.45	100	76.45	100	76.45	100	76.45	100

Fig. 4.5 New city axis: land use change from 1990 to 2008 (Strohschön et al. 2013)

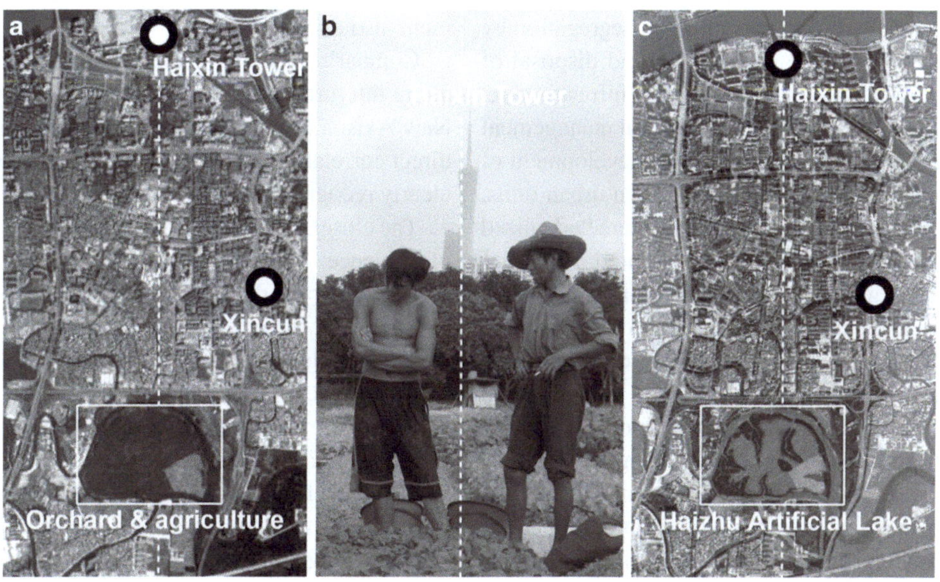

Fig. 4.6 Land use change from orchard and agriculture (**a** and **b**) to an artificial lake (**c**) (**a** and **c** are based on Google Earth) (Strohschön et al. 2013)

of three parks and nine ecological zones, a 3.19-km-long dyke, and three floodgates. It is mainly planned for floodwater storage in the rainy season and water supply of the nearby creeks which will be connected with the lake and to further improve the local ecological environment in the dry seasons.

The axis can be seen as the basis for a maximum urban expansion and densification by replacing old small-scaled structures by megastructures with maximum density ranges on the one hand and as a large inner-city project creating new urban greenbelts and water surfaces on the other hand. The recognizably increasing aggravation of the structural differences and the neglect of the outside areas of the city lead to an intensification of the problems on microscale:

A new form of socio-spatial structure arises that is mainly related to:

- The transformation of China's socioeconomic system
- The internationalization of Guangzhou's urban functions
- The general development of services and hi-tech industries in the region (Gaubatz 1999)

It will lead in further steps of city development to new dimensions of urban segregation by means of adequate supply with and disposal of water, adequate living space, and infrastructure. Due to China's dualistic urban-rural management system, the social and economic development of especially the heterogeneous grown urban units, such as, for example, existent or future "urbanized villages," has not been integrated into city planning and development yet, although they belong to the urban agglomeration of Guangzhou. The ongoing land use change, observable during the last 6 years, can be seen as an important indicator for what will happen medium term to the historic town center of Guangzhou. Today, new areas with substantially changed use and function occur, which follow the program of the overarching master plan. The example of Liede village close to the city axis shows one of the first complete disappearances of an urban village in favor of future development in Guangzhou (Wehrhahn and Bercht 2008). Further direct as well as indirect effects of the total mega-urban development on the environment, settlements, and humans are to be expected in the near future, which will have to be integrated in city and water management planning (see again Fig. 4.1).

4.4.3 Transformation of Urban Units and Effects on the Local Water Household

4.4.3.1 Densification and Water

Due to the morphologic investigations between 2007 and 2009, coherence can be postulated between spatial situation and thus the existing current development pressure and the change of the units' morphology: The morphology of the areas, which today are located in the direct sphere of influence of the urban development, exhibits a particularly high degree of spatial densification and surface sealing. In structural terms, they seem to follow similar rules during the transformation phases. Investigations of 67 small-scale investigation areas demonstrate that regarding the spatial structure on macroscale, the transformation from rural or peri-urban to urban area forces various vulnerabilities for humans and the environment also on microscale.

Comparing the four characteristic urban units along the rural-urban transect of the Guangzhou New Axis, a downward gradient, which exists in direct correlation with the distance to the axis, is clearly recognizable:

The closer the position in relation to the sphere of influence:

- The larger the development pressure
- The larger the pressure on inhabitants
- The stronger and more urgent the demand of living space
- The worse the open space supply due to enormous surface sealing and densification
- The higher the degree of informal design and use of the existent space, and finally
- The stronger the deterioration of surface water quality with total coliform bacteria, but not
- The stronger the pollution of groundwater with total coliform bacteria as even higher contamination was measured in the area of the old city center

The city area increasingly becomes a place to go for migrants due to the proximity to potential jobs. The per capita need for living space in the described areas rises synchronously with the need of the whole area. In 1997, the average floor space per household was 7.5 m^2 per person, while a huge number of people from the middle and low-income groups were still living in accommodation with less than 5 m^2 per capita (Chi-Man Hui and Seabrooke 2000). Between 1980 and 2000, urban per capita living space increased from less than 4–10 m^2. For 2010, it was aimed to achieve 15.5 m^2 of urban living space per capita (Hugentobler et al. 2002). This will inevitably lead to a higher grade of surface sealing due to the need of horizontal construction processes leading also to a reduction of leaching areas, to a densification because of vertical development, and to rising water consumption and amounts of sewage, respectively, as a result of enhancing living conditions.

While the degree of structural extent in the areas next to the town center, as for instance in Xincun and Liede, nearly goes beyond the spatial capacities and necessary expansions must yield to the narrow transition areas, the periphery of Yuangangcun and Shibi still has sufficient surface for expansion. Xincun is characterized by an already high and constantly growing spatial density and has a population density exceeding more than 20 times the average value of Guangzhou (cp. Guangzhou Municipal Statistic Office 2007). The number of migrants living in Xincun rose from 50% in 1993/1994 to 83% in 2007, while the predominant part of the local inhabitants moves away (Wehrhahn et al. 2008).

In the already urbanized units next to the town center, this intracellular re-densification is nearly ten times higher than in the peri-urban areas. The simple settlement structure found in Xincun and Liede built in the 1990s is/was arranged strictly, and buildings with five to eight storeys placed at distances of only 0.5–1 m between produce a very high basic density leading to bad lightning and ventilation. However, these kinds of buildings can also be found increasingly in the rural village structures Shibi and Yuangangcun, here also as an indication of social ascent. In Shibi and Yuangangcun, each having an average building density of 60%, less growth and structural density can be detected. There are still areas available for a surface-moderate expansion outward. Old structures in the settlement core are left, while at the settlement's border, buildings are constructed with higher standards such as connection to public water supply and the possibility of placing household appliances like washing machines and air conditioners due to more living space. Therefore, a more fragmented settlement structure arises, consisting of a mixture of abandoned surfaces, wasteland, agriculture, new housing estates, and traditional buildings, making the qualitative gaps between insufficient and adequate water infrastructures as obvious as the gaps between environmental conditions and living conditions.

In particular, the border and transitional areas of urban units seem to play a decisive role in the context of urbanization as potential building and development areas often still exist there. Thus, border areas can be seen as an indicator for the ongoing dynamic changes of the mega-urban development as they are permeable against inside and outside development pressure. Based on the analysis of the urban units' border areas, general explanation criteria can be gathered about the spatio-structural changes of the entire urban unit.

In comparison with the neighboring areas, Xincun, for example, can be described as an urban system with a relatively high degree of permeability and reciprocal effects between "inside" and "outside" between the unit and its surroundings. In contrast to this, a substantially more closed settlement system can be found in rural structures like in Shibi and Yuangangcun. Here, the transition areas between agricultural and settlement area do not develop rigidly but flowing: Buildings are constructed at the settlement border. In case of space requirement, the agglomeration can expand over the original settlement borders toward the agricultural areas. In Xincun, as in other urban or urbanized structures of this kind, various layers along the units' outskirts appear, depending on the position: Mainly old and informal dwellings being inhabited by

Table 4.2 Concentration of total coliform bacteria in surface water

Urban unit	Sample 1 [in MPN/100 ml]	Sample 2 [in MPN/100 ml]	Sample 3 [in MPN/100 ml]
Liede	1.3×10^7	5.4×10^8	1.3×10^7
Xincun	1.1×10^7	7.9×10^6	1.3×10^5
Shibi	1.7×10^4	1.3×10^5	1.3×10^5
Yuangangcun	2.2×10^6	3.5×10^6	4.9×10^5

disadvantaged groups such as migrants who have found temporary shelter and who shape their environment to fit their needs are located on such outskirts. The housing structures are of marginal conditions having not only the lowest standards such as deficient sanitation facilities accompanied by open sewers and basic living conditions or creative demands.

Manifold adaptations of simple structures as substitutes for living or storage space could be found that cope with the increasing and changing demands. Thus, somehow recurring characteristics of design and use of transitional areas could be identified, such as the installation of little vegetable gardens with sealed floor around in order to keep the need for water as low as possible. A second typical example is the "outsourcing" of normally internal functions, such as kitchen or sanitary facilities like washbasins. In the examined units, many improvised cooking and/or sanitary facilities but also simple storage areas can be found in the transition areas; however, having also the highest functional requirement as a comparatively slightest prospect of being able to remain prevails there: The risk is high that in the near future, these people and buildings must give way to new development projects or will be forced to relocate due to the sale of land, however, without adequate compensation or alternatives.

4.4.3.2 Water Infrastructure and Water Quality

Related to the quality of surface waters, all four investigation areas showed several signs of water-related vulnerability such as deficient access to public sewage disposal, poor drainage in open sewers, or informal and thus insecure dumps of various sizes containing mixed waste materials such as garbage, paper, or even batteries.

The dumps, whose contaminants seep away uncontrollably, are located either in house niches or adjacent to surface waters that are used for fish cultivation purposes serving the local and even national markets or within agricultural areas (in the case of Xincun: were) traversed mostly by feeders as in Shibi and Yuangangcun. Agricultural activities in Shibi and Yuangangcun, animal husbandry in the area of Shibi, and untreated domestic sewage discharges in all surveyed units are leading to severe surface water contamination by microorganisms and organic pollutants.

The measuring results of in situ parameters of flowing and stagnant surface water showed values of pH between 7.06 and 7.64, electrical conductivity ranging from 424 (Shibi) to 1,789 μS/cm (Xincun), oxygen content of 2.3 (Liede) to 6.56 mg/L (Xincun), oxygen saturation of 28 (Liede) to 76.4% (Shibi), and a redox potential from −214 (Liede) to 345 mV (Yuangangcun). Analyses of total coliform bacteria resulted in extreme water contamination from 1.7×10^4 MPN/100 ml in Shibi during the dry season in 2007 up to 1.3×10^7 MPN/100 ml in Liede which was measured at the beginning of the rainy season in 2009. However, in consideration of the measuring point's position to the axis, it became clear that the maximum concentration of total coliform bacteria was found in Liede which is in the new CBD, respectively, within the center of the New Axis (see Table 4.2). At present, this is not very surprising, as the measuring point was chosen at the southern end of Liede Creek which flows through the northern urban area from north to south, accumulating domestic sewage.

According to the environmental quality standards for surface water GB 3838-2002 (see Sect. 4.2.2), the highest water quality class V only allows a maximum concentration regarding total coliform bacteria of 40 000 MPN/L. Thus, the

Table 4.3 Concentration of total coliform bacteria in groundwater

Urban unit	Sample 1 [in MPN/100 ml]	Sample 2 [in MPN/100 ml]	Sample 3 [in MPN/100 ml]
Xincun	3.3×10^4	1.7×10^4	3.3×10^4
Shibi	3.3×10^2	4.9×10^3	4.9×10^4
Yuangangcun	9.2×10^5	3.5×10^5	7.0×10^5

critical value of this standard for surface water quality was seriously exceeded at all measuring points. The measured microbiological pollution by animal and/or human waste of the tested surface water might cause gastrointestinal infections or sickness as the water is/was also used for irrigating purposes in three of the four units (Shibi, Yuangangcun, and Xincun) – even if coliform bacteria are not a direct health threat to human beings (cp. U.S. Environmental Protection Agency 2009).

Other potential surface water contamination and related health risks might arise from the situation in regard to the availability of limited technological filtering or treatment resources on the one hand and the use of liquid fertilizer on large-scale agricultural areas on the other hand, as personal communication in Shibi revealed. In the 1980s alone, use of chemical fertilizers in the Pearl River Delta increased by 40% according to Hugentobler and Lütolf (2006).

Several residents believe that the consumption of groundwater tainted by fertilizer-polluted run-off is a major cause of stomach cancer. Due to the fact that local-level statistical data concerning cancer mortality rates is not open to the public in China, the correlation mentioned between mortality rate and polluted water could not be verified. However, studies from the World Resources Institute (1998) indicate (1) that China's water resources are more and more polluted by toxic contaminants such as synthetic nitrogen fertilizers, (2) that China has the highest stomach cancer death rate in the world, (3) that stomach cancers are three to seven times higher in polluted rural areas compared to cleaner ones, and (4) that stomach cancers in China are caused in part by water pollution. These results might be transferred to Shibi and thus support the residents' statements.

The interactions between land use, humans, and groundwater systems become visible in three of the four units when considering people using groundwater as part of their daily diet from privately and publicly available wells. In urban Xincun as well as in peri-urban Yuangangcun and Shibi, places around wells are often used for washing clothes whereby cleaning agents easily get into the groundwater. Sampling results, however, showed no significant pollution burden with heavy metals like cadmium or copper, but instead concentrations of total coliform bacteria in the range from 3.3×10^2 in Shibi, sampled in the dry season of 2007, up to 9.2×10^5 MPN/100 ml in Yuangangcun, which was tested at the end of the rainy season in 2008 (cp. Table 4.3).

According to the Chinese Quality Standard for Groundwater GB/T 14848-93, which is also divided into five categories (Nantong Municipal Environmental Protection Bureau 2006), all measuring values are assigned to class five indicating water with maximum pollution and total coliform bacteria concentration > 100 MPN/L. The results of in situ parameters showed a pH between 5.42 (Shibi) and 7.3 (Yuangangcun), electrical conductivity in the range from 38 (Shibi) to 1,165 μS/cm (Xincun), oxygen content of 1.69 (Yuangangcun) to 6.67 mg/L (Shibi, Xincun), oxygen saturation of 75.6 (Xincun) to 76.5% (Shibi), and a redox potential from 180 (Xincun) to 325 mV (Shibi). Groundwater data of Liede, the unit where surface water quality with regard to total coliform bacteria was worst, are not available as no well could be found.

Because of the groundwater's "good taste," several residents – locals as well as migrants – consume the water regularly without previously boiling it; some of them store the water for a few days in order to suspend the solids, as personal communication revealed. Others know about potential pollution or dislike the taste and use the water only for irrigation or sanitary purposes. Different knowledge about poor water quality and its potential effects on human health but also diverse potential risk exposures lead to different perceptions and appraisals by the inhabitants. Although the contamination of surface water is highly visible due to the grey to dark color, putrid smell of the river water, and also by packaging materials, pet bottles, cans, and other disposables in many river parts and even in ponds used for fish cultivation, the interviewees vary in their appraisals:

While many locals of Shibi complain that local factories do not apply mechanisms and processes to treat their wastewater and concerning the river's dirty and smelly condition, many working migrants do not worry about the water pollution at all because it is their aim to make as much money as they can to go back home some day as auto-photography and interviews revealed. It becomes evident that personal and situational factors always determine whether the relationship between man and the environment is appraised as vulnerable. Environmental conditions or a deficiency in resources makes a person vulnerable only when the deficit refers to something that really matters. In comparison to the migrant who is not emotionally attached to Shibi, the villager feels a reduction of his quality of life as he has experienced the changes of the natural environment. From this point of view, the river no longer stands for fishery, water supply, or a recreation area. The villager's commitment to benefit from the river in various ways is constrained by the harmful river's contamination. The different perceptions and appraisals of locals and migrants regarding water quality deterioration were also reported in other units.

Water infrastructure regarding domestic sewage disposal is very similar in all four units, and the proximity to the new city axis does not have any influences to restructuring the sewage disposal yet: A combination of open and covered wastewater gutters, a main conduit leading into the nearby agricultural area, and/or a creek and canalized areas can be found.

The existence of open wastewater gutters causes significant negative impact in terms of hygiene factors such as domestic waste in the gutters, optical impairment, and bad smell especially in the summer, rats, and other vermin-transmitting etiologic agents to humans, especially to children playing outdoors. In order to avoid the bad smell and unpleasant sight, many of the open wastewater gutters and small canals have recently been covered. Enclosures that are installed at the changeover from open wastewater gutter to canalization shall prevent raw particles from trespassing into canalization and preclude rats coming out of the canalization could be observed in urban as well as in peri-urban units not depending on the axis' proximity. However, both measures – coverage and installation of enclosures – are only improvised and do not address the problem of a deficient water management at its source.

In an interunit comparison and in relation to the axis' proximity, no systematic between the houses' age and the form of sewage disposal is recognizable. While most domestic sewage from people living in new houses is often discharged in open wastewater gutters like in Shibi, Xincun's newer houses, which were built in the 1990s, are mostly connected to the public sewerage. Nonetheless, most effluent enters the water courses. This fact makes many inhabitants of Guangzhou feel as if they have been treated unfairly since they have to pay 0.70 CNY per m^3 for wastewater treatment of domestic sewage (cp. KFW n.d).

In this context, the effects of water management and of land use change on human performance and the appraisal of water quality are an important addition to hydrological research: According to the villager's perceptions, the construction of the Guangzhou South Railway Station in Shibi accelerates the processes of changing living conditions and the impairment of environmental structures. The development of the village from 2004 to date shows that Shibi has rapidly been changing from a traditional Chinese village characterized by a rural way of life to an urbanized village with predominantly urban land use structures. Due to the sale of most of the farmland, many villagers have lost their basic source of income and additionally fear resettlement due to the pending demolition of some parts of the village. The interviewees, however, did not recognize or know about any severe impacts of the megaproject on certain water-related processes and structures such as variations of underground or surface water quality or decrease in drinking water quality. As one of the main concerns of Shibi's locals is to lose the assured place to live and to be relocated, it became clear that the personal meaning of certain conditions is fundamental for the experience of social vulnerability. The uncertainty of not knowing what is going to happen to one's life in the near future increases the level of vulnerability, as villagers have a significant goal at stake but have no guarantee of achieving it and no available resources with which to try. Fear, uncertainty, and regarding the situation as not totally contingent upon one's actions increase the subjective level of vulnerability. Research results show that vulnerability is determined by the relationship between the individual's pattern of commitments and resources for warding off threats to those commitments.

In this sense, vulnerability can be thought of as a potential threat that is transformed into an active threat when that which is considered of importance is jeopardized. Due to diverse individual goals, commitment patterns, and personal factors, even most severe crises are thus differently appraised with respect to the experience of vulnerability. In this case and in the scope of the threatening situation to individual's livelihoods, water contaminations are considered as less important.

The distress of the deterioration of water quantity and quality will be heightened in the near future due to the fact that multifaceted urbanization processes of various scales will take place in several units and their adjacent areas along the axis: Large amounts of farmland and settlement areas have already been sold by the village committees of Liede, Xincun, and Shibi to the government. In the north of Xincun, for example, many new high-rises were built evolving the impression of moving closer to and thus exerting pressure on the village. Also inside the village borders, fundamental changes relating humans and environment happen: Instead of vegetables, buildings are to "grow" on the former agricultural areas. First constructional measures such as leveling of the agricultural area or filling up of a creek and ponds formerly used for fish cultivation have started in 2008, reducing the city's water area, but also the villagers agricultural or rental income.

While land use changes within the central core of the axis will primarily happen due to the demolition of low-standard houses and small agricultural areas, the large-scale land use changes will especially take place at the southern city fringe like in Yuangangcun and Shibi: Especially those large-scale land use changes and the growth of population based on in-migration have far-reaching impact on the city's water household such as a decrease of unsealed surfaces allowing infiltration and natural groundwater recharge on the one hand, and impact on the water infrastructure with rising consumption and disposal patterns, the need for access to clean drinking water, and adequate sanitation on the other hand. As treatment capacities are still insufficient, seepage of wastewater will thus be an important but critical source of the so-called urban groundwater recharge (cp. Putra and Baier 2009), which endangers the health of Guangzhou's ecosystems.

4.5 Discussion

One prominent aim of this chapter is to enrich the transdisciplinary debate on vulnerability from a natural as well as from a social science perspective by discussing changing land use and its effects on water resources in a Chinese megacity. The benefit natural and social scholars gain from exchanging and commonly applying theoretical and methodical backgrounds is evident, especially with regard to the analysis of complex and multifaceted water problems in the scope of rapid urban transformation processes.

4.5.1 Mega-Urban Restructuring, Urban Water Household, and Human Vulnerability

With respect to the mega-urban restructuring, the administration of Guangzhou Municipality is making great efforts to realize its target of the five modernizations (cp. Sect. 4.4.2). Especially the goals "adjustment in the middle" and "exploration in the south" relate to the new city axis. Besides the generally changed spatial characteristics, the analysis in Guangzhou also showed the emergence of new forms of urban space on the microscale. These are on the one hand an expression of the drastic changes of the urban system and give reference to the personal and social identity of the inhabitants and to their needs (Watson and Bentley 2007).

On the other hand, the examples of coping strategies like using groundwater as drinking water due to tap water contamination show the "ability of adaptability" within a constantly changing mega-urban living situation. In particular, the heterogeneous structure of the examined units offers a large range of coping strategies in the form of re-densification, which is, however, limited by the maximum density range. The ongoing densification tendency always enables a multiplicity of individual solutions: A kind of micro-urbanism arises, which can mostly be found in old, grown areas. The vacuum of the national regulations and laws thus permits a large variety of structural extensions, of low-cost housing and of informal sources of income. These are transferred from generation to generation and produce in this way a living system, which is up to a certain degree able to help itself. Organizational and structural self-help generates an organic, self-renewing urbanism, which changes the social and cultural character of the entire city.

While urban restructuring in terms of mega-urban building constructions is visible especially along the new city axis, the reaching of the municipality's target of turning Guangzhou into an international eco-metropolis by, for example, construction, respectively, expansion of four sewage treatment plants, enhancement of water supply, river improvement programs, and creation of green areas is still far away as the research results and in particular the measuring values of the groundwater and surface water samples have shown.

This mirrors the fact that millions of Guangzhou's residents and a large number of migrants without access to improved sanitation living in urban as well as in peri-urban areas of the city still pose a severe threat to both surface and groundwater resources by uncontrolled disposal of human excreta. The need of changing this situation, however, is in the municipality's area of responsibility. In terms of rising population, this means on the one hand that an increasing number of people must be provided with safe drinking water and sanitation services and on the other hand that the need for more living space and changing land use patterns, ecological and social vulnerability, might get worse.

However, many ecological improvements to enhance the city's image – also in the sphere of the axis – have already taken place until the Asian Games in 2010. Chinese tradition like urban garden design finds its way back in the cityscape: On the one hand, green spaces are being integrated into new apartment complexes for upgrading purposes and as a soft location factor in terms of selling point. On the other hand, they are integrated in the cityscape as public open spaces such as in an area of about 26,400 m^2 in the northern city axis or as a major greenbelt for recreational purposes between the "Haixin Tower" on the north end and the Pearl River on the south end forming the southern axis.

While creeks, especially in the old city, had to give way to road and building constructions in the past, surface water represents a water source utilized by millions of people for sewage disposal today, and it is and above all will be integrated in the current master plan to become reservoirs for water storage, flood protection, and recreational area purposes in the future. Artificial water surfaces like lakes, waterfalls, or fountains are already and will continue to be integrated as modern design elements in the cityscape – currently forming considerable contrast to severely polluted creeks and ponds.

4.5.2 Spatial Adaptability: Micro-Urbanism and Its Effects on Water Resources

Various layers of urban fabric form a kind of typical mega-urban tissue within a period of several years as it could be observed in many units. The dynamic states of the city and the pervasive relationship between its elements and their effects

Fig. 4.7 Temporary constructions: solitaires or additional space on/at existing building (**a**), unused areas as storage surfaces and floor spaces (**b**), and composition areas for privatization of public space by, for example, placing flowerpots and stones or constructing seat walls and awnings (**c**) (Strohschön et al. 2013)

on water resources quantity and quality have led to an interdisciplinary research design evaluating these ongoing changes.

Up to a certain degree, humans adapt to the existing spatial offer in a passive way. But there is a growing active adaptation of the existing situation for individual demands in terms of alternative use and design. Altogether, a high degree of adaptability could be observed in the examined urban units as Fig. 4.7 shows, identifying the following three types of the micro-urbanization based on recurrent self-organization characteristics. Thus, Fig. 4.7a, for example, shows a small, purely functional motivated standard, while Fig. 4.7c is evidence of a somehow differentiated use extending, but not replacing the original use.

In addition to these active uses, self-regulating measures also occur, such as groundwater use for daily diet or informal building construction with the lowest water infrastructural standards in terms of a lack of toilets and thereby open sewage disposal, influencing ground and surface water quality. These have been set up since the number of migrants and the urban density started rising in Guangzhou. Function, shape, design, and quality of all measures depend on the sociocultural background, the needs, and the economic options of the users.

4.5.3 Can the Transdisciplinary Concept of Urban Units Serve as a Basis for Research?

The problems and challenges of the mega-urban development can be read off due to the systematic of the urban units and their characteristics such as density, surface sealing, and water infrastructure. The selected urban units, each characterized by their highly dynamic and transforming structures and processes that have proceeded within a comparatively short period of time, generate to be demonstrative examples for an interdisciplinary in-depth analysis and discussion on social and ecological vulnerability in the context of complex and fast-growing (mega-)urban development processes.

The authors consider the mega-urban city to be a lively system of various functions and interdependencies which are submitted to permanent transformation processes as the comparison of satellite images in regard to changes of the settlement area and water surfaces field studies and interviews could reveal. The examination of the morphologic species offers the possibility identifying these interactions on the microscale to get a detailed understanding of the mega-urban context and the reciprocal influences on the urban water household. Thus, the urban units are not to be seen as a rigid system, but as a basis for the analysis of complex structures such as formal and informal living spaces each with its characteristic infrastructures, types of water supply or sewage disposal, and dynamic processes like densification or changing channels of supply.

This approach can help to reveal formal and informal characteristics and puts special emphasis on the development of built and open space, in particular on the transitional areas. Apart from population density and constellation, also the development and the quality of the spatial density, the contemporaneously increasing surface sealing, and decreasing of open space can be seen as indicators for the progressive development process. In the same way, self-regulating processes such as micro-urbanization phenomena show the degree of ecological and human vulnerability on the one hand, and the variety of already existent coping strategies like compensating for bad tap water quality by buying plastic bottles or using groundwater on the other hand.

In comparison to hydrological and morphological analyses, as well as to objective sampling of water quality and quantity factors, a fundamental proposition of the social vulnerability perspective is the personal meaning of the humans-environment relationship that, in turn, depends on the perception and appraisal processes by means of which that meaning is constructed. This implies that not all potential stressors actually cause vulnerability from a personal point of view (for theoretical background, see Bercht and Wehrhahn 2010). What is appraised as harmful or threatening by one person may not be so appraised by another. Research results indicated that people exposed to the same environmental conditions appraise certain risks differently. This is based on varying personal factors and access to appropriate resources.

However, a social vulnerability perspective does not argue against the importance of ecological vulnerability factors and the influence of meso- and macroscale factors (e.g., institutions, water systems). Within this transdisciplinary research framework, the social perspective approaches the effects from an individualistic perspective to better capture people's way of thinking and their emotional states and hence explains different levels of vulnerability and coping behavior. However, in this context, a quantitative analysis of water-related structures and processes is required to verify, for instance, if subjective appraisals with regard to poor drinking water quality agree with corresponding water samples that prove contamination.

The essential point is the difficulty of understanding vulnerability from the standpoint of the environment or person per se. A transdisciplinary approach is needed in which the two subsystems, humans and environment, are considered as one entity. In this chapter, insights and theoretical backgrounds from natural sciences are thus combined with those from social science to concretize the impacts of structural changes along Guangzhou's city axis on environmental as well as on social factors.

4.5.4 Prospects on the Future City Development

How would it look tomorrow, if the speed at which the city develops continues as before? Will all of the units be absorbed by the city after they exceeded a certain limit of densification or if they do not fit optically into the modern cityscape anymore because of, for example, inadequate sewage disposal by open sewers? Will the deterioration of water quality and environmental pollution further increase?

Without adequate steering mechanisms within the range of water resources and surface management, the units, for example, will necessarily self-collapse or will be demolished and will be subjected to the overarching modern urban development projects. Following examples from European city development practice, a user and thus need-oriented regulation of land consumption in medium- to long-term perspective may lead to a discharge of the urban system, as for instance by supply of additional (temporary) land areas either on occasional unused fallow lands or on future development surfaces.

The examined areas are characterized by different permeability. It can be expected that the different settlement units will continue to densify, the closer and more directly they are related to the sphere of the urban development's influence. It can, however, also be assumed that more units will be restructured in the sphere of the axis and in this context will get a modern water infrastructure such as an extensively canalization and thus an optical upgrading leading to an objective better standard of living. But, several of these urban renewal processes will take place after processes of forced relocation, loss of living space, loss of sources of income due to a loss of rentable living space, and maybe a loss of social networks.

4.6 Conclusions

Mega-urban restructuring as demonstrated with regard to Guangzhou's new city axis has decisive influences on the local water household, on technical supply and disposal systems, on traditional urban units as much as on peoples' living conditions, and at least their need of evolving coping strategies. Restructuring processes planned on the macroscale which in this case is Guangzhou City do also lead to restructuring or local influences on the microscale, here within the urban units. The analyses of microscale land use changes within the context of mesoscale restructuring processes have shown various hotspots of vulnerability with regard to the water resources, settlement structures, and humans:

The *shape, use, and spatial structure* of especially the open space will continue to change drastically within the different urban units during the next years. However, the land use change must take place in an appropriate way, meeting the needs of a growing megacity which is on the way from local agriculture to a global production region. In general, this development is bound to a reduction of the public open space in favor of bigger residential capacities and public traffic infrastructure. Many urban elements will be changed, removed, or substituted.

Regarding the *local water resources,* the vulnerability of megacity Guangzhou is noticeable everywhere, especially in terms of water quality. The interactions of high-speed population growth leading to an increasing use of water, land use, peoples' actions, and surface and groundwater systems become obvious due to high concentration of total coliform bacteria up to 1.3×10^7 MPN/100 ml in surface waters and values up to 9.2×10^5 MPN/100 ml in groundwater as the results of the investigation substantiate. The strain on ground and surface water due to coliform bacteria is heavily present in Liede and Xincun located in the new CBD and in the central area of Guangzhou but is also present in the peri-urban areas Yuangangcun and Shibi. A downward gradient relating to the new city axis becomes obvious, with the exception of groundwater contamination by total coliform bacteria as the highest concentration was measured within the old city which leads to the presumption of a concentration of leaking water and wastewater pipes of the old houses. It can be assumed that coliform bacteria are primarily of excrement origin, since large amounts of domestic wastewater are piped into the watercourses untreated and are accumulated by flowing through the urban area.

Countless other examples like open wastewater ditches or eutrophied feeders show the water supply's vulnerability. Open wastewater gutters reduce living environments and pose health risks to humans, especially to children. Activities like improvised covering of those ditches in order to reduce bad smell and sight express the population's need for better living and environmental standards. In many cases, a lack of modern sewage disposal is accompanied by a deficient public water supply: There are still people living at the peri-urban fringe of the city, facing the problem of not being connected to the public supply network and thus having to buy water bottles or use groundwater, whose samples have also been polluted by total coliform bacteria in particular in many cases; others who are provided with tap water do not want to use it directly as drinking water, and thus either boil it or use private suppliers or publicly available water vending machines (Wehrhahn et al. 2008).

Based on the *macroscale consideration* and description of the overarching impacts emerging from the City Administration's Master Plan (cf. Sect. 4.4.2), the reciprocal influences between the city's development projects and the *microscale case study areas* are focused on in this chapter. The various urban unit types and their ecological, economic, and social characteristics as much as their changes act as reference areas and thus as a pointer structure for the entire megacity. The application of a morphological typology on microscale is a good starting point for future proposals for comparability in a first step and for a standardization of urban and water planning measures in a second step, and can thus improve the understanding of the relationship between water quality and landscape pattern. Using this approach in a transdisciplinary research group, it becomes possible to reduce the confusing system of the megacity into significant main elements which are characteristic and informative for long-term developmental analysis and sustainable urban and water management planning and governability.

Against this background and the research experiences of the last 6 years, it can be substantiated that information drawn from the transdisciplinary investigations related to the allocation of surface water, the interaction between sanitation systems and surface water and groundwater, and the relating water quality, as much as the spatial and psychological processes involved, is an essential input to sustainable land use and water resources protection and planning.

Acknowledgements The authors would like to thank the German Research Foundation, as this research material is based upon work within the Grant Priority Program 1233: Megacities – Megachallenge: Informal Dynamics of Global Change.

Note

Meanwhile, a revised form of the manuskript was submitted to the international Journal of Environmental Research. It has been accepted for publication and will presumably appear in March 2013 (cf. Strohschön et al. 2013).

References

Albeverio S, Andrey D, Giordano P, Vancheri A (2008) The dynamics of complex urban systems – an interdisciplinary approach. Physica-Verlag, Heidelberg

Banister D, Watson S, Wood C (1997) Sustainable cities: transport, energy and urban form. Environ Plan B Plan Des 24:125–143

Bercht AL, Wehrhahn R (2010) A psychological-geographical approach to vulnerability: the example of a Chinese urban development project from the perspective of the transactional stress model. Environ Plan A 42(7):1705–1722

Biswas AK, Tortajada C (2004) The water challenges of megacities. The Stockholm International Water Institute, Technical report, pp 12–13

Breil P, Lafont M, Fletcher TD, Roy A (2008) Aquatic ecosystems. In: Fletcher TD, Deletíc A (eds) Data requirements for integrated urban water management, Urban water series. UNESCO/Taylor & Francis, Leiden

Bureau of Foreign Trade and Economic Cooperation of Guangzhou Municipality (2009) Liede commercial project kicks off. Article 27 Aug 2009. http://www.investguangzhou.gov.cn/web/eng/jsp/content/content_detail.jsp?catEncode=001006001001&contentId=26420. Accessed 18 Mar 2010

Chen M (2009) Guangzhou to build second artificial lake in June. http://www.lifeofguangzhou.com/node_10/node_37/node_85/2009/03/14/123699672561710.shtml. Accessed 24 Feb 2010

Chen X, Chen Y, Lai G, Zeng C (2006) Optimal allocation of water resources in Guangzhou City, South China. J Environ Sci Health A 41:1405–1419

Chi-Man Hui E, Seabrooke B (2000) The housing allowance scheme in Guangzhou. Habitat Int 24:19–29

China Daily (2004) Guangzhou plans new train station. http://english.peopledaily.com.cn/200412/31/eng20041231_169292.html. Accessed 20 Apr 2009

Gaubatz P (1999) China's urban transformation: patterns and processes of morphological change in Beijing, Shanghai and Guangzhou. Urban Stud 36(9):1495–1521. doi:10.1080/0042098992890. http://usj.sagepub.com/cgi/content/abstract/36/9/1495

Goudie A (1990) The human impact on the natural environment, 3rd edn. The MIT Press, Cambridge

Guangdong News (2010) Guangzhou south railway station to open on Jan 30. http://www.newsgd.com/news/GDNews/content/2010-01/05/content_7807746.htm. Accessed 15 Mar 2010

Guangzhou Municipal Statistics Bureau (2007) Guangzhou statistical yearbook 2007. No 19. China Statistics Press, Guangzhou

He M (2005) The water household as an example of ecology in the Pearl River Delta. In: Ipsen D (ed) The genesis of urban landscape: the Pearl River Delta in South China. Faculty of Architecture, Kassel, Work report 161, pp 77–84

Hiwasaki L, Arico S (2007) Integrating the social sciences into ecohydrology: facilitating an interdisciplinary approach to solve issues surrounding water, environment and people. Ecohydrol Hydrobiol 7(1):3–9

Hong Kong Trade Development Council (2009) Selling in China – Guangzhou. http://www.hktdc.com/info/mi/a/bgdscn/en/1X069VKV/1/Guide-To-Selling-In-China/Selling-In-China-Guangzhou.htm. Accessed 15 Mar 2010

Hugentobler M, Lütolf T (2006) Zhu village: urban renewal in the city of Guangzhou. ETH, Zürich

Hugentobler M, Jia B, Maovenzadeh F, Hanaki K (2002) AGS future cities: Guangzhou – a partnership for sustainable urban development. disP – Plan Rev 151(4):51–58

Huang J, Keyton D (2010) Guangzhou population closes to 15 million. http://www.lifeofguangzhou.com/node_10/node_37/node_85/2010/06/18/127683154477641.shtml. Accessed 11 Mar 2011

Hwang J (2006) Key projects of urban utilities and landscaping bureau in 2003. http://www.lifeofguangzhou.com/node_10/node_38/node_46/node_47/2006/03/23/11430965651149.shtml. Accessed 21 Jan 2010

Ito C (2005) Urbanization and water pollution in China. Discussion papers 0513, The Australian National University, Canberra

Jia X, Richards JA (1994) Efficient maximum likelihood classification for imaging spectrometer data sets. IEEE Trans Geosci Remote Sens 32(2):274–281

Jin H (2007) Aspects of Guangdong Province. Cartographic Publishing House of Guangdong Province, Guangzhou

KfW (n.d.) China, PR: Guangzhou sewage disposal. http://www.kfw-entwicklungsbank.de/EN_Home/Ex-post_Evaluation_at_KfW/Ex-post_evaluation_reports/PDF-Dokumente_A-D/Kurz_Guangzhou_Abwasser_engl.pdf. Accessed 14 Jan 2010

Kraas F, Sterly H (2009) Land use change in megacities and challenges for water management. In: Baier K, Strohschön R (eds) Proceedings of megacities – interactions between land use and water management. Mitteilungen zur Ingenieurgeologie und Hydrogeologie 99, Mainz, Aachen

Larkham PJ, Jones AN (1991) A glossary of urban form, Historical geography research series no. 26. Institute of British Geographers, London

Lin W, Zhang L, Du L, Yang HL, Zhang Y, Li J (2009) Quantification of land use/ land cover changes in Pearl River Delta and its impact on regional climate in summer using numerical modeling. Reg Environ Chang 9:75–82

Liu S, Li X, Zhang M (2003) Scenario analysis on urbanization and rural-urban migration in China. Interim report IR-03-036. Chinagro project: report of WP1.2, Beijing

Liu D, Chen X, Lou Z (2010) A model for the optimal allocation of water resources in a saltwater intrusion area: a case study in Pearl River Delta in China. Water Resour Manag 24(1):63–81

Marshall S (2005) Streets and patterns. Spon Press, London/New York

Ministry of Environmental Protection of the People's Republic of China (2002) Environmental quality standards for surface water, GB 3838-2002. http://english.mep.gov.cn/standards_reports/standards/water_environment/quality_standard/200710/W020061027509896672057.pdf (in Chinese). Accessed 14 Jan 2010

Moudon AV (1994) Getting to know the built landscape: typomorphology. In: Franck K, Schneekloth L (eds) Ordering space. Types in architecture and design. Van Nostrand Reinhold, New York

Nanfang Daily (2008) Bai Yun Xin Cheng Gui Hua 6 Fang An Jin Qi Gong Bu. http://www.nanfangdaily.com.cn/epaper/nfrb/content/20080411/ArticelGC01003FM.htm (in Chinese). Accessed 14 July 2009

Nantong Municipal Environmental Protection Bureau (2006) Quality standard for groundwater, GB/T14848-93 (in Chinese). http://www.nthb.cn/standard/standard02/20030414085238.html. Accessed 14 Jan 2010

Pan A, Zhou H, He C, Wang F (2006) Urban transportation – traffic planning and management in Guangzhou (in Chinese). Chinese Construction Industry, Peking

People's Government of Guangzhou Municipality (2007) Administrative regions and population. http://www.gz.gov.cn/vfs/subsite/JGIN7QPB-AZE4-2MTO-EA6G-R281E8V2SFJH/category/category07.jsp?catId=5713&PageNo=5. Accessed 20 Jan 2010

People's Government of Guangzhou Municipality (2008) Orchard turned into Wetland Park. http://big5.gz.gov.cn:82/gate/big5/www.gz.gov.cn/vfs/subsite/JGIN7QPB-AZE4-2MTO-EA6G-R281E8V2SFJH/content/content_visitor.jsp?contentId=645435&catId=5944. Accessed 14 Jan 2010

People's Daily Online (2009) China's urbanization level reaches 45.68 percent. http://english.people.com.cn/90001/90776/90882/6637891.html. Accessed 15 Mar 2010

Putra DPE (2007) The impact of urbanization on groundwater quality – a case study in Yogyakarta City – Indonesia. Mitteilungen zur Ingenieurgeologie und Hydrogeologie 96, Mainz, Aachen

Putra DPE, Baier K (2009) Der Einfluss ungesteuerter Urbanisierung auf die Grundwasserressourcen am Beispiel der indonesischen Millionenstadt Yogyakarta. Cybergeo, Environnement, Nature, Paysage, article 469

Rogers P (1994) Hydrology and water quality. In: Meyer WB, Turner BL (eds) Changes in land use and land cover: a global perspective. University Press, Cambridge

Rowe C, Koetter F (1978) Collage City. Basel, Boston, Berlin 1984, 3rd edn, 1988

Settle JJ, Briggs SA (1987) Fast maximum likelihood classification of remotely-sensed imagery. Int J Remote Sens 8(5):723–734

Shao M, Tang X, Zhang Y, Li W (2006) City clusters in China: air and surface water pollution. Front Ecol Environ 4(7):353–361

Stead D, Marshall S (2001) The relationships between urban form and travel patterns: an international review and evaluation. Eur J Transp Infrastruct Res 1(2):113–141

Strauch G, Musolff A, Leschik S, Oswald S, Osenbrück K, Krieg R, Schirmer M, Reinstorf F (2009) The Watershed-Sewershed approach – Experience from urban surface-groundwater system in Germany. In: Baier K, Strohschön R (eds) Proceedings of megacities – interactions between land use and water management. Mitteilungen zur Ingenieurgeologie und Hydrogeologie 99, Mainz, Aachen

Strohschön R, Wiethoff K, Baier K, Lu L, Bercht AL, Wehrhahn R, Azzam R (2013) Land use and water quality in Guangzhou, China: a survey of Ecological and Social Vulnerability in four urban units of the rapidly developing megacity. Int J Environ Res (forthcoming)

Tong STY, Chen W (2002) Modeling the relationship between land use and surface water quality. J Environ Manag 66:377–393

United Nations (UN) (2006) World urbanization prospects – the 2005 revision. United Nations, New York

US Environmental Protection Agency (2009) Drinking water contaminants. National primary drinking water regulations. List of drinking water contaminants and their MCLs. http://www.epa.gov/safewater/contaminants/index.html#listmcl. Accessed 1 July 2009

Watson GB, Bentley I (2007) Identity by design. Elsevier, Oxford

Wehrhahn R, Bercht AL (2008) Konsequenzen der Weltmarktintegration für die mega-urbane Entwicklung in China. Das Beispiel Guangzhou/Perlflussdelta. Geographie und Schule 173:19–27

Wehrhahn R, Bercht AL, Krause CL, Azzam R, Kluge F, Strohschön R, Wiethoff K, Baier K (2008) Urban restructuring and social and water-related vulnerability in mega-cities - the example of the urban village of Xincún, Guangzhou (China). Die Erde 139(3):227–249

Weng Q, Yang S (2003) An approach to evaluation of sustainability for Guangzhou's urban ecosystem. Int J Sustain Dev World Ecol 10:69–81

Wenig Q (2001) Modelling urban growth effects on surface runoff with the integration of remote sensing and GIS. Environ Manag 6:737–748

Whitehand JWR (ed) (1981) The urban landscape: historical development and management: papers by Conzen MRG. IBG special publication no 13. Academic, London

World Resources Institute (1998) World resources 1998–99. Environmental change and human health. A joint publication by the World Resources Institute, the United Nations Environment Programme, the United Nations Development Programme, and The World Bank, Washington, DC

Wu F, Xu J, Yeh AGO (2007) Urban development in post-reform china. State, market, and space. Routledge, London

www.stats.gov.cn (2009) Tong Ji Nian Jian. http://data.gzstats.gov.cn/gzstat1/chaxun/njsj.jsp (in Chinese). Accessed 19 Apr 2010

Xinhua News Agency (2006) Guangzhou to upgrade water supply system. www.china.org.cn/English/government/189120.htm. Accessed 12 Dec 2007

Xu J, Yeh AGO (2003) City profile. Cities 20(5):361–374

Xu J, Yeh AGO (2005) City repositioning and competitiveness building in regional development: new development strategies in Guangzhou, China. Int J Urban Reg Res 29(2):283–308

Yu XJ, Ng CN (2007) Spatial and temporal dynamics of urban sprawl along two urban-rural transects: a case study of Guangzhou, China. Landsc Urban Plan 79:96–109

Zhao Q, Lin H, Jiang L, Chen F, Cheng S (2009) A study of ground deformation in the Guangzhou urban area with persistent scatterer interferometry. Sensors 9:503–518

Zhu Y (2009) Guangzhou expected to add four man-made lakes. http://www.lifeofguangzhou.com/node_10/node_37/node_85/2009/01/19/123233013959031.shtml. Accessed 24 Feb 2010

Zhu Z, Deng Q, Zhou H, Ouyang T, Kuang Y, Huang N, Qiao Y (2002) Water pollution and degradation in Pearl River Delta. South China. Ambio – J Human Environ 31(3):226–230

Food Security in Dhaka: Between Global Risks and Local Vulnerabilities

5

Markus Keck, Benjamin Etzold, Hans-Georg Bohle, and Wolfgang-Peter Zingel

Abstract

Food security is a central issue in the numerous megacities of the global south. However, basic knowledge lacks about how food supply and distribution currently work in these agglomerations. Drawing on recent research in Dhaka, Bangladesh, in this contribution, the supply and distribution channels of rice are outlined. It is clarified how rice is produced, how the goods pass from the fields to the city and how they are distributed within the city. The contribution concludes by discussing the current situation of food insecurity of Dhaka's poor citizens.

Keywords

Food security • Megacity • Rice supply • Distribution • Accessibility

Megacities can be regarded as new focal points of an emerging global risk society. A main challenge in these agglomerations is feeding the urban population. Hunger is a chronic feature of life for the poor population living in marginal settlements. Nevertheless, little is known of how food supply and food distribution in megacities work. This will be examined in the present contribution, based on a case study conducted in the capital of Bangladesh.

Hardly any other city in the world has grown as rapidly during the last decades as the capital of Bangladesh. In 1951 Dhaka had a population of 336,000 (Islam 2005, p. 13), today there are roughly 14.2 million (Islam 2010, p. 6): 42 times more inhabitants than before! Thus, Dhaka has joined the ever-growing list of megacities and now stands at number nine among the largest

M. Keck (✉) • B. Etzold • H.-G. Bohle
Geography Department, University of Bonn,
Meckenheimer Allee 166, 53115, Bonn, Germany
e-mail: markus.keck@giub.uni-bonn.de;
etzold@giub.uni-bonn.de; bohle@giub.uni-bonn.de

W.-P. Zingel
Department of Economics, South Asia Institute,
University of Heidelberg, Im Neuenheimer Feld 330,
69120, Heidelberg, Germany
e-mail: h93@urz.uni-heidelberg.de

F. Kraas et al. (eds.), *Megacities: Our Global Urban Future*, International Year of Planet Earth,
DOI 10.1007/978-90-481-3417-5_5, © Springer Science+Business Media B.V. 2014

cities of the world (cf. UN 2008). Like other megacities, Dhaka is marked by extreme contrasts in close proximity, hence challenging city planners and politicians in unprecedented ways. Dhaka is the seat of government, the centre of political power, and – with most of the garment industry concentrated in and around the metropolis – the centre of one of the main industries that keep the economy running. With its vast university campus and a number of private universities, as well as museums, exhibitions and art performances, Dhaka is also the cultural heart of the country.

Despite Dhaka's centrality, however, 30–40% of the population of the city live in marginal, partly illegal, settlements and under conditions of extreme poverty (cf. World Bank 2007). Providing basic amenities like drinking water and sewerage systems for all inhabitants is beyond the capacities and will of the city's administration (cf. Siddiqui et al. 2000, 2004). As Castells (1996, p. 403) has emphasised, size alone is not the dominant risk factor in a megacity. Along with other megacities, Dhaka is experiencing the clash of a multitude of actors with very divergent interests, unequal positions in the social arena and different possibilities of acting. Life in Dhaka offers great potential for innovation and financial wealth for some of the city's inhabitants; for many others it primarily means risk and a daily fight for survival (cf. Kraas 2003, 2007; Etzold et al. 2009).

The food system of megacities has been studied very little, save a few exceptional cases (Pryer and Crook 1988; Gertel 1995; Koc et al. 1999; FAO 2001; Bohle and Adhikari 2002; Keck et al. 2008; Gertel 2010). The present contribution aims at showing how the supply of Dhaka with rice and its distribution within the city functions. How much food is necessary to meet the daily demand of the population in a megacity like Dhaka? Where do all the food supplies come from? What is the structure of the supply chain and how is it functioning? Which city dwellers are most vulnerable in terms of insecurity of food supply? And how is the access to food negotiated from day to day?

5.1 Dimensions of a Megacity: Demand for Food

After its official founding in 1610 by *Islam Khan Chishti*, Dhaka prospered as the capital of Bengal Province within the Mughal Empire. Following the annexation of Bengal by the British in the eighteenth century, however, it lost its importance and decreased in size, while Calcutta became the seat of colonial power. The number of inhabitants fell from 200,000 (some sources even speak of 900,000) to only 40,000 in the early nineteenth century; the first population census in 1872 counted 70,000 inhabitants. With the partition of Bengal in 1905, Dhaka became the capital of the new province of East Bengal. This enhanced status was short-lived, as the British soon reunited Bengal; the whole region suffered from the transfer of power to Delhi in 1911.

As a regional centre of trade and education (Dhaka University was founded in 1921), Dhaka had slowly reached a population of around 250,000 by the time colonial rule ended in 1947. It became the capital of the Pakistani province of East Bengal and later of East Pakistan. The city expanded rapidly, the number of its inhabitants increased sixfold during Pakistani times. After Bangladesh had gained independence in 1971, growth accelerated with annual growth rates reaching 10% a year in the mid 1970s (Siddiqui et al. 2000, pp. 2ff; Islam 2005, pp. 6ff.). Today, more than 14 million people live in Dhaka (Islam 2010).

Without a steady stream of food supplies from outside, life in the megacity would not be possible. Gigantic quantities of cereals, fish, vegetables, meat, fruit and spice, amounting to more than 9,000 tons, have to be brought into Dhaka City (cf. Fig. 5.2) every day (Keck 2012, p. 130). On average the citizens of Dhaka meet their caloric demand to 65% by eating cereals; with 378 g per person and day, cereals (i.e. mostly rice) are the most important staple foods for Dhaka's and Bangladesh's population (cf. BBS 2007). In this way the old Bengali proverb still holds true that the people in Bangladesh were made of rice and fish (*Machhe-Bhat-Bengali*). Even today a good meal in Dhaka has to consist of rice with lentil sauce (*dhal*) and – for all who can afford it – fish or meat.

Every year the people of Dhaka consume almost two million tons of rice (BBS 2007). Moving these quantities into and within the city is a considerable logistical challenge. The transport into the city requires a large number of trucks; if they were put in a queue, it would extend for 1,600 km. The rapid growth of the city's population must also be taken into account. At the present growth rate of 3.5% (World Bank 2007, p. 22), around half a million people must be added every year. This means that an additional Dhaka the size of city in 1950 has to be supplied with the respective quantities of rice every year.

Because of the ongoing rural-urban migration this trend will not change too fast. Important reasons for migration to the cities are the lack of employment in rural areas and the loss of arable land because of floods and erosion. The migrants dream of finding a job in the "city of hopes" and of getting access to basic public services, especially to education and healthcare (ibid., p. 23).

5.2 Rice Production and Import in Bangladesh

Although 11% of Dhaka's working population are still employed in agriculture (ibid., p. 14), food is mainly produced outside the city. The country's rice granaries are located in the northwest and north, in regions such as Rajshahi, Bogra, Dinajpur, Mymensingh and Rangpur (see Fig. 5.1). There, numerous small-scale farmers (*krishok*) with cultivation areas of less than 2.5 acres (1 ha) each grow rice with up to three harvests a year. The seedlings of *aman* are usually broadcasted in March and April, transplanted during monsoon in June and July and harvested in November and December. Winter rice, *boro*, is planted in December and January and harvested in May, just before the beginning of the monsoon rains. *Aus* is planted in March and April and harvested in July and August (Ahmed 2001, p. 2; Dorosh et al. 2004, p. 14). In the fiscal year 2006/2007, 55% of the total rice production stemmed from the *boro* season, whereas *aus* lost its importance and plays only a marginal role now with hardly 6% (FPMU 2009, p. 1).

After being harvested, rice starts its journey along the supply chain. The peasants thresh the sheaves on the farm and sell the paddy (*dhan*) to middlemen. Grain merchants purchase larger quantities of rice, transport them to the markets in district towns (*mokam*) and sell them to millers. In the rice mills, the paddy is preboiled. After the grain has been dried in the sun on large concrete yards (*chatals*), it is milled. First, the husk (*tush*) is separated from the grain; in a second step, the brown membranes (*kura*) around the grain are peeled off. Whereas the husk is used as fuel for boiling the paddy, the thin membranes are fed to the livestock. After this two-day procedure, the rice (*cal*) is put into 50–85-kg sacks, loaded onto trucks and transported to Dhaka. In the course of a day, dozens of grocers, hawkers and restaurant owners roam Dhaka's markets and bargain over prices. Once the price is fixed, the rice sacks are loaded onto rickshaws, push carts and pickups to be delivered to the kitchen markets, retail shops, canteens and street food stalls all over Dhaka. Here, the journey of the grain ends, after the cooked and prepared rice (*bhat*) is consumed by the city dwellers.

Bangladesh has been a net importer of rice throughout the last decades. Since around 1999, 1 year after severe flooding devastated the country, the rice production has been able to, at least in theory, meet the national demand. New technologies such as high-yielding varieties and mechanised irrigation systems, an improved infrastructure in terms of roads, telecommunications and electricity networks and market development through new financial service providers helped to make Bangladesh more or less self-sufficient in rice production (Dorosh et al. 2004, p. 13f.). However, production is still volatile. Extreme weather events, such as floods and cyclones, both devastating Bangladesh in 2007, are likely to cause severe crop losses.

The government of Bangladesh tries to counteract the effects of poor harvests by maintaining

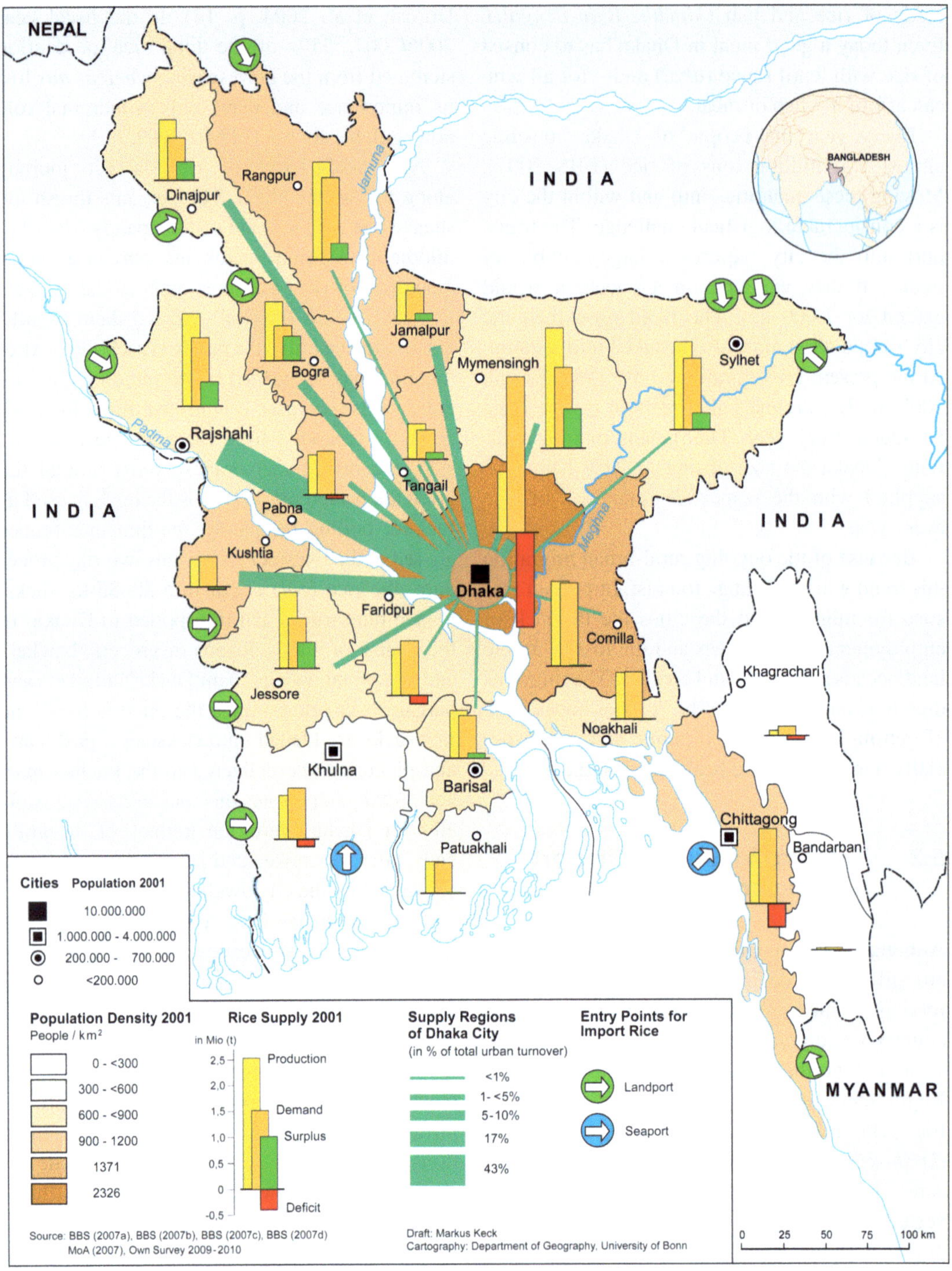

Fig. 5.1 Rice production in Bangladesh and supply of the Megacity Dhaka

public grain reserves, importing rice and wheat flour and selling food grain at subsidised rates, often supported by foreign food-aid programmes. Since the beginning of market liberalisation in 1994, private food imports have started to play an important role (Zingel 2006). Bangladesh's market liberalisation coincided with India's removal of its export restrictions. Once markets had been opened, India started to dominate the rice imports to Bangladesh, thus becoming more important than Thailand (Dorosh and Murshid 2004, p. 109). Rice imports increased from 233,300 tons on average during the 10 years from 1984 to 1993 to 838,100 tons during the 10 years from 1994 to 2003 (FAO 2008a).

Buffer stocks, government procurement at home and abroad and opening the borders even for private traders helped to protect the urban consumer against minor production shortfalls. The sudden rise of world market prices, however, limited the government's ability to come to the rescue of the urban consumers. In the second half of 2007, the world market price of rice doubled within a few months. As the global market for rice is much smaller than that for wheat or maize, small cutbacks of the leading rice-exporting countries Thailand, India, Vietnam, USA, Pakistan and China led to significantly increasing world rice prices (FAO 2008b, p. 7). Experts identified several factors as being responsible for rising prices such as the increasing world energy prices (of oil and gas) that raised the prices of major agricultural inputs, like fertiliser and water (via diesel for pumps and tractors), an increasing demand for rice in China and India, more land used to grow fuel crops, the weak US dollar and massive price speculation in agricultural commodities. As a consequence of the beginning world economic crisis, food prices went down again from mid-2008 onwards. These events have shown how vulnerable Bangladesh still is – and even more so the urban poor – as long as the country has to rely on the world market in times of production shortfalls (Zingel et al. 2011, p. 305).

5.3 Entry Points and Main Markets for Rice in Dhaka

Dhaka is not a junction of the national railroad network, and even inland waterways play a marginal role as far as the transport of rice is concerned. At least 80% of the rice consumed in Dhaka reaches the megacity by truck. Aged vehicles and poorly maintained roads make transportation risky; long waiting times at ferries and bridges add to the inefficiency of the system and, thus, to the consumer's costs.

Dhaka's wholesale markets are mainly located along three entry axes (see Fig. 5.2): Rice from northern regions arrives via the Dhaka-Mymensingh Highway and enters the capital at Tongi; rice from the northwest as well as imported rice from India is transported on the Dhaka-Aricha Highway and enters the capital at Gabtuli; at Jatrabari, rice coming from Bangladesh's major sea port, located in the southeast of the country, enters the city by the Dhaka-Chittagong Highway. Nearly 50% of the quantity of rice consumed in the megacity is traded at the seven big rice wholesale markets, that is, Babubazaar/Badamtuli, Krishi Market, Kochukhet, Mirpur 1, Jatrabari North, Malibag and Mirpur 11 (see Fig. 5.2). In order to avoid even more traffic congestion during daytime, trucks are only allowed to enter the city at night, between 8:00 p.m. and 6:00 a.m.; thus, the megacity's stocks are replenished at night (see Text Box 5.1).

Modern means of telecommunication connect Dhaka's wholesalers (*arotdar*) to the price mechanisms of the world market and help integrating the market; today, price information can be gathered from all over the country within a few minutes by mobile telephone. Wholesalers respond to harvests, price changes and interruptions in transport immediately, thus, securing the megacity's rice supply (and realising profit at the same time). Our investigations revealed that rice and fish wholesalers in Dhaka operate only a small number of exchanges via formal market ties, while the majority of transactions is organized through informal business relations. These latter relations are characterized by the sharing of more detailed information and a higher degree of mutual knowledge among business partners. This leads gradually to an ever growing versatility of the relationship, which brings up flexibility, options for action, and an atmosphere of mutual trust. This trust, in turn, leads to joint problem solving activities and mutual support among the traders and their suppliers in times of adversity and crisis. Even though short-term efforts to bail

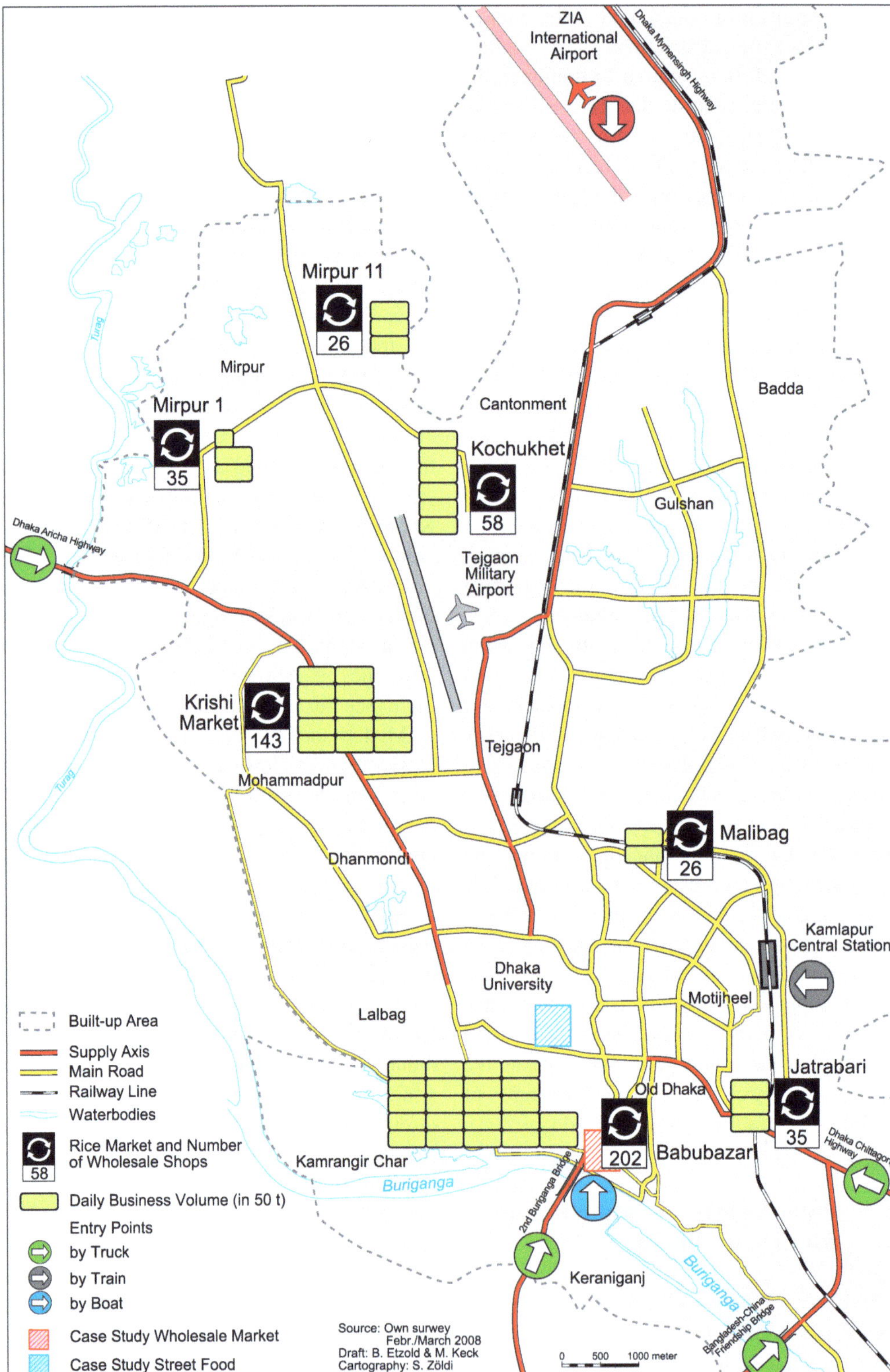

Fig. 5.2 Rice wholesale markets in Dhaka and daily turnover

Text Box 5.1 Babubazaar. Everyday Business at Dhaka's Biggest Rice Wholesale Market

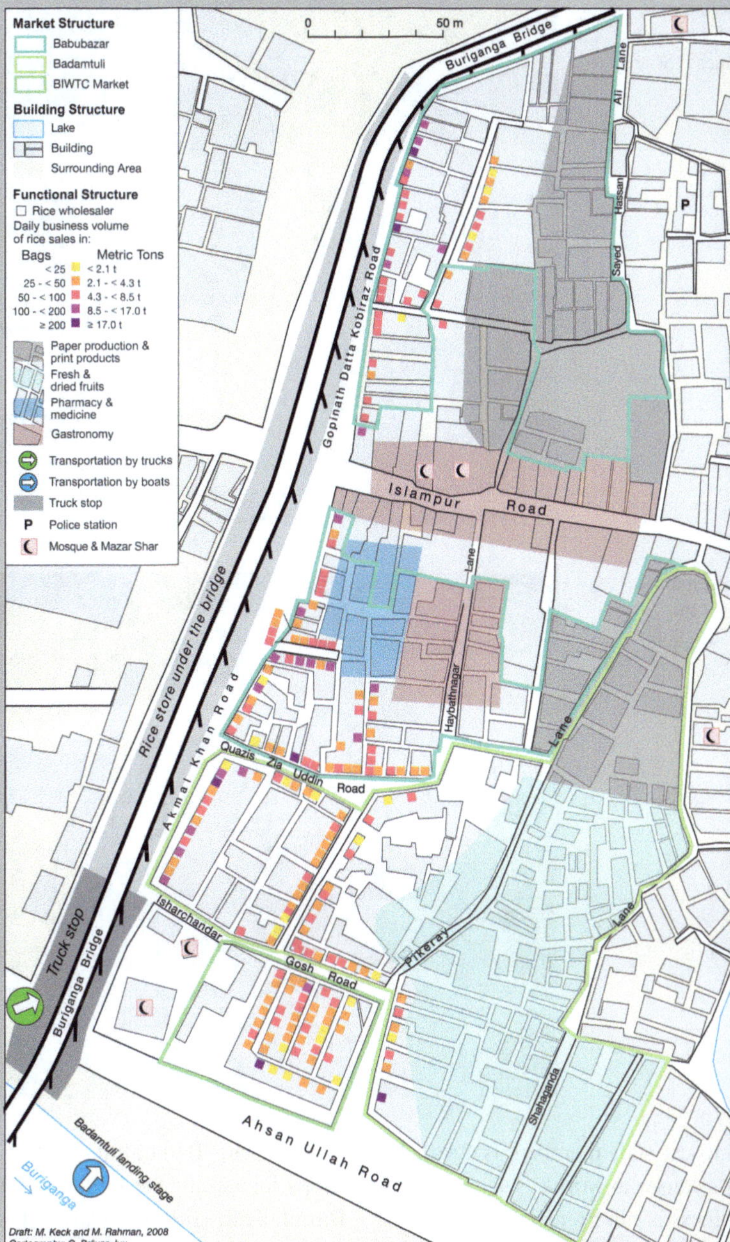

Fig. 5.3 Babubazaar, the largest wholesale market for rice in Dhaka

Mohammad Shaiful Khan has been in the rice business for 26 years; 8 years ago, he opened up his own shop. His father also dealt in rice. Shaiful is proud to be one of more than 200 wholesalers at Babubazaar, the biggest, oldest and most important rice market of the capital (see Fig. 5.3).

(continued)

Text Box 5.1 (continued)

Photo 5.1 Mohammad Shaiful Khan and his rice wholesale business (Source: Markus Keck)

Photo 5.2 Informal rice warehouse under the Buriganga bridge (Source: Markus Keck)

One of his closest staff members is Iqbal, his elder brother's son, who works as shop accountant. The business is run by them together. In a couple of years, Iqbal is going to become the manager (see Photo 5.1).

During the day, Shaiful stays in his tiny showroom, where he displays samples of the varieties he has to offer. His storage is not at the same place. After the 2nd Buriganga Bridge was built in 2001, the merchants of Babubazaar, Badamtuli Market and of the shops located in the building complex of the Bangladesh Inland Water Transportation Corporation (BIWTC) reached an agreement with local authorities to use the space under the bridge as a warehouse (see Photo 5.2). Shaiful is happy to have access to one of the plots, even though he is aware of the fact that Dhaka City Corporation can evict him from this informal warehouse at any time.

(continued)

Text Box 5.1 (continued)

Photo 5.3 Restocking of rice warehouse at Babubazaar at night (Source: Markus Keck)

Every morning at around 8:00 a.m., he opens his shop and welcomes his customers, who usually come before lunch. The majority of them are floating customers; even though with some of them long-term business relationships exist. In the afternoon the number of customers decreases, so that Iqbal has time to do the bookkeeping from 4:00 p.m. onwards. He documents the number of rice sacks, the varieties and prices per kg and works out which varieties of rice are selling well and which are not. Later, he contacts his suppliers in the countryside to place orders on the basis of these notes.

Two permanently employed workers are responsible for carrying the heavy rice sacks weighing up to 85 kg from the storage to the showroom and to the customers' vehicles. In contrast to Shaiful, their work is not finished at 6:00 p.m. After dinner, they both stay in the showroom the whole night, sleeping on the ground while waiting for the shelves to be restocked (see Photo 5.3). Around 10:00 p.m. the first trucks arrive at the market with the rice sacks that were ordered the day before. Sometimes the workers have to wait until 2:00 a.m. until they can start unloading. Then Babubazaar awakes for the second time that the day. Trucks queue up and dozens of day labourers hope to find work.

Usually the permanent labourers employ day labourers informally in order to operate the unloading smoothly. For a rate of 0.04 Euro per sack, the workers carry the heavy load on their heads to the store. Each of them is given a small wooden stick at the truck's loading platform which he drops in front of the shop when unloading the sack. Thus, the supplier guarantees that every wholesaler receives the number of sacks he has ordered. These informal workers with low salaries, no legal contracts and no coverage of expenses in case of illness or injury – in other words, the most vulnerable city dwellers – have to be considered as the fundamental pillar of Dhaka's entire food system. Under enormous physical strain, they literally carry the rice supply of the megacity on their shoulders.

a partner out of his misery are usually asymmetric, eventually, they cause both business partners stable and long-term returns. As such, informal business networks add to the merchants' adaptive capacities and help Dhaka's food supply to be guaranteed (Keck et al. 2012, p. 55).

5.4 Urban Poverty and Access to Food in Dhaka

While the supply of major food staples is guaranteed in Dhaka, it must be emphasised that the food security of the growing urban population largely depends on how these people are integrated in the labour market. How people get access to food is crucial, because only people with a regular income can afford to buy food in sufficient quantity and quality, as long as food prices do not rise (c.f. Bohle and Adhikari 2002). The majority of the population can actually find work in Dhaka, although very often only under highly precarious working conditions. In 2000, about 84% of men and 33% of women in working age – that is, older than 10 years in Bangladesh – were integrated in the megacity's labour market (World Bank 2007, p. 13). The majority of these workers, that is, 69%, are employed in the service sector, in particular in retail, in transport and in catering. The numerous garment factories and the manufacturing industry employ about 20% of the workforce. Official sources state that 49% of all jobs are classified as informal (ibid., p.14).

Thus, the crucial role of the so-called informal economy in Dhaka must be emphasised again and again. Whole economic sectors of the megacity, such as food trading and food distribution, are largely organised informally and rest almost completely (and literally) on the shoulders of day labourers, employees without secure working contracts and unregistered street vendors (cf. Etzold et al. 2009). Above all, the urban poor who work in the so-called 3D jobs (dirty, dangerous and demanding) are subject to poor working conditions and severe health risks. Working as a waste collector, as a labourer in the plastic recycling and processing industry (Kulke and Staffeld 2009), as a rickshaw puller (Begum and Sen 2004) or as a day labourer on construction sites (Daniels 2004) or on Dhaka's wholesale food markets is extremely demanding, both physically and psychologically.

Furthermore, the majority of these workers are highly vulnerable to food insecurity. The poorest households have to spend the largest portion of their income on food. According to the poverty line that has been drawn on the basis of food consumption, 28% of the population of Dhaka are poor, while 12% are considered to be extremely poor (World Bank 2007, p. 3). The poorest quintile of the megacity's inhabitants spends 62% of their income on food, on average. In stark contrast, the richest quintile only needs to spend 32% of its income for a sufficient supply of food (ibid., p.9). People with casual income or no income at all can afford only two full rice meals a day. A highly vulnerable minority of old or disabled persons, divorced or widowed women or street children have to live on only one meal per day. According to estimations, about half of the population living in slums is chronically undernourished (c.f. Pryer 2003).

5.5 Food on the Streets

In Dhaka most people do not go home for their lunch due to the high costs in terms of time and transport. They therefore depend on food being directly delivered to their workplace. As the great majority of the population prefers to eat a good, home-prepared meal, there is a large demand for supply services – just as in other Asian megacities. Shortly before lunchtime, thousands of rickshaw pullers and carriers deliver food to their customers' workplaces in metal carriers (tiffins) (see Photo 5.4). Thus, the working husband can enjoy fresh food that was prepared by his wife. Depending on the distance, this service only costs 1–3 Euros per month.

For those who cannot use these services, there are numerous small restaurants and canteens that offer rice meals in different price categories. However, considering the great demand, the number of food stalls, where good-quality food is available at a decent price, is rather small. Street vendors who sell rice meals, snacks, bakery products, fresh fruit, drinks and sweets often fill this niche.

Small tea and food stalls on pavements and more mobile vendors selling from baskets, pushcarts or rickshaws at the roadside are a common sight in Dhaka. This ready-to-eat "street food" (Tinker 1997, p. 15) is available everywhere, and the poorer parts of the population can also afford it.

Photo 5.4 A rickshaw puller brings food to his customers (Source: Benjamin Etzold)

Therefore, street food plays an important role for food security in Dhaka. In particular, physically hard-working day labourers and rickshaw pullers depend on small but nutritious snacks in between. Consuming fatty snacks, white bread, biscuits, bananas and sweet tea and chewing betel leaves drive out hunger during the day. Despite being tempted by light snacks such as pastries filled with meat or vegetables (*singhara*, *samosa*) or taco shells filled with chickpea-potato mash, cucumber, onion and chillies (*fuchka*, *chatpoti*), the middle class and the elites of Dhaka usually refrain from eating food from the streets. In particular, full rice meals sold outside are considered to be unhygienic and unhealthy.

Selling street food is an important self-employment opportunity for the urban poor. The street vendors make up a significant share of Dhaka's informal economy. With their flexibility they contribute in a perfect way to the mega-city's food supply (c.f. Etzold et al. 2009). However, in contrast to many other developing countries where the street food trade is largely organised and maintained by women (c.f. Tinker 1997; Nirathron 2006), the majority of street food vendors in Dhaka are male. This is mainly due to traditional Muslim values, according to which women should not appear in public (*purdah*). Nevertheless, women play a crucial role in securing the city's food supply, because they cook the meals at home, and also often help to prepare the snacks that are sold by the men on the street.

Street food vendors earn between one and three Euros per day, depending on the products sold, the vending site and the number of customers. This is quite a substantial income compared to the wages of factory workers, day labourers and rickshaw pullers, but the vendors face considerable risks every day. Since their food stalls are put up illegally and since there are no licenses for food hawkers, they sometimes have to deal with being evicted from their vending sites and their equipment being confiscated by the police or by security officers working for companies or public institutions (see Text Box 5.2).

How the street food trade is organised and controlled can be demonstrated vividly using the example of street food vending in front of Dhaka's University Hospital (see Fig. 5.4). In the course of the day, the number and type of vendors as well as their products vary significantly. In the early morning hours, mobile and semi-mobile vendors sell tea, rice cakes (*pitha*), boiled eggs as

Text Box 5.2 Rice Cakes and Restrictions – A Street Food Vendor's Daily Struggle for Survival

Photo 5.5 Mohammad Selim, Nazira and Shazia and their street food shop (Source: Benjamin Etzold)

Mohammod Selim is 25 years old. Together with his mother Nazira and his sister Shazia, he operates two small street food shops on the campus of Dhaka University (see Photo 5.5). After the death of his father, Selim took over the street food business when he was only 13 years old. For 5 Eurocent a piece, they sell *pitha*, small rice cakes that can be served either sweet or spicy. This snack is particularly popular with students, university and hospital staff as well as with rickshaw pullers and taxicab drivers.

Selling *pithas* from the early morning to late at night, Selim has a business volume of approximatly 5 Euro per day. Even though his family manages to make ends meet, Selim stated that it was getting more and more difficult to secure their livelihood. Regularly, the police evict the street food vendors as the university and hospital administrations disapprove of their activities. They argue that food sold on the street is illegal, unhygienic and undesirable. Since the military-backed caretaker government took power in Bangladesh in January 2007, the police have taken more rigorous action against the hawkers on Dhaka's streets. Once, the police confiscated Selim's stove, his pots and a water drum four times within 10 days. Because of these events, he lost about 80 Euros. In order to continue with his business, he used up his complete savings, thus diminishing his meagre resources even more. When his elder sister got married 5 years ago, he also had to take up a high loan from a money lender to pay the dowry. Since then, his family has been living close to their vending site in a makeshift shelter made of plastic sheets and rags. Every further police raid gets him deeper into debt, thus aggravating his struggle for survival. But making ends meet is difficult enough even without these police interventions: extreme weather events such as heat waves in the early summer or heavy rain in the monsoon season can have serious health consequences. If Selim falls ill, he will not be able to sell his rice cakes, and the only source of income for his family will collapse.

The rapid increase in food prices starting in 2007, in particular for rice, edible oil, vegetables, spices, meat and fish, became an additional burden. In early 2008, the price for rice of medium quality almost doubled from 25 to 50 Eurocent per kg within a few months. Since then, Selim's family can hardly afford to buy vegetables, meat or fish for their meals.

(continued)

Text Box 5.2 (continued)

On some days, when they earn far too little, they hardly manage to cook one full meal per day. But while Selim has to cope somehow with the increased food prices, he is not inactive when it comes to evictions by the police. Together with some other street vendors at his site, he is in contact with an informant who warns the hawkers about upcoming police raids, of course only for a little tip. Furthermore, like other vendors, Selim has become more flexible. Instead of using his large table, he takes only the really necessary equipment to the site and sells from the ground. In the case of a police raid, he is now able to leave his vending site quickly and without any further losses.

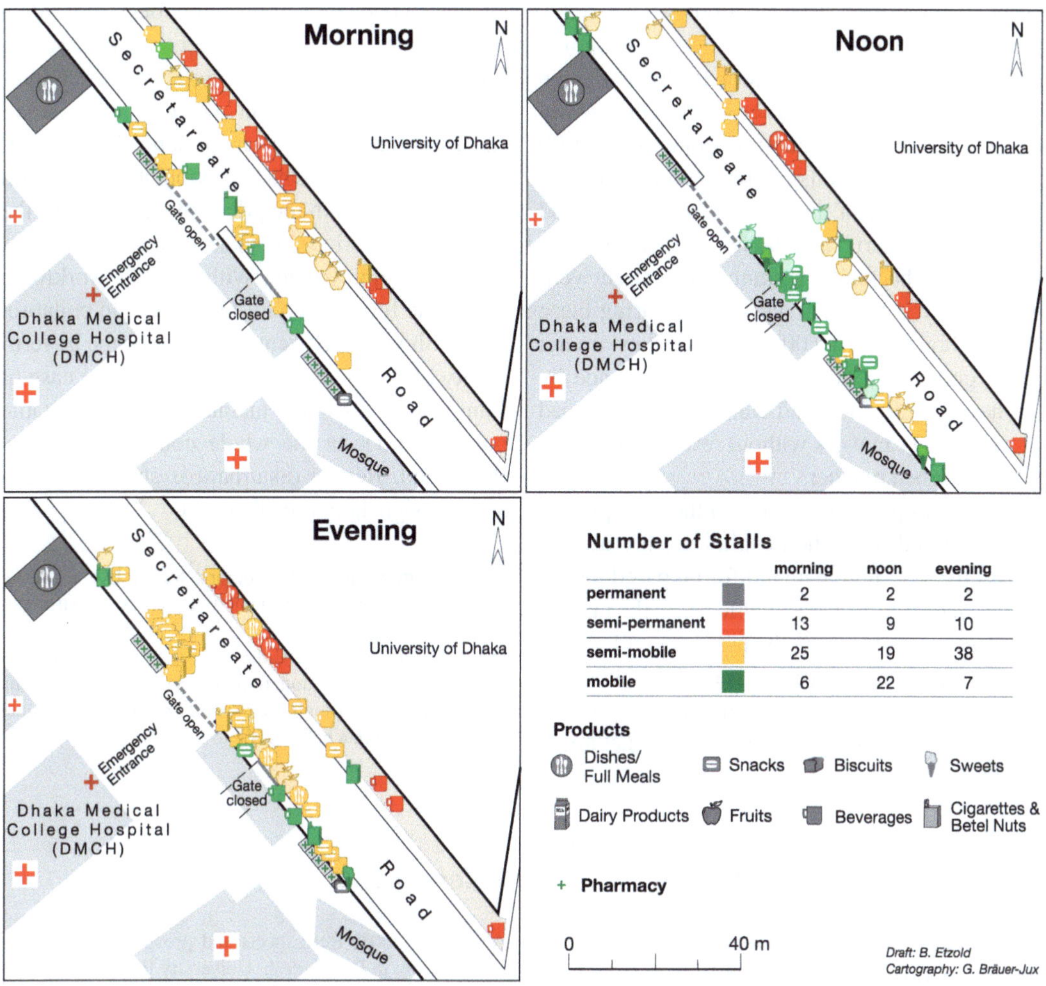

Fig. 5.4 Mobility patterns of street food vendors in front of Dhaka's Medical College Hospital

well as cigarettes and betel from pushcarts and small tables right in front of the hospital's main entrance, where all the hospital's visitors, and thus potential customers, pass by. Shortly before the hospital opens, the vendors are chased away from the gate by the hospital's security guards. From 9:00 a.m. to 5:00 p.m., only mobile vendors can do business there, as they are able to run away quickly with their light trays, baskets and flasks, if necessary.

On the other side of the road, the situation is quite different. There, the small food stalls and tea shops are not on the premises of the hospital, and thus their business cannot be disturbed by the hospital authorities. But evictions of the semi-permanent vendors from these illegally occupied places are quite common, too. In this case, the police clear the site by order of the Dhaka City Corporation, the city's administration. After 5:00 p.m., when the hospital director has left his office, the street is back in the hands of the street food vendors. Pushcarts and tables are installed in front of the hospital's entrance. The street vendors make most of their daily profit during these evening hours until 10:00 p.m. when they are not under constant control/observation/vigilance by the hospital direction and can work undisturbed or when they can work without restrictions and in an undisturbed manner.

As the hospital's canteen and the only permanent food stall near the hospital are not able to satisfy the very high demand for prepared meals, snacks and fruits at a reasonable price, the numerous street vendors secure the provision of food for the hospital staff, its patients and their visitors; but obviously neither the hospital administration nor the Dhaka City Corporation want to acknowledge this fact.

5.6 Summary and Outlook

The United Nations estimates that more than 614 million people will live in 59 agglomerations with more than five million inhabitants each by the year 2015. More than two-thirds of these megacities can be found in the countries of the global south (UN 2008, p. 220). In each of these cities, a sufficient supply of food and an affordable access to it have to be secured. As the example of rice supply in Dhaka shows, a great number of actors are engaged in the chains of production and trade and in the intra-urban distribution system. Global factors like the sudden and unprecedented price increase during the years 2007 and 2008 are as important as national factors like the quality of the transport and communication systems. The urban poor and other marginalised groups are also put at risk by local security forces like the police and other security guards who exert property rights directly and indirectly.

The distribution of food within the city of Dhaka is characterised by a high degree of informal organisation. Day labourers are responsible for refilling the rice stocks in the wholesale markets; non-licensed street vendors offer small snacks to the megacity's inhabitants. With a continuing growth of Dhaka's population and with further increases in the prices of staple foods, the city administration has to reconsider the role of informal employment. Without this workforce the whole food supply of Dhaka would collapse. And without a sufficient food supply, the governability of the megacity Dhaka, and in the case of Bangladesh also the maintenance of a stable political order in the whole country, would be threatened. Recent disturbances should be seen as clear signals in this direction.

Acknowledgments The authors would like to thank Prof. Dr. Shafique uz Zaman, Economics Department, Dhaka University, as well as our research assistants in Dhaka, Heidelberg and Bonn for their support, and the German Research Foundation (DFG) for financing this research project through their special priority programme "Megacities-Megachallenges".

References

Ahmed R (2001) Retrospects and prospects of the rice-economy in Bangladesh. University Press, Dhaka

BBS (2007) Report of the household income and expenditure survey 2005. Bangladesh Bureau of Statistics, Dhaka

Begum S, Sen B (2004) Unsustainable livelihoods, health shocks and urban chronic poverty. Rickshaw pullers as a case study. ICPRC working paper 46. Dhaka

Bohle HG, Adhikari J (2002) The metropolitan food system of Kathmandu – conceptual considerations and empirical evidence. Die Erde 133(4):401–421
Castells M (1996) The rise of the network society, The information age vol I: economy, society and culture. Blackwell, Malden/Oxford
Daniels PW (2004) Urban challenges: the formal and informal economies in mega-cities. Cities 21(6):501–511
Dorosh P, Murshid KAS (2004) Trade liberalization and national food security: rice trade between Bangladesh and India. In: Dorosh P et al (eds) The 1998 floods and beyond. Towards comprehensive food security in Bangladesh. University Press, Dhaka, pp 101–126
Dorosh P et al (2004) Overview of the Bangladesh foodgrain. In: Dorosh P et al (eds) The 1998 floods and beyond. Towards comprehensive food security in Bangladesh. University Press, Dhaka, pp 13–52
Etzold B, Keck M, Bohle HG, Zingel WP (2009) Informality as agency. Negotiating food security in Dhaka. Die Erde 140(1):3–24
FAO (2001) Feeding Asian cities. In: Proceedings of the regional seminar 2001. Food and Agriculture Organization of the United Nations, Rome www.fao.org/DOCREP/003/X6982E/X6982E00.HTM
FAO (2008a) Food balance 2008. Food and Agriculture Organization of the United Nations, Rome. www.faostat.fao.org
FAO (2008b) Rice market monitor 2008. Food and Agriculture Organization of the United Nations, Rome. www.fao.org/es/esc/en/15/70/highlight_71.html
FPMU (2009) Bangladesh food situation report 72, January–March 2008. Food Planning and Monitoring Unit, Ministry of Food and Disaster Management, Dhaka
Gertel J (ed) (1995) The metropolitan food system of Cairo. Freiburg Studies in Development Geography, Saarbrücken
Gertel J (2010) Globalisierte Nahrungskrisen. Bruchzone Kairo. Transcript, Bielefeld
Islam N (2005) Dhaka now: contemporary urban development Dhaka. University Press, Dhaka
Islam N (2010) Dhaka in 2050. CUS Bulletin on Urbanization and Development 58–59:5–10
Keck M (2012) Market governance and social resilience. The organization of food wholesaling in Dhaka, Bangladesh. Unpublished dissertation thesis, University of Bonn
Keck M, Etzold B, Bohle HG, Zingel WP (2008) Reis für die Megacity. Nahrungsversorgung von Dhaka zwischen globalen Risiken und lokalen Verwundbarkeiten. Geographische Rundschau 60(11):28–37
Keck M, Bohle HG, Zingel WP (2012) Dealing with insecurity. Informal business relations and risk governance among food wholesalers in Dhaka, Bangladesh. Zeitschrift für Wirtschaftsgeographie 56(1+2):43–57
Koc M et al (1999) For hunger-proof cities. Sustainable Urban Food Systems, Ottawa
Kraas F (2003) Megacities as global risk areas. Petermanns Geographische Mitteilungen 147(4):6–15
Kraas F (2007) Megacities and global change in East, Southeast and South Asia. Asien 103(2):9–22
Kulke E, Staffeld R (2009) Informal production systems – the role of the informal economy in the plastic recycling and processing industry in Dhaka. Die Erde 140(1):25–43
Nirathron N (2006) Fighting poverty from the street: a survey of street food vendors in Bangkok. International Labour Office, Bangkok
Pryer J (2003) Poverty and vulnerability in Dhaka slums: the urban livelihood study. Dhaka
Pryer J, Crook N (1988) Cities of hunger: urban malnutrition in developing countries. Oxfam, Oxford
Siddiqui K et al (2000) Overcoming the governance crisis in Dhaka City. University Press, Dhaka
Siddiqui K et al (2004) Megacity governance in South Asia: a comparative study. University Press, Dhaka
Tinker I (1997) Street foods: urban food and employment in developing countries. Oxford University Press, New York/Oxford
UN (2008) World urbanization prospects. The 2007 revision. United Nations, Department of Economic and Social Affairs, Population Division. http://www.un.org/esa/population/publications/wup2007/2007WUP_Highlights_web.pdf
World Bank (2007). Improving living conditions of the urban poor. The World Bank Office, Dhaka. http://siteresources.worldbank.org/BANGLADESHEXTN/Resources/295759-1182963268987/dhakaurbanreport.pdf
Zingel WP (2006) Food security in South Asia. In: Ehlers E, Krafft T (eds) Earth system science in the anthropocene: emerging issues and problems. Springer, Heidelberg, pp 229–246
Zingel WP, Keck M, Etzold B, Bohle HG (2011) Urban food security and health status of the poor in Dhaka, Bangladesh. In: Krämer A, Khan MMH, Kraas F (eds) Health in megacities and urban areas. Springer, Heidelberg, pp 301–319

6 Land Development Strategies in Megacities: Guiding Land Use and Land Rights in the Context of Urban Sprawl and Informality

Babette Wehrmann

Abstract

In terms of land use and land rights, megacities are characterised by uncontrolled, informal developments that seem to be outside of the control of public management. After analysing the situation, it has become obvious that at the international level, there are sufficient methods and tools that could be applied to guide land use and ownership. What are lacking, however, are ideas, strategies, and options on how to deal with power structures and how to negotiate with powerful actors who – at least partly – act illegally and possess a lot of power.

Keywords

Land use • Land-use planning • Land tenure • Tenure security • Intermediate tenure • Informality • Legitimacy • Corruption • Governance • Power structures • Intuition

6.1 Key Issues Linked to Current Land Use and Land Development

Megacities are expanding spatially, mostly in an uncontrolled fashion and at an unprecedented speed. Within the already built-up area, agglomerations are becoming increasingly fragmented due to gated communities and the resulting patchwork developments and physical barriers within the city. At the periphery, agricultural land, hazardous areas, environmentally sensitive areas, and village lands are being converted into (peri-)urban areas – most often without any comprehensive long-term planning that facilitates a sustainable use of the land. Due to the conversion of land, generally accompanied by land transactions, land values increase. The municipalities, however, which often lack resources (e.g. for fringe land development) do not benefit from this process. The profits go to private actors (investors, land developers, real estate managers, banks, etc.).

They are also the ones – rather than the state – who mainly influence and control urban land development. Little is known about the formal and informal actors and institutions guiding and controlling land development and the dynamic

B. Wehrmann (✉)
Freelance consultant, Munich, Germany
e-mail: Babette.Wehrmann@land-net.de

DOI 10.1007/978-90-481-3417-5_6,

processes that are linked to it. It can only be assumed that urban land development is – especially fringe land development – steered (a) by informal actors who quite often "cooperate" with representatives of the formal sector (Wehrmann 1999 on Johannesburg), (b) by the population itself, whereby the rich have a greater influence than the poor, and (c) at least partly and mainly indirectly by international/global actors (investors/bank funds). Detailed knowledge of who is ruling megacities' (land) development, however, is still lacking, and it can be assumed that it is neither easy nor safe to conduct more profound investigations on this matter.

Awareness is high among experts from institutions such as the World Bank, UN Habitat, and GTZ (e.g. Farvacque and McAuslan 1992) that the existing instruments of urban (land) management, which have mainly been imported from Europe where they were developed for nineteenth/twentieth century cities, do not work in present-day megacities in developing countries. However, so far, there has only been relatively little research on this issue, and only few approaches exist which are rather piecemeal and were spontaneously developed for a given situation rather than in the context of comprehensive citywide approaches, concepts, or strategies. The limited (if at all existing) power of the public sector constitutes the main problem.

There are many case studies from all over the world dealing with the shortcomings of urban land management, mainly focusing on urban land-use planning, access to land for the poor, housing, land conflicts, politically motivated and increasingly also market-driven eviction processes, tenure insecurity, etc. These case studies mainly identify problems connected to slums or analyse informal settlement development, some of them identifying well-functioning informal approaches towards urban land management (Lopes de Souza 2003; Lupala 2002). A few studies also look at the link between formal and informal land management (Kombe and Kreibich 2000). Systematic research on the spatial and social consequences of uncontrolled urban land development includes issues such as fragmentation and social exclusion (Lopes de Souza 2001; Mertins 2003; Kundu 2000; Scholz 2002), Lopes de Souza (1995) being one of the very few looking at power structures and the influence of criminal actors (drug traffic) on the spatial development of the city.

More recently, land tenure issues (Durand-Lasserve and Royston 2002; Fernandes and Varley 1998; Payne 2002a; UN Habitat 2003, 2004a, b) and inadequate planning and building standards (Payne and Majala 2004) have become research objectives. However, none of the researchers have come up with an overall model of formal-informal urban land management. This would require the inclusion of topics such as urban power structures and formal/informal governance structures. Some research has been conducted on these issues (Stratmann 1999 on mega-urban lifestyles; Rüland 1996 on metropolitan management and politics), but they are more general and do not focus specifically on land management and settlement development.

It can therefore be summarised that there is an awareness of the fact/problem that peri-urban land management and fringe land development which mainly occur in form of urban sprawl are organised almost entirely informally. There is a common presumption that there are powerful people and powerful interests involved, but that little is known about how the underlying processes work, how efficient they are, how harmful and/or beneficial to urban development they are, etc. Further, there is a lack of a conceptual frame connecting the different topics and approaches mentioned above such as land management, tenure (in)security, network analysis, power structures, and governance.

Given the insufficient knowledge about land development dynamics and processes, it is not surprising that dealing with these issues leads to more question marks than it provides answers. Still, there are some promising approaches and a number of lessons to be learned.

6.2 Key Issues Connected to Land Tenure and Land Rights in Today's Megacities

Megacities are very diverse when it comes to the distribution of land ownership. Due to their very specific historical development and differences between the current political and economic systems in different countries, megacities are characterised by different land tenure systems and combinations of property regimes such as state land, leases on state land, private property, collectively owned property, and usage rights on land under customary ownership (common property).

In addition to various forms of formal land rights, in many megacities, there are numerous informal categories of land rights. The crucial issue concerning these informal land rights is whether they are legitimate or not. Many (so-called) informal or illegal (as opposed to statutory law) land rights are actually accepted within (local) society. Only those informal actions and rights that are not legitimate at all (neither in the context of customary rights nor as commonly accepted ways of muddling through so-called legitimate behaviour (extra-legal rules)) can be considered negative for overall development as they only serve the interests of a small group of people. An example for this is the well-organised land mafia in Johannesburg where some developers illegally developed state land, organised the occupation of land against payment, and collected monthly rents as well as protection fees (Reeves 1998).

The relevance of legitimacy: formal and informal actions and rights regarding land markets

6.3 Regulating Land Use and Securing Land Rights

Uncontrolled urban growth and urban sprawl are often seen as something terrifying. People believe that the situation has got out of control. They complain that there is no land-use planning and no guidance for future land use. It is often said that we need new forms of land-use planning for megacities because they are so big and involve so much space and so many people with different interests. It could, however, also simply that the old-fashioned master planning has never been very efficient and is outdated, not only for megacities. Even in Europe where cities are generally much smaller, we have observed a decline in urban land-use planning and regional planning since the 1980s, leaving the floor (or giving it back) to architects (Helbrecht 1991).

Considering the still ongoing search for urban development strategies (DISP 2005) and the only recent re-emergence of (hardly ever implemented) visions for urban development in Europe, mainly due to the discussions on sustainability (Becker et al. 1999), it is becoming obvious that the absence of effective land-use planning is not a particular problem of megacities. However, in the case of megacities, the consequences of ineffective land-use planning are greater because of the scope of the agglomera-

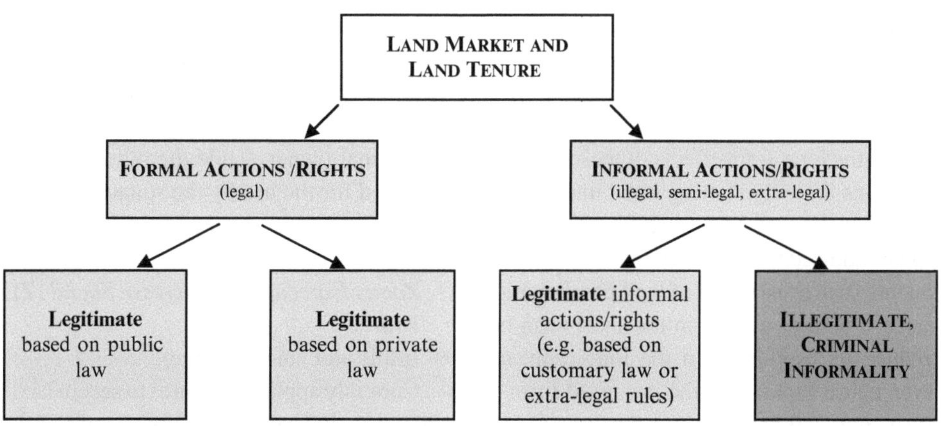

Source: Wehrmann (2005)

tion, the dynamics of change, and the (in) efficiency of the public sector, as well as a lack of transparency and accountability of the public as well as the private sectors. In megacities, the issue is probably less a lack of knowledge on and tools for land-use planning but rather a lack of power to implement measures, as informal land developers are more powerful than the public sector, and it is hardly possible to control or even influence them. What seem to be needed, therefore, are creative and negotiable deals and compromises.

The situation is slightly different when it comes to securing property rights. Only recently, a broader range of tenure options providing short- and long-term security has been identified. The spectrum ranges from political statements guaranteeing a protection from eviction and the provision of services without legal tenure, over intermediate tenure solutions, to the provision of titles, creating a registered freehold (Payne 2002b; UN Habitat 2004a). It has been agreed upon that there is a wide range of alternative options to achieve tenure security, freehold only being one of them (UN Habitat 1999). Examples are delayed freehold, registered leasehold, public rental, private rental, shared equity (a combination of delayed freehold and rental), co-operative tenure, customary ownership, religious tenure systems, and nonformal tenure systems (UN Habitat 2004b).

The question if titles matter more than rights or vice versa (Payne 2001) can be answered for the moment: secure land rights are more important. A surprisingly broad range of intermediate tenure options has been identified within the last 5 years (see below) and is now being promoted by UN Habitat (2004a, b). However, so far, there is little evidence of the long-term effects of these options, as evaluations are only starting to be undertaken. A recent evaluation on land proclamations in the Philippines showed that "intermediate land tenure instruments such as issuance of land proclamation can increase tenure security in the short-term but may reduce its usefulness in the long-term and may even cause backlash if the promised form of tenure (in the Philippines' case the issuance of individual title) will not be achieved in a reasonable time" (Antonio 2007: 150).

Although knowledge and instruments are available today, in terms of land-use planning as well as concerning the provision of secured land tenure options, neither of the above mentioned is implemented broadly. Apart from shortcomings in finance, human capacity, technology, and adequate laws and by-laws, there is a lack of coordination and cooperation between land-use planning and land administration departments and other departments. This is not a problem specific to megacities. However, megacities encounter one other problem – power and power relations. In megacities, there is a lot of speculation on future land values as well as continuous rent from land. As a consequence, land markets and land administration that are supposed to facilitate land transfers and secure property rights are subject to high levels of corruption. Finding solutions to problems linked to land-use planning and land administration in megacities therefore requires a strong focus on governance issues and explicit anticorruption actions.

Currently applied land management tools are

- Simple zoning instead of master planning.
- Copying informal land development through reverse development: This means establishing settlements directly after parcelling and then developing the neighbourhood, instead of the traditional way of settling people as the last step. This makes land development faster and cheaper and can be compared to sites-without-services projects.
- Land readjustment: In Cambodia, for example, land readjustment is done on private properties which have been occupied by squatters. As a result, part of the land can be used by the original owner, while the other part is re-parcelled for the use by the squatters.
- Creating special zones for low-cost housing, such as social concessions in Cambodia or *Zonas Especiais de Interesse Social* (ZEIS) in Brazil.
- Infills and land recycling.

Currently applied measures to secure land tenure and to supply land for all segments of society are

- The identification and promotion of intermediate land tenure options: Examples are temporary land rentals in Bangkok (private land owners who are waiting for the market price of their land to rise even higher before developing it rent the land on a short- to medium-term lease), temporary occupation licences in Kenya (on public land for an annual land rent to avoid illegal occupation), *anticredito* in Bolivia (2 years use right on private land against a deposit that enables the land owner to make investments without formal bank credits, thereby avoiding the payment of high interest rates), and land proclamations in the Philippines which assure squatters on public land that they will not be evicted and social services will be improved (UN Habitat 2004b).
- Regularisation programmes for informal settlements including legalisation, planning, services, land administration, funding, and housing (UN Habitat 2003).
- Recognising and legitimising the diversity of land-delivery systems and land development actors, including those that are informal, based on the belief that they are an asset and not a liability (UN Habitat 2003).
- Decentralising land management responsibilities to local authorities who should play a major role in land regularisation, land policy development, the establishment of land-delivery mechanisms, the development of land-use guidelines and building regulations, processing land records, provision of infrastructure, the use of land tax, the management of state land, the introduction of decentralised simple land information systems, and the discussion of a new role played by customary authorities – all these in the light of participation and inclusiveness (UN Habitat 2003).
- Good governance in land administration.

While appropriate technical solutions for securing property rights have become accessible and affordable, widespread corruption, an incoherent and inconsistent legal framework, a weak judiciary, and poor public sector management have been recognised as the key obstacles hindering change. Here is where good governance comes in. Good governance in land administration aims to protect the property rights of individuals and enterprises as well as of the state by introducing such principles as transparency, accountability, rule of law, equity, participation, and effectiveness into land-related public sector management (Zakout et al. 2006).

Experience shows that it is much easier to address petty corruption (bureaucratic corruption) than grand corruption (state capture). While few examples exist on how to improve state land management in terms of reducing corruption (e.g. through the introduction of a state land inventory), there are a number of recently applied measures to combat petty corruption, such as improved services to enhance transparency and accountability, counter offices to increase transparency, client surveys and hotlines, web-based land information systems, people's participation in demarcation and adjudication, and public displays of newly established cadastral maps for public verification before they are legally approved.

In conclusion, it can be stated that dealing with land use and land rights in megacities is quite a hot issue because of the powerful interests involved. Therefore, guiding land use in a sustainable way and securing access to land for all segments of society require an accountable public sector that is legitimate and – through this legitimacy and the support of the public – powerful enough to negotiate with local, national, and international elites, quite a number of them conducting illicit practices (some of them being drug dealers, etc.).

Future research as well as policy advice and practical support from the side of development cooperation should rather focus on unconventional approaches to negotiation, governance issues, and public-private partnership with questionable actors than on new ways of land-use planning or further innovative tenure approaches. It is not the "hard tools" that are lacking but rather the "soft tools." Maybe, what we need to proceed successfully in these days is intuition rather than additional elaborated methods and tools.

References

Antonio DR (2007) Intermediate land tenure instruments for the urban poor: concepts and realities. The case of land proclamation in the Philippines. Unpublished master thesis, Munich

Becker H et al (ed) (1999) Ohne Leitbild? Städtebau in Deutschland und Europa. Karl Krämer, Stuttgart

DISP 160 (1/2005) Urban Sprawl: Strategien und Instrumente einer nachhaltigen Flächenhaushaltspolitik. Zürich

Durand-Lasserve A, Royston L (2002) Holding their ground: secure land tenure for the urban poor in developing countries. Earthscan, London

Farvacque C, McAuslan P (1992) Reforming urban land policies and institutions in developing countries, UPM policy paper 5. The World Bank, Washington, DC

Fernandes E, Varley A (1998) Illegal cities: law and urban change in developing countries. Zed Books, London/New York

Helbrecht I (1991) Das Ende der Gestaltbarkeit? Zu Funktionswandel und Zukunftsperspektiven räumlicher Planung. Bibliotheks- und Informationssystem der Universität Oldenburg, Oldenburg

Kombe WJ, Kreibich V (2000) Informal land management in Tanzania. SPRING-Center, Dortmund

Kundu A (2000) Inequality, mobility and urbanization. China and India. Indian Council of Social Science Research and Manak Publications, New Delhi

Lopes de Souza MJ (1995) Die Fragmentierte Metropole: Der Drogenhandel und seine Territorialität in Rio de Janeiro. In: Giese E et al (eds) Geographische Zeitschrift, Stuttgart, pp 238–249

Lopes de Souza MJ (2001) Metropolitan deconcentration, socio-political fragmentation and extended suburbanization: Brazilian urbanisation in the 1980s and 1990s. Geoforum 32:437–447

Lopes de Souza MJ (2003) Alternative urban planning and management in Brazil: instructive examples for other countries in the south? In: Harrison P et al (eds) Confronting fragmentation, housing and urban development in a democratising society. University of Cape Town Press, Cape Town, pp 190–208

Lupala A (2002) Peri-urban land management for rapid urbanization: the case of Dar es Salaam. Springer, Dortmund

Mertins G (2003) Jüngere Sozialräumlich-strukturelle Transformationen in den Metropolen und Megastädten Lateinamerikas. Petermanns Geographische Mitteilungen 147(4):46–55

Payne G (2001) Urban land tenure options: titles or rights? Habitat Int 25:415–429

Payne G (ed) (2002a) Land rights and innovation: improving tenure security for the urban poor. ITDG, London

Payne G (2002b) Introduction. In: Payne G (ed) Land rights and innovation: improving tenure security for the urban poor. ITDG, London, pp 3–22

Payne G, Majala M (2004) The urban housing manual: making regulatory frameworks work for the poor. Earthscan, London

Reeves J (1998) The move from slums to suburbs. In: Saturday star, Johannesburg, 21 Feb 1998

Rüland J (ed) (1996) The dynamics of metropolitan management in Southeast Asia. Institute of Southeast Asian Studies, Singapore

Scholz F (2002) Die Theorie der "Fragmentierenden Entwicklung. Geograpische Rundschau 54(10):6–11

Stratmann B (1999) Stadtentwicklung in Globalen Zeiten. Lokale Strategien, Städtische Lebensqualität und Globalisierung. Birkhäuser, Basel

UN Habitat (1999) Global campaign for secure tenure: implementing the habitat agenda: adequate shelter for all. UN Habitat, Nairobi

UN Habitat (2003) Handbook on best practices – security of tenure and access to land. UN Habitat, Nairobi

UN Habitat (2004a) Pro poor land management – integrating slums into city planning approaches. UN Habitat, Nairobi

UN Habitat (2004b) Urban land for all. UN Habitat, Nairobi

Wehrmann B (1999) Zum Umgang mit informellen Institutionen des städtischen Bodenmanagements in Afrika – ausgehend von Beispielen in Mauretanien, Senegal und Südafrika (unpublished master thesis). Marburg

Wehrmann B (2005) Landkonflikte im urbanen und peri-urbanen Raum von Großstädten in Entwicklungsländern: mit Beispielen aus Accra und Phnom Penh. Urban and Peri-urban land conflicts in developing countries. Lit, Berlin

Zakout W, Wehrmann B, Törhönen M (2006) Good governance in land administration – principles and good practices. World Bank and FAO, Washington, DC

7 The Influence of Foreign Direct Investment on Land Use Changes and Regional Planning in Developing-World Megacities: A Bangalore Case Study

Margaret Pugh O'Mara and Karen C. Seto

Abstract

Economic reforms and trade policy since the 1980s, combined with concurrent technological changes, have opened up parts of the developing world to unprecedented levels of foreign direct investment. This infusion has transformed regional economies, cultures, political systems, and the local environment. This chapter discusses how foreign direct investment in Bangalore, India, has served not simply to fuel rapid growth in urban population and urban extent but also has strongly affected regional planning and infrastructure policy. Bangalore's and India's political history plays an instrumental role, directly or indirectly creating incentives for industry and middle-class workers to decentralise into self-contained landscapes at the urban periphery. We argue that policy and planning approaches must understand and consider the legacies of local and national policies, measure how and why private capital is reshaping urban space, and incorporate private-sector actors into sustainable development discussions.

Keywords

Globalisation • Land use • Infrastructure • Regional planning • High technology

M.P. O'Mara (✉)
Department of History, University of Washington, Box 353560, Seattle, WA, USA 98195-3560
e-mail: momara@u.washington.edu

K.C. Seto
School of Forestry and Environmental Studies, Yale University, 380 Edwards Street, New Haven, CT, USA 06511
e-mail: karen.seto@yale.edu

7.1 Introduction

The South Indian metropolis of Bangalore has become known worldwide as a hub of the "global knowledge economy", whose name has become synonymous in the international media with the outsourcing of white-collar jobs from North America and Western Europe. A major centre of

F. Kraas et al. (eds.), *Megacities: Our Global Urban Future*, International Year of Planet Earth,
DOI 10.1007/978-90-481-3417-5_7, © Springer Science+Business Media B.V. 2014

back-office operations and customer support services since the mid-1990s, Bangalore is increasingly a place where more sophisticated technology functions and management activities occur/are performed/take place.

With a population swollen by the return of educated migrants from elsewhere in India and returning Indian nationals who began their careers in Silicon Valley and other established high-tech clusters, Bangalore is in many respects an archetypal "world city" (Friedmann 1986; Sassen 2000) and "technopolis" (Hall and Markusen 1984) whose post-industrial economy exemplifies Castells' "space of flows" (1993). This essay seeks to historicise and analyse how Bangalore's ascent to the top tier of the world's technology regions has played out in urban space.

This approach reveals how Bangalore's historical growth patterns and politics laid the groundwork for current land use trends towards extreme peripheral development and piecemeal and privatised infrastructure provision. A place that has long prided itself on its reputation as a "garden city" of low-rise bungalows and abundant parks and water bodies, Bangalore provides a particularly vivid example of how the rapid influx of foreign direct investment (FDI) not only enlarges the metropolitan economy but also reshapes metropolitan form.

In this rapidly growing metropolitan area, FDI has shaped the expansion of urban areas in three ways. First, international real estate developers and property management firms have become a major presence, contributing to the development of high-rent industrial and residential facilities built in a geographically indistinguishable "global modern" style that could be found anywhere from Silicon Valley to Singapore. Second, the space, infrastructure, and human resource demands of multinational firms – particularly software industries – have required new facilities with particular amenities, like broadband Internet, uninterrupted electrical power, air conditioning and computational cooling facilities, and security. Third, the influx of FDI has changed the landscape of consumption in Bangalore, as educated workers flow into the city from extended stints in the United States and Europe, and rising worker incomes increase the demand for higher end, and often imported, consumer goods and services.

All of these changes have played out in the context of a weak regional planning infrastructure that historically has focused on the central city. The agricultural hinterland has been a distinct and separate space, governed by its own powerful brand of politics. These political realities, combined with rapidity of growth and liberality of economic regulations, allowed FDI to accelerate a process of planning privatisation and economic segregation. In Bangalore, foreign capital is, effectively, trumping public regional planning (Fig. 7.1).

7.2 Growth and Planning in Bangalore Before Market Liberalisation

Standing on the rooftop of one of Bangalore's highest buildings (13-storey tall), one surveys an urban landscape of parks and low-rise residential neighbourhoods, punctuated by large government and institutional buildings, set amid verdant parks and landscaped grounds. Far in the distance, towards the edge of the city, there are clusters of modern high-rise buildings. Beyond those, there are small farms and villages. From this height, the landscape is surprisingly bucolic and unexpectedly horizontal. If the high rises on the horizon are the "new" Bangalore of high technology and outsourcing, the parks and government buildings are evidence of the "old" Bangalore, a reflection of the "garden city", a sprawling community of gardens and bungalows that has been controlled by a single city government only since 1949. The role of public agencies in shaping Bangalore's urban growth has been conditioned by the city's long history of decentralisation, incremental development, and fragmented governance.

Situated on a plateau of nearly 1,000 m and enjoying cooler and drier weather than much of the rest of South India, the earliest city of Bangalore was a sixteenth-century fort and market settlement that developed around numerous water reservoirs or "tanks" (Srinivas 2001; Nair 2005). The climate and amenities attracted British

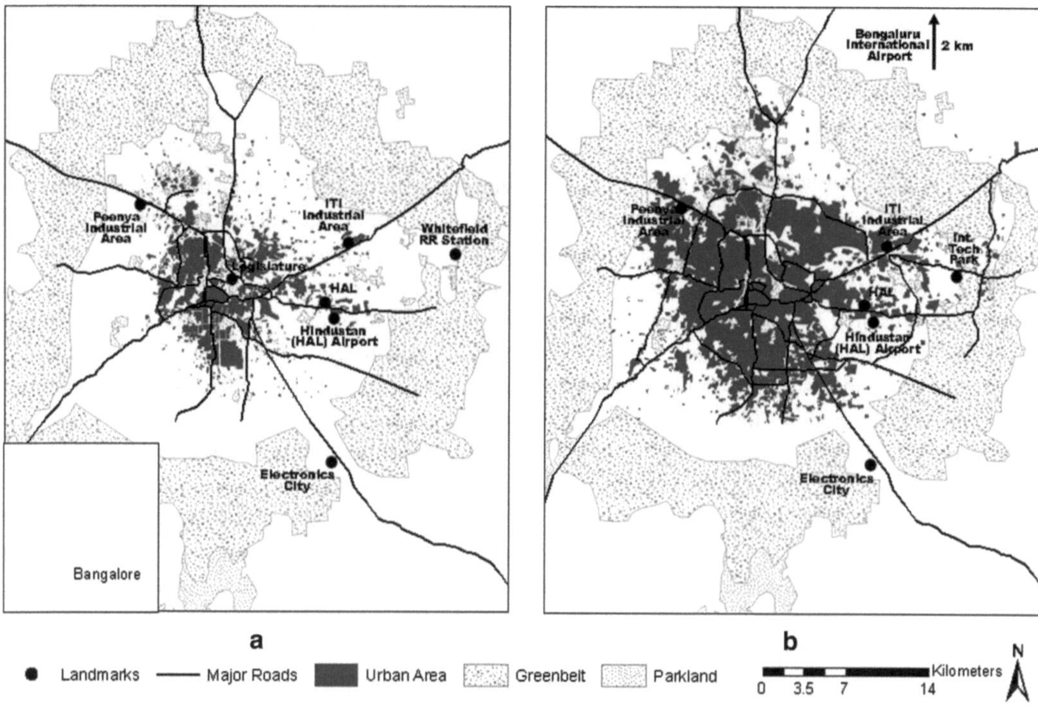

Fig. 7.1 Bangalore 1980 and 2000

colonial officials, who established a cantonment four miles away from the Bangalore city in 1809. The two settlements maintained separate town governments even after a large city park was established between the two, and new neighbourhoods developed around and beyond them.

Although the layout of the British-controlled cantonment reflected the Western embrace of the gridded street plan and a clear delineation between city and country, Bangalore as a whole grew more organically and incrementally, incorporating agricultural lands without necessarily urbanising them. Suburbs came early to Bangalore, and the city expanded through a series of suburban extensions that established new residential neighbourhoods at the periphery (Heitzman 2004; Nair 2005). By the turn of the twentieth century, Bangalore was a self-proclaimed "garden city" with large parks in the centre, broad avenues radiating out, and self-contained residential neighbourhoods at the urban edge; its 1901 population density was only 12 people per acre (Prakasa Rao 1983, p. 189).

After Indian Independence in 1947, the unification of market town and cantonment into the Bangalore City Corporation, the establishment of Bangalore as the capital of Karnataka State, and the expansion of nationalised industry in the city catalysed a series of comprehensive regional plans between the 1950s and the 1990s. While the particulars of these plans changed over the decades, they were consistent in advocating a planning approach that decentralised large industry, established satellite towns with self-sustaining infrastructure, and accommodated the growing middle class.

In the 1940s, Bangalore's population grew significantly as a result of the establishment and subsequent nationalisation of Hindustan Aeronautics Limited (HAL) and Indian Telephone Industries (ITI) in the 1940s, followed by Bharat Electronics Limited (BEL) and Hindustan Machine Tools (HMT) in the 1950s (Nair 2005, p. 127). These firms employed tens of thousands and attracted a significant middle-class population of public servants and white-collar technicians to the city. Jawaharlal Nehru proclaimed

Year	Population	Sq km	Persons/sq km
1981	2,921,751	365.7	7,991
1991	4,130,288	445.9	9,263
2001	5,686,844	531.0	10,710

Fig. 7.2 Population growth and density

Bangalore "India's city of the future" (Nehru 1958), and his government led a further push to enlarge the presence of research- and engineering-driven industry in the city and expand its higher education infrastructure and headquarter research-intensive bodies, like India's national space agency, in Bangalore (Fig. 7.2).

The establishment of formal planning entities and the arrival of large-scale industry to Bangalore both occurred at a moment when city planners worldwide were embracing ideas of comprehensive regional planning, urban renewal, population dispersal, and industrial decentralisation. The answer to overtaxed urban infrastructure, ageing housing, and congested roadways was expansion outwards, posited many planners. In Great Britain, this solution took the form of publicly planned and state-subsidised new towns; in the United States, it spurred large-scale rebuilding of central neighbourhoods, an interstate highway system, and public subsidy of real estate development at the suburban fringe (Jackson 1985; Fishman 1987; Clapson 1998).

In Bangalore, where rapid influx of people and jobs put new strains on the city's infrastructure, comprehensive plans for the city embraced this pro-dispersal ethos in a series of plans by a changing succession of public agencies. The Bangalore Development Committee Plan of 1953 proposed five satellite towns on the city's outskirts, centred upon the major industrial employers and featuring worker housing. The Outline Development Plan for the Bangalore Region 10 years later echoed this decentralisation scheme.

The Bangalore Development Authority was established in 1976 with the mandate of developing the city, planning for infrastructure, and servicing the needs of the population. The Comprehensive Development Plan of 1976 reflected a growing worry that growth needed to be controlled and industry contained. It proposed satellite communities be created a good 50 km outside the city, with self-contained infrastructure. The Comprehensive Development Plan of 1985 took this growth control one step further by zoning a green belt around the far edges of the city. Each plan became greater in its physical reach, reflecting a concurrent annexation of the city limits and an enlargement of the definition of the metropolitan area (Bangalore Development Committee 1954; Bangalore Metropolitan Planning Board 1963, 1976; Bangalore Development Authority 1984, 1995).

The result was a city that grew outwards rather than upwards and that decentralised its infrastructure rather than improving centralised systems. Inadequate infrastructure has since been a hallmark of these policies. Facilities for HAL, ITI, HMT, and BEL "leapfrogged" 13 km from the city centre, leaving the areas in-between to be developed in their wake (Prakasa Rao 1983, p. 193). The chief constituents of the city planning effort were the nationalised industries and the middle-class managers they employed; as in other nations, the working class and poor were not well served by urban dispersal initiatives (Teaford 1990; Nair 2005).

By the early 1970s, Bangalore had high degrees of spatial segregation between higher and lower economic classes, castes, and occupational categories. Peripheral neighbourhoods tended to be less dense and more affluent. Prakasa Rao's 1974 survey found that in the urban core, 8% of residents were high income and 37% were low income; in the periphery, 25% were high income and 27% were low income (Prakasa Rao 1983). Between nodes of exurban development, interstitial areas remained largely agricultural, but as the city expanded, these were zoned as industrial and residential and soon turned over to private uses. With growth control policies weak to non-existent, and ample political incentive for government officials to broker smaller-scale land use deals with constituents, the green belt was easily encroached upon by incremental private real estate development, both residential and industrial.

India's highly atomised patterns of property ownership – where land was often organised in small parcels held by multiple owners – confounded public and private efforts to secure large tracts of land in the centre city. As in other places, it was relatively easier to obtain land and build on farmland at the urban edge, yet the process was more easily effected through a private transaction between landowner and buyer rather than through government land acquisition measures (O'Mara 2005; Varma 2006). Bangalore's own past history of expansion via suburban "layout" and industrial satellites further encouraged this tendency, as did the common practice of state and local elected officials engaging in personal real estate deals (Varma 2006). The city continued to have a relatively large proportion of open spaces and retained its character as a place of trees and bungalows, but the dispersed and multi-nodal development laid the groundwork for more severe land use and infrastructure problems to come.

As the numerous and successive plans indicate, it is much easier to write a regional plan than to actually execute it. Bangalore is hardly alone in this (Rusk 1993; Orfield 1997). Every plan had to appease multiple stakeholders, extending the process; the 1963 plan did not win approval until nearly 10 years into its proposed 15-year development process. The rapid pace of Bangalore's industrial growth between 1940 and 1990 confounded the ability of government to predict and plan for population increases, employer demands, and infrastructure needs.

Bangalore's long history as a colonial and post-colonial space with multiple political identities fragmented local government, and piecemeal development certainly contributed to the difficulties it encountered in implementing strong regional land use practices. Bangalore had long been a sprawling city; industrial and residential suburbanisation was in keeping with its horizontal past and self-identification as a city of parks and gardens. The city had no political tradition or economic imperative to prepare it for comprehensive and resource-efficient city and regional planning. This legacy would allow a coming influx of foreign capital to have a significant impact on land use and urban space.

7.3 Economic Reform and Trade Liberalisation

The urban and economic landscapes of Bangalore have been transformed since the central government initiated comprehensive policy reforms in 1991. Reforms were introduced in June 1991 on the heels of a severe payment crisis due primarily to fiscal extravagance. In 1990–1991, the central government fiscal deficit had reached a peak of 8.4% of GDP. The Indian government had introduced reforms throughout the 1980s, but the reforms of 1991 were different in that they centred on the global economy and the recognition that integration with the world economy through trade and investment inflows would provide opportunities for economic growth.

The government embarked on an ambitious reform programme to reform through two main mechanisms: decentralisation policies and market-oriented reforms. By Indian standards, the reforms were far-reaching and included industrial, trade, and agricultural policies, although the most ambitious and successful reforms were in industrial policy. Most of the central government controls on investments have been reduced or fully dismantled.

7.3.1 Industrial Policy

Prior to reforms, industrial policy was characterised by state control over private investment and a burdensome bureaucracy that created obstacles to efficient business development. Government controls limited industries and sectors for private investment. Industries identified as strategic, including telecommunications, iron and steel, mining, energy generation and distribution, and transportation, were off-limits to private investment. Reforms have significantly reduced the number of industries restricted to public investment.

Reforms have also changed the policy of limiting the scale of certain industries. These small-scale industries (SSI) were kept artificially small through limitations on investments in capital and technology. These industries included some with the highest export potential, such as food processing, shoes, garments, and leather products. Ceilings on investment capacity limited the growth and comparative advantage of these units, making them uncompetitive in the world market. Reforms have increased the investment ceiling for many of these industries, and government officials are hoping that the reforms will unleash small-scale entrepreneurship and spark domestic growth, similar to the town and village enterprises in China.

Industrial licencing has been reduced significantly. The legacy of Nehruvian development was the complex system of licences, bureaucratic controls, and regulations of the "Licence Raj". Prereform, the cost of doing business in India was substantial in terms of time and financial resources. The "Licence Raj" has been nearly abolished such that doing business in India has become significantly easier. Even with the dismantling of the "Licence Raj", starting a business in India remains a long bureaucratic process, requiring 11 procedures and taking 35 days, compared to 6.2 procedures and 16.6 days for OECD countries (United Nations Conference on Trade and Development 2007).

Recent policy developments at the national level have further reduced the disincentives for investment by exempting certain areas and industries from tax and regulatory requirements. The Special Economic Zones (SEZ) programme established in 2005 offered private firms and developers streamlined development procedures as well as a 100% tax exemption on export income for 5 years. Because developers have to package together very large parcels of land in order to apply for SEZ designation, these industrial zones are, by and large, located at the periphery of cities or in rural areas (Special Economic Zones Act of 2005).

Although the policy reforms initiated by the central government have been wide-reaching, they have not been accompanied by similar reforms at the state level. In some states, private investors must still overcome institutional hurdles such as licencing. As such, the investment climate across states can vary considerably (Stern 2002). Some states such as Tamil Nadu and Karnataka, home of Chennai and Bangalore, respectively, have policies that have created a hospitable home for private investment. The government of Tamil Nadu is strategically developing research parks to attract both domestic and foreign investments. Master-planned research communities and research parks in Tamil Nadu such as Mahindra World City aim to compete domestically with Electronics City and Whitefield in Bangalore.

These communities also aim to be attractive for global firms by offering amenities and services that meet international standards. In contrast, poorer states such as Uttar Pradesh and Bihar have fewer strategies for attracting foreign investment and also receive significantly less investment, both domestic and foreign. The reforms at the national level have helped to make industrial production less bureaucratic, but without parallel policies at the state level, large variations in economic growth and increasing economic inequities across states will be inevitable.

Even where policy reforms at both state and central government levels have created investor-friendly environments, barriers to private investments still remain. For example, in Bangalore, as in many Indian cities, the provision of urban infrastructure lags behind economic development. Cities are often plagued with unreliable water quality, poor public transportation systems, and inadequate water and waste treatment facilities. The lack of an efficient transportation infrastructure in Bangalore has led some senior management of foreign firms to question the investment capacity of the city. Roads are often in poor condition, and infrastructure planning and delivery can be burdened by politics or insufficient public funds.

In some regions of the city, private firms have filled the infrastructure gap by providing their own emergency, education, and transportation services. Although these stopgap measures are one way to solve the infrastructure problem, the privatisation of public services has led to inefficient delivery of services. Reforms have driven a "fast-track" urban development strategy to provide efficient urban services through community participation and increased transparency and accountability. The Jawaharlal Nehru National Urban Renewal Mission (JNNURM) is aimed to provide assistance to states and local governing bodies from the central government. The thrust of the JNNURM programme is to strengthen local urban bodies by devolving administrative functions and financial powers and by building capacity of the local communities. Bangalore is one of the 63 cities that have been identified for support.

7.3.2 Trade Policy

Prior to reforms, trade policy was characterised by high import tariffs and restrictive import licencing. The two hallmarks of trade policy reforms are the elimination of import licencing and a reduction in tariffs. As a strategy to protect domestic industries and to manage balance of payments, imports of manufactured goods were severely restricted prior to reforms. Manufactured consumer products were banned, and intermediate goods and raw materials faced high import tariffs, making their importation nearly impossible. Raw materials could be imported, but if there were domestic producers of a particular manufactured product, its importation would be restricted through import licences. Reforms have abandoned licencing requirements for raw and intermediate goods, but only relaxed them for manufactured goods, in part because of the strength of the domestic industries that benefit from high restrictions. It was not until 2001, nearly a decade after reforms began, that restrictions on the importation of manufactured consumer products were finally lifted. Although import tariffs are lower than prereform period, they are still high compared to international standards, and there is room for further reductions.

Similarly, high import tariffs were established to protect domestic manufacturers. Prior to reforms, imports were subject to high tariffs, up to 400%. Reforms have reduced tariffs, but they are still high by international standards, and the country has had several complaints brought against it under World Trade Organization regulations. High tariffs may protect domestic industries in the short run, but they may also slow progress towards a transition to an internationally competitive economy. Domestic producers of commodities that are highly protected are likely to be uncompetitive in the world market. Tariffs may actually hurt domestic industries by not providing incentives for improvements in efficiency and quality. Indeed, despite the reforms, trade policy still has an anti-export bias.

7.3.3 Foreign Direct Investment

For many countries undergoing neoliberal reforms, opening up the domestic market for foreign investment has been a critical component to the growth strategy. One of the most significant impacts of the reforms has been the increase in foreign direct investment. Prior to reforms, foreign investment was discouraged. Annual average inflows of FDI were USD 200 million in 1987–1990 compared to 4.1 billion in 2001–2004 (United Nations Conference on Trade and Development 2007). The post-reform period has been characterised by a considerable increase in FDI inflows and a change in the type of FDI, as well as its composition. FDI is now permitted in many sectors, including financial services and telecommunications, and manufacturing. Since the mid-1990s, FDI has shifted from the agriculture and manufacturing sectors to services. There are high expectations among government officials that an increase in FDI will spark economic growth in India in the same way that it has in China, with the implicit assumption that FDI and growth are positively correlated (Fig. 7.3).

Although FDI inflows have increased, they are still low and tend to be concentrated in several sectors primarily focused on the domestic market. As a percentage of gross fixed capital formation, FDI inflows only accounted for 3.5% in 2005, compared to 9.2% for China and 12.8% for all developing countries during the same period (United Nations Conference on Trade and Development 2007). Similarly, FDI stocks as a percentage of GDP was only 5.8% in 2005,

Year	Nominal FDI ($ million)	FDI (2006 $ million)	% of GNP
1991-92	165	244	0.13
1992-93	393	566	0.28
1993-94	654	916	0.43
1994-95	1374	1869	0.81
1995-96	2141	2826	1.14
1996-97	2770	3546	1.32
1997-98	3682	4639	1.65
1998-99	3083	3823	1.28
1999-00	2439	2951	0.6
2000-01	2908	3402	0.66
2001-02	4222	4813	0.88
2002-03	3134	3510	0.62
2003-04	2634	2897	0.47
2004-05	3754	4017	0.61
2005-06	5549	5715	0.79
2006-07	15726	15726	1.57

sources: FDI: Indian Ministry of Commerce and Industry, Department of Industrial Policy and Promotion Website; GNP: Indian Ministry of Finance, Economic Survey, 2007.

Sectors Attracting Highest FDI Equity Inflows

Sector	2003-04	2004-05	2005-06	2006-07	1991-2007	% of Total Inflows
Electrical Equipment	532	721	1451	2733	8227	18.77
Services Sector	269	469	581	4749	7840	17.84
Telecommunications	116	129	680	521	3892	8.7
Transportation	308	179	222	466	3644	8.04
Fuels	113	166	94	250	2831	6.31
Chemicals	20	198	447	206	2348	4.95
Construction	47	152	151	985	1420	3.33
Drugs and Pharmaceuticals	109	292	172	215	1222	2.75
Food Processing	111	38	42	98	1277	2.68
Cement and Gypsum	10	0	452	243	989	2.26

US$ millions

source:Department of Industrial Policy & Promotion, Ministry of Commerce and Industry, India, Fact Sheet on Foreign Direct Investment (FDI)

Fig. 7.3 FDI increase, leading sectors

compared to 14.3% for China and 26.3% for all developing countries during the same period (United Nations Conference on Trade and Development 2007). However, this is a considerable increase from the 0.3% of gross fixed capital formation during 1987–1990.

The services sector has attracted the bulk of FDI inflows in recent years. Most notably, "back-office" work such as call centres has been transferred to Indian cities like Bangalore and Hyderabad. However, there is evidence that the type and structure of FDI has important implications for the economic development of the country (Asian Development Bank 2004). The quality of FDI is just as if not more important than the quantity of FDI. Export-oriented FDI usually has more impact on GDP growth than domestic-oriented FDI because they can facilitate technology transfer to the host country and create positive economic externalities that benefit local enterprises and workers (Enderwick 2005).

While there is no doubt that FDI inflows to the country have increased, there is disagreement as to whether FDI has actually increased exports (Banga 2006). Local officials and foreign investors acknowledge that these service industries will migrate away from India as cities begin to lose their international comparative advantage in low wages. As such, domestic investors and local policymakers are interested in developing Indian cities into full service R&D hubs for a wide range of industries. Currently, foreign firms often transfer high-level management from overseas rather than recruit from within India, citing a lack of business expertise and managerial competence.

The combination of reforms at the central and state levels, strong local educational institutions, and cultural ties to the non-resident Indian (NRI) population overseas helped Bangalore gain a domestic competitive advantage in attracting foreign direct investment. Approximately one third of engineers in Silicon Valley are of Indian descent, and 7% of Silicon Valley's high-tech companies are led by an Indian. It is not surprising that Bangalore became the geographic target for IT-related foreign investments.

7.4 Growth of India's Silicon Valley

Entering the 1990s, Bangalore already had a reputation as India's hub of research-intensive industry and the home of its nascent technology sector. In the late 1970s, the state of Karnataka had embarked on a major economic development push to grow the electronics industry in Bangalore, building upon its existing large employment base in aerospace and telecommunications and taking advantage of its educated workforce. Existing research-based institutions, governmental and educational, gave Bangalore a high concentration of trained workers. The combination of this with state economic development support, national industrial subsidy, and trade restriction allowed local technology companies to thrive. Bangalore-based Western India Vegetable Products Limited (Wipro) began to expand its operations into computing in the mid-1970s; in 1983, it was joined by IT services company Infosys, who moved its headquarters to Bangalore.

The first foreign entrant into the Bangalore IT market arrived in 1986, when US-based Texas Instruments (TI) opened a back-office facility in the city; Digital Computers and Hewlett-Packard had joined TI in Bangalore by the end of the 1980s. But with market liberalisation still years away, all of these companies had to find an Indian partner in these endeavours due to Indian restrictions on foreign ownership of land and business assets. Faced with these limitations, other foreign companies chose to take the relatively easier path of locating offshore operations in Taiwan, Singapore, or the Philippines (Bresnahan and Gambardella 2005).

The situation changed dramatically once Indian economic reform opened up national markets to private and foreign capital. Significantly, market liberalisation occurred just at the moment that technological changes, particularly in telecommunications, made the physical distance between India and the rest of the world matter less and the high skill level of its technical workforce matter more. The huge technological demands created by the Y2K adjustment of computers and software, combined with the economic downturn in the US technology sector beginning in 2001, served to enlarge the client base of IT services firms like Infosys and Wipro. The search for cheaper skilled labour brought US and European computer hardware and software giants – like Compaq, IBM, Motorola, and Sun Microsystems – to Bangalore in the 1990s.

These large firms were followed in turn by small- to medium-sized foreign technology companies, drawn in part by the presence of their larger competitors, what one Silicon Valley executive termed "the lemming effect" in a conversation with the authors. The changes linked Indian firms and workers to a worldwide network and had an explosive effect upon the growth of Bangalore's technology sector. Educated, white-collar, non-resident Indians (NRIs) returned to India and migrated to Bangalore to work as managers and technicians. The number of jobs in computer hardware, software, and IT services in Bangalore grew from about 2,600 in 1991 to an estimated 160,000 in 2004 (Heitzman 2001; Morphy 2004). The growth of the technology sector helped to propel Bangalore's population further upwards.

7.5 The Built Environment of the Technology Industry

A transformation of the industrial built environment accompanied the growth of Bangalore's IT economy. The historical development of technology clusters of the United States (Silicon Valley in particular) strongly associated the white-collar technology industry with a very specific kind of architecture and landscape design. Technology itself was a driver of this – as the fabrication of sophisticated hardware and software required features like air conditioning and high telecommunication capacity – but so was an American cultural predilection to provide technical workers with high-amenity workspaces as well as public policies that favoured industrial de-densification.

In the mid-twentieth century, the high-tech aesthetic meant low-rise modernist buildings set amid landscaped grounds. By the 1990s, it had translated into sleek and glass-sheathed buildings of about 4–10 storeys, developed singly or as part of a more extensive campus. As it globalised, the IT industry continued to be associated with facilities that were modern and spacious, contained the best and newest infrastructure, provided employee amenities like cafeterias or jogging paths, and that had an aesthetic homogeneity that had little relation to cultural context (O'Mara 2005).

The research park environment has been so strongly associated with white-collar technology industry in the popular imagination that public-sector-driven efforts to grow regional high-tech clusters have often centred upon the physical development of research parks, sometimes to the exclusion of all other strategies (O'Mara 2007). Yet in developing regions with outmoded or unreliable utility and road infrastructure, provision of these facilities became essential. This was especially true for Bangalore, where in the years prior to the IT boom, the city was already expansive, with development driven more by walking and public transportation than by transportation corridors. Land development was a key element of the State of Karnataka's initiative to build out Bangalore's computer industry in the 1970s and 1980s, and in 1984, the state government established an "Electronics City" on a 300-acre plot 20 km south of the centre city.

In 1980, urban development in the Bangalore metropolitan area covered 75 km^2, approximately the size of Hong Kong Island (Survey of India 1980). Drawing both from the American research park idea and from the campuses of the nationalised industries, Electronic City provided its tenants with self-contained electrical and other utility services and direct access to Bangalore's centre via Hosur Road, a broad thoroughfare running through largely undeveloped farmland. The privatisation of infrastructure services also filled an important gap as the public sector was unable to provide electricity and other urban services. Infosys and Wipro were two of its first tenants.

Despite the presence of Electronic City, however, the Bangalore of the early 1990s had few class-A commercial buildings to accommodate the influx of new IT firms. It continued to have highly inconsistent provision of public utilities. Its residential neighbourhoods lacked the amenities that NRIs had enjoyed in London, Boston, or San Jose. Despite decades of public planning effort, the region continued to have vastly overburdened transportation networks and a fragmented and weakly regulated land use strategy.

Bangalore's regional population of six million was nearly double the size of what city officials had predicted 20 years earlier; the unexpected influx had left city planners "unable to cope" (Prakasa Rao 1983; Schifferes 2007). Economic liberalisation had opened up India to the global high-tech economy, but it was accompanied by little guidance or public funding for remaking the urban built environment to accommodate this growth (Sheshagiri 1999). Multiple administrative offices with overlapping and competing functions, low inter-institutional cooperation, and weak local bodies exacerbated the problem of a lagging infrastructure.

By the turn of the twenty-first century, these circumstances, occurring in an environment of extraordinarily rapid growth, led to IT firms and workers becoming among the most visible of those private-sector entities "opting out" of the public infrastructure and planning system in order to obtain the commercial and residential facilities they needed. Bangalore commercial real estate development became not only a matter of leasing adequate square footage, but also providing necessary electrical, water, and sewer services independent of the city grid (Varma 2006). Some developments went so far as to provide private ambulance services to their tenants (Murali 2006). By 2005, Bangalore's urban footprint had expanded to approximately 225 km^2 (TKK Healthcare Limited 2005).

As in other globalising megacities, real estate development and property management in Bangalore became dominated by large firms who had the resources and personnel to tackle these multilayered land development demands. While managed locally by Indians with deep knowledge of regional real estate markets and necessary political connections, many were foreign-based firms. Singaporean developers became a particularly visible presence, as did American and European multinationals who entered the Indian market after being hired by their existing corporate clients to find and manage new facilities. Property management giant Jones Lang LaSalle came to India in 1998 to facilitate the offshore move of Silicon Valley-based software firm Oracle; by 2006, Jones Lang LaSalle had 1,000 employees in the country (Jones Lang LaSalle 2007).

The Bangalore metropolitan region became dotted with office buildings, research parks, and industrial "campuses" that not only offered office facilities nearly indistinguishable from buildings found in California's Silicon Valley or Massachusetts' Route 128 but also featured electric generators, independent water supplies, and private road networks. The built environment of the technology industry fell into three general categories. The first category consisted of single building or closely clustered developments near the city centre or near its airport. These were often built for a specific sole foreign tenant, whose name was emblazoned on the building's upper storeys.

The second category was the research park, developed by a single public or private entity for the use of multiple large and small tenants. Electronic City remained one of Bangalore's more significant developments of this kind, but it was now rivalled by several joint public-private research park developments in the eastern suburb of Whitefield. While Electronic City was home both to major Indian technology concerns and foreign market entrants, Whitefield had a distinctly international flavour – perhaps appropriate for a suburb originally settled in the nineteenth century by British expatriates (Nair 2005). Singaporean developer Ascendas played a major role in developing this part of Bangalore's technology landscape, and its research parks were home to some of the small- to mid-size foreign technology enterprises who opened Bangalore facilities during the US tech downturn of the early 2000s.

The third category of development was the research campus, a single-tenant development set in expansive grounds and located at the far periphery of the city. The General Electric Company's research arm occupied such a facility on the eastern outskirts of Bangalore, in a largely rural landscape of farms and villages.

7.6 Changes in Urban Space

The industrial complexes and research parks of Bangalore's Silicon Valley were the leading edge of a broader economic and cultural rearrangement of urban space and consumer culture. Economic growth, foreign investment, and the privatisation of infrastructure provision in Bangalore accelerated decentralisation, shifted the industrial centre of gravity in the city and region, and created a new consumer market demanding particular urban amenities. In many respects, these trends are not simply representative of the urban characteristics inherent in high-tech clusters, but they signal a fundamentally new global economic order in which cityscapes – and city planning processes – are profoundly influenced by multinational corporations and their middle-class workforce.

Maps of Bangalore's urban extent over time show how the rapid escalation in number of jobs and people played itself out in urban space. Bangalore had long trended towards lower-density settlement than other Indian cities (Prakasa

Rao 1983). Over time, legal annexations made the urban area larger, and growth trends spread settlement outwards, but the pattern of development remained remarkably consistent: a denser centre city, a widely dispersed, and only partially urbanised fringe. The consequence of piecemeal control of land and private delivery of utilities, combined with a strong market preference for modern structures and low floor-area ratios, was a hopscotch pattern of development and increased inefficiencies in infrastructure provision.

One result of these inefficiencies was severe traffic congestion. Providing adequate transportation and public transit systems had long been a struggle for Bangalore; the rapid growth and decentralisation of the high-tech era – accompanied by rising affluence and increased private vehicle ownership – made the situation critical. Bangalore lacked a unified regional public transit system. Although its road mileage had tripled between 1980 and 2005, when the region had approximately 3,000 km of roadways, increased traffic had slowed average speeds to a 5-mph crawl during peak hours (Kāmat 1991; World Bank 2005). Unlike in North America, where roads exert a strong pull on development patterns, in Bangalore, proximity to existing urbanisation such as services, jobs, and housing is a greater driver of urban form (Reilly et al. 2009).

Therefore, Bangaloreans were not simply driving in more congested environments, but the decentralisation of employment and residence meant they were travelling longer distances. Further aggravating traffic congestion, Bangalore lacks a dense city core. With houses and jobs distributed throughout the city, traffic patterns are unpredictable without clear commute patterns. Foreign and multinational firms' solution to this traffic dilemma was, once again, to opt out of a subpar system and provide parallel transit and transportation services for their employees. Many IT firms, particularly those running call centres where employees worked throughout the night, provided extensive private bus service for their workers.

Although local officials publicly despaired about Bangalore's traffic woes, the city's zoning policy helped give firms incentives to move further outwards and rely on private cars rather than mass transit. Companies in the central-city zone could only offer a certain number of parking spaces without having to pay the city a fee; a ratio of one space per 50 employees was typical. At the periphery, corporate tenants were able to offer many more parking spaces, up to one space per eight employees (Varma 2006; Murali 2006).

Another aspect of urban change in the high-tech era was a reorientation of the direction in which the city grew outwards. In the era of nationalised industry, the parks that housed Bangalore's aerospace and telecommunications industries radiated out to the northwest as well as to the northeast. Major industries like HMT and educational institutions like the Indian Institute of Science (IISc) established themselves in the north-western quadrant, making the thoroughfare of Tumkur Road a chief concentration of industry in the 1970s (Bangalore Metropolitan Planning Board 1976, p. 57). Pragmatic resource concerns factored into industrial location choices, as water supplies were more plentiful on the west side.

The countertrend began with the establishment of Electronic City on the far southern edge of the city, far from the city centre and from all the previous industrial campuses. As Electronic City grew, middle-class housing layouts were built in the vicinity and infill development began to turn Hosur Road into a crowded thoroughfare. After economic liberalisation, Bangalore's trajectory of growth was further altered, as foreign firms chose to locate their facilities in the vicinity of Bangalore's airport. The airport was directly east of the city centre, and the resultant clustering of multinationals along the Airport Road became the most visible manifestation of Bangalore's arrival in the global economy.

The firms, managers, and markets targeted by Electronic City in the 1980s were all Indian; in the era of nationalised industry, proximity to the airport was less important than driving proximity to neighbouring cities. In the 1990s and 2000s, nearness to the airport was essential, particularly given Bangalore's traffic congestion. The orientation of growth may shift further in the future as a result of Bangalore's new international airport, which opened in 2008 in the far northern outskirts of the city, 40 km away from

Fig. 7.4 Executive housing

the central business district. Because the current geography of IT and other export-oriented businesses skews south and east of the city, international business travellers must endure rides of up to 3 h through Bangalore's congested streets in order to get to the airport. Within months of the airport's opening, traffic woes spurred the launch of a helicopter taxi service to ferry the more affluent of these travellers between the airport and the city.

Residential developments catering to returning NRIs, expatriates, and the growing middle class created similarly self-contained environments in the vicinity of Bangalore's high-tech clusters. The Indian real development firms Prestige Corp. and Heritage Corp. dominated this market, offering high-rise apartments and single-family home developments with names like Whispering Pines, Palm Meadows, and the Acropolis. In contrast to some industrialised nations, increased levels of affluence did not necessarily trigger decreased levels of residential density. Bangaloreans had long been attached to villa and bungalow living, thus creating a significant local demand for suburban developments of single-family homes. However, the new migrants to the city from elsewhere in India and the world tended to favour the more cosmopolitan atmosphere of the luxury high-rise apartment complex (Varma 2006).

With their many modern amenities, both types of developments were a departure from earlier, more modest forms of housing in Bangalore. And, with little land available in the centre city and escalating traffic congestion, the middle-class housing developments appeared in the intermediate suburban neighbourhoods or in the metropolitan periphery. Whether high-rise apartment or gated community, the new white-collar residential developments of Bangalore provide self-contained electric, water, and sewer services, as well as offering the promise of security from the increasingly chaotic city outside. As historian Janaki Nair writes, "in a city like Bangalore, which lacks several kinds of infrastructure facilities, what builders place on offer is the dream of successful planning itself" (Nair 2005, p. 162) (Fig. 7.4).

The increased privatisation of space in Bangalore during the high-tech era, combined with the failures of an overstressed and underfunded local government to address very real infrastructure problems, led to regional service provision itself becoming increasingly privatised by the turn of the millennium. Foreign corporations Volvo and Coca Cola invested significant sums building roads and research parks in the vicinity of their Bangalore operations. Indian IT leader Infosys funded the police department. Real estate developers Brigade and Prestige each supported road improvements and extensions. Karnataka state leaders trumpeted "public-private partnerships" as key in planning and infrastructure provision (Ghosh 2005).

While the growth of India's Silicon Valley made the Bangalore of the 1990s profoundly different from the city of a generation before, there are persuasive continuities. Twenty-first century tendencies towards privatisation, architectural standardisation, and self-contained new development were not only a result of the entry of multinational firms into the land market but also an outgrowth of the earlier era of industrial planning. New tech parks and gated communities are latter-day satellite townships, akin to the communities outlined in Bangalore's regional plans of the 1950s and 1960s.

7.7 Implications for the Future

Reforms have facilitated economic growth and opened up opportunities for urban development. However, economic dynamism and open markets have presented new challenges to urban growth and sustainability. The growth of Bangalore has been driven largely by an influx of foreign capital concentrated in high-tech and IT-related services industries. It has benefitted a minority of workers – the educated middle class, many of whom are not native Bangaloreans – and it has created zones of segregation between the rich "new" landscape of high technology and the poorer, "old" Bangalore.

Bangalore also has staked much of its future on multinational companies with fickle loyalties. They came to the city in good part because business and labour costs were cheaper, and they may just as easily depart for other cities in India and the rest of the world that can offer similarly well-educated workers without the infrastructure and transportation hassles of "India's Silicon Valley." This increases the vulnerability of Bangalore to international economic woes. There are already indications that the current global economic recession has affected Bangalore, with a slowdown in the real estate sector and some manufacturing companies looking elsewhere for less expensive labour.

If Bangalore's domestic and international competitive advantage declines, it is unclear how the economy will transition as international capital moves to the next target of investment opportunity. Despite reforms at the central and state levels, the city is laden with bureaucratic complication which creates obstacles to efficient planning, provision of urban services, and investment utilisation. Bangalore's impressive rate of urban growth over the last two decades would challenge even the most effective urban municipalities. In Bangalore, it has underscored the city's inefficiencies and politics.

National policy has condoned and encouraged peripheral development and privatisation of infrastructure in Bangalore. Policy priorities at the national level since 1990 have focused on giving the software industry as smooth a path as possible to investment and profit. By 1999, the director general of India's National Informatics Centre was declaring (somewhat hyperbolically) that the nation had "removed all remaining red-tape and has virtually transformed it into a red carpet. The immediate result has been that virtually every state is now providing a Silicon Valley kind of environment with state-of-the art modern hi-tech infrastructure and plug and play facilities, so that Indian and overseas companies can immediately set up their software operations" (Sheshagiri 1999).

There are few indications that national policy is going to move in a different direction in the near future. The SEZ programme provides a rich package of government incentives to domestic and foreign firms who are opting out of the public infrastructure system and locating in peripheral satellite settlements that are physically and politically separate from central cities. Rather than helping cities like Bangalore make systemic improvements at the city core and giving incentives for foreign capital to locate there, national economic policies continue to echo the pro-dispersal policies of several generations before.

Yet, in spite of the challenges, there are exciting opportunities for the city and its inhabitants. Compared to other rapidly urbanising regions in Asia, where urbanisation and economic growth are often synonymous with westernisation, in Bangalore, there is a distinct South Indian cultural influence to development. Western fast food chains have penetrated the domestic market, but their presence is limited, especially compared with other rapidly growing Asian cities of comparable size. International companies with branches in Bangalore appear interested and concerned with sustainability issues. This could stem from the fact that senior management often are NRIs who are temporarily stationed in Bangalore. Corporate interest in sustainability could be encouraged by senior management with cultural connections to Bangalore and a vested interest in the well-being of the city.

There is also evidence that urbanisation in Bangalore is occurring at a human scale and is being developed through Indian interpretations. Walkable places with pedestrian-friendly streets are common. Returning high-tech executives are choosing high-rise apartments in part because they present the kind of vibrant, social living environment that they missed during their stays in the low-density suburban landscapes of California and elsewhere. New housing developments include multi-generational living spaces which belong to cultural tradition in India. The planned mass-transit metro system that is expected to be completed in 2011 will bring much relief to a congested city.

When it comes to urban liveability, foreign direct investment in Bangalore has seemed, so far, to be more of a bane than a boon. Population growth, traffic congestion, and peripheral development appear to have all degraded the quality of life. Yet, the choice of some returning expatriates to maintain denser, more energy-efficient patterns of living and working, combined with the presence of multinational firms that may feel more corporate pressure to "build green," signals that foreign firms could become part of the solution to this region's infrastructure and planning challenges. Could foreign firms be given incentives by the government to move to the centre rather than the periphery? Could corporate executives and developers help planners integrate economic growth strategies with sustainable urban development goals? Could multinational firms be engaged in the effort to integrate more of Bangalore's citizens in the new technology economy, rather than segregating themselves from these populations?

There is evidence that these things are possible – both in the way that some of these actors are choosing to live now and in their stated aspirations for a more liveable urban environment of the future. Although the global economic downturn that began in 2008 will reduce the amount of capital available to Bangalore's city builders, slower economic growth rates may provide planners and citizens the breathing room they need to institute a more socially and environmentally sustainable urbanisation. It is up to local actors to decide how best to structure such collaborations, but it becomes clear from our observations that, in a megacity context where public planning may be at odds with the priorities of multinational corporations (and those of national economic policy), the multinationals prove to be far more powerful. Rather than allowing FDI to trump public planning efforts, these private actors must be integrated into the planning process.

Bangalore has long been a "garden city" and a place where a low-rise built environment has contributed to its quality of life, rather than detracting from it. The challenge for the greater Bangalore metropolitan area is to preserve the cultural and spatial characteristics that have made Bangalore desirable and liveable over the centuries, while finding a new paradigm for urban and regional planning that prevents further physical and infrastructural fragmentation of the city. The multinational companies and internationalised executives who have transformed Bangalore over the past 15 years may be instrumental to this solution.

Acknowledgments The authors wish to thank Michael K. Reilly for his assistance in producing the maps in this chapter. We also thank the Stanford Woods Institute for the Environment for its support of this project through a Interdisciplinary Faculty Ventures grant.

References

Asian Development Bank (2004) Asian development outlook 2004. Part 3: foreign direct investment in developing Asia

Banga R (2006) The export-diversifying impact of Japanese and US foreign direct investments in the Indian manufacturing sector. J Int Bus Stud 37(4):558–568

Bangalore Development Authority (1984) Comprehensive development plan report. Bangalore Development Authority, Bangalore

Bangalore Development Authority (1995) Comprehensive development plan (revised) Bangalore report. Bangalore Development Authority, Bangalore

Bangalore Development Committee (1954) Report of the Bangalore development committee. Government Press, Bangalore

Bangalore Metropolitan Planning Board (1963) *Outline* development plan for the Bangalore metropolitan region. Government Press, Bangalore

Bangalore Metropolitan Planning Board (1976) Report on the comprehensive development plan (CDP) of Bangalore. Government Press, Bangalore

Bresnahan T, Gambardella A (2005) Building high-tech clusters: silicon valley and beyond. Cambridge University Press, Cambridge

Castells M (1993) The information age: economy, society and culture, vol 1, The rise of the network society. Blackwell, Cambridge

Clapson M (1998) Invincible green suburbs, brave new towns: social change and urban dispersal in post-war England. Manchester University Press, Manchester

Enderwick P (2005) Attracting "desirable" FDI: theory and evidence. Transnatl Corp 14(2):93–119

Fishman R (1987) Bourgeois utopias: the rise and fall of suburbia. Basic Books, New York

Friedmann J (1986) The world city hypothesis. Dev Chang 17:69–83

Ghosh A (2005) Public-private or a private public? Promised partnership of the Bangalore agenda task force. Econ Political Weekly 40(47):4914

Hall PG, Markusen A (1984) Silicon landscapes. Taylor and Francis, New York

Heitzman J (2001) Becoming silicon valley. Seminar (India): 503

Heitzman J (2004) Network city: planning the information society in Bangalore. Oxford University Press, New Delhi

Jackson KT (1985) Crabgrass frontier: the suburbanization of the United States. Oxford University Press, New York

Jones Lang LaSalle (2007). http://www.jllm.co.in/en-GB/company/. Accessed 21 Sept 2007

Kāmat S (1991) Karnataka state gazetteer. Office of the Chief Editor, Karnataka Gazetteer, Bangalore

Morphy E (2004) Bangalore Besting Silicon Valley? NewsFactor, July 29, 2004. http://www.newsfactor.com/story.xhtml?story_id=111003KJS4RC. Accessed 21 Sept 2007

Murali (2006) Interview with the authors, Bangalore, 9 Oct 2006

Nair J (2005) The promise of the metropolis: Bangalore's twentieth century. Oxford University Press, New Delhi

Nehru J (1958) Jawaharlal Nehru's speeches (1958–1968). Publications Division, Ministry of Information and Broadcasting, Govt. of India, New Delhi

O'Mara MP (2005) Cities of knowledge: cold war science and the search for the next silicon valley. Princeton University Press, Princeton

O'Mara MP (2007) Landscapes of knowledge: history and the evolving geography of high technology. Places 19(1)

Orfield M (1997) Metropolitics: a regional agenda for community and stability. Brookings Institution Press, Washington, DC

Prakasa Rao VLS (1983) Urbanization in India: spatial dimensions. Concept, New Delhi

Reilly MK, O'Mara M, Seto KC (2009) From Bangalore to the Bay Area: comparing transportation and activity accessibility as drivers of urban growth. Landsc Urban Plan 92(1):24–33

Report on the Comprehensive Development Plan for Bangalore (1976). Government Press, Bangalore

Rusk D (1993) Cities without suburbs, Woodrow Wilson center special studies. Woodrow Wilson Center Press, Washington, DC

Sassen S (2000) Cities in a world economy, 2nd edn. Pine Forge Press, Thousand Oaks

Schifferes S (2007) Bangalore's boomtown blues. BBC News, 29 Jan2007 http://news.bbc.co.uk/2/hi/business/6288325.stm. Accessed 21 Sept 2007

Sheshagiri N (1999) The informatics policy of India. Inf Syst Front 1(1):107–116

Special Economic Zones Act of 2005 (2005) No. 28 of 2005, 23 June 2005. Government of India, New Delhi

Srinivas S (2001) Landscapes of urban memory: the sacred and the civic in India's high-tech city, Globalization and community 9. University of Minnesota Press, Minneapolis

Stern N (2002) Building a climate for investment growth and poverty reduction in India. In: A Strategy for Development (ed) Speech presented at the export-import Bank of India, Mumbai, March 22, 2001. World Bank Publications, pp 51–68

Survey of India (1980, 1990). Bangalore. New Delhi

Teaford JC (1990) The rough road to renaissance: urban revitalization in America, 1940–1985. Creating the North American landscape. Johns Hopkins University Press, Baltimore

TKK Healthcare Limited (2005) Bangalore map. Bangalore

United Nations Conference on Trade and Development (2007) World investment report 2006. United Nations, New York/Geneva

Varma K (2006) Interview with the authors, 6 Oct 2006, Bangalore

World Bank Energy and Infrastructure Unit, South Asia Region (2005) Towards a discussion of support to Urban transport in India. The World Bank, Washington, DC

Part II

Economic, Social and Infrastructure Transformation

8 Inner-City Development in Megacities Between Degradation and Renewal: The Case of São Paulo

Martin Coy and Tobias Töpfer

Abstract

The development of inner cities in Latin America during the twentieth century is characterised by three stages: growth and "beautification", modernization according to the Athens Charter as well as degradation and renewal. During the first and second stages, the city centre of São Paulo has experienced a strong verticalisation and an orientation towards individual transport. Later on, due to several limitations, new inner-city fragments emerged southwest of the central area. In this third stage, the old city centre suffered manifold degradations in general and of the public space in particular. The latter has lost significance in favour of the private space. Different actors of the "formal city", the "informal city" as well as the public sector are engaged to change the city centre. Based on their different interests, there are several conflicts concerning the treatment of the city centre. Against this background, constantly changing city governments took various renewal measures, focusing either on sectoral upgrading or on holistic, integrated rehabilitation. The possible future development of the inner city of São Paulo might follow one of three scenarios and depends not only on the local but also on the superordinate framework conditions.

Keywords

São Paulo • Urban development • Fragmentation • Public space • Inner-city degradation • Inner-city renewal

8.1 Stages of Inner-City Development in Latin America

In the last couple of years, inner-city renewal has become an increasingly important topic in Latin American city politics. This is to be interpreted as a consequence of profound functional

M. Coy (✉) • T. Töpfer
Institute of Geography, University of Innsbruck,
Innrain 52, A-6020 Innsbruck, Austria
e-mail: Martin.Coy@uibk.ac.at;
Tobias.Toepfer@uibk.ac.at

DOI 10.1007/978-90-481-3417-5_8, © Springer Science+Business Media B.V. 2014

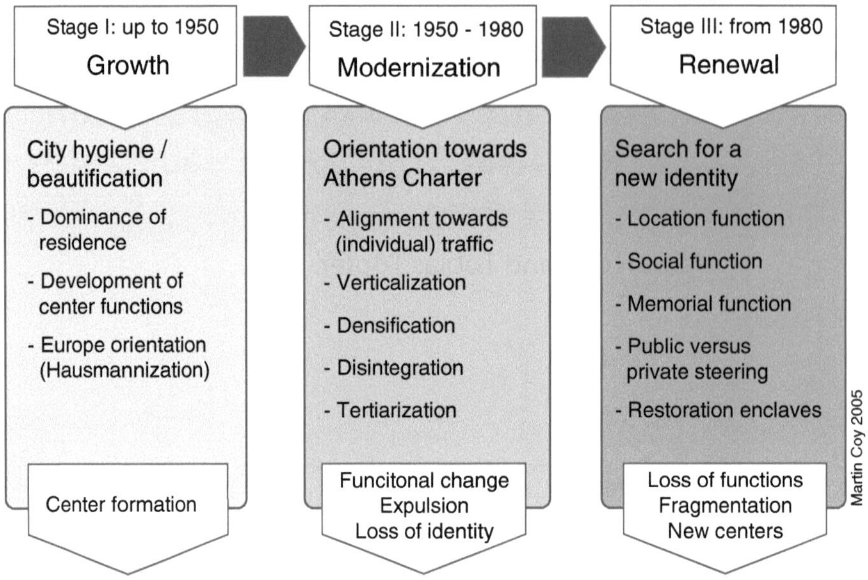

Fig. 8.1 Stages of inner-city development in Latin America

and socio-spatial degradation processes which are directly related to the increasing fragmentation of Latin American cities. It is therefore often the city centres that, as a result of explosive urban growth, modernisation attempts and socio-economic splitting processes, are marked particularly by physiognomic, social and functional signs of decay (cf. on the subject of correlations between city decay and city renewal generally, Lichtenberger 1990). Consequently, city centres often become the venue of widely varying conflicts of interests and competition for land.

Simultaneously, urban researchers, planners, politicians and segments of the public rediscover the inner cities as an "emblematic place", that is, a place of high importance for the collective memory of a city and for its identity. It is in its public areas and representative buildings where the "symbolic" content of a city is concentrated. Due to diverging interests in many places, inner-city renewal nonetheless represents a highly conflict-laden arena, both in the city discourse and in specific urban policy, in which groups of actors operate, who assign varying objectives to the process of renewal.

The development of the city centres in Latin America during the twentieth century can simplistically be divided into three stages (cf. Fig. 8.1): A first phase, lasting approximately up to the middle of the twentieth century, was above all characterised by growth processes. Although the city centres were still places of residence for the various social classes, over time the typical city functions emerged. Mostly under the influence of European city planners and architects, the city centres were reorganised in accordance with city hygiene and visually "improved" by the design of splendid boulevards as well as with the construction of buildings of prestige according to the spirit of the times. They became, in a manner of speaking, "hausmannised", for which Buenos Aires, Rio de Janeiro and also São Paulo are good examples (cf. on São Paulo Porta 2004).

The second period, lasting until the 1980s, can be defined as a phase of modernisation, and planning is essentially oriented towards the principles of the Athens Charter. The city centre was aligned by means of road corridors for individual traffic, services were expanded at the expense of residential functions, and, in terms of construction, an excessive verticalisation determined the increasingly specialised and segregated city centres (cf. Bähr and Mertins 1995).

In the last 30 years, nearly everywhere city centres have had to once again define their position within the fragmented city. In this context, the functions of the inner city, especially in relation to the newly emerging city centre structures, and above all, the control mechanisms in place, are being questioned. This has resulted in a third phase, one characterised by the loss of function and, alternatively, by increased efforts geared towards the renewal of city centres. In the meantime, in most Latin American countries, endeavours are concentrating on the preservation, rehabilitation and renewal of historically still intact city centres as well as those largely superimposed by modernity. Measures of historic-city renovation and inner-city restoration can be accomplished in principle in different ways. In Latin American cities, the impetus comes mostly from above, in the sense of a top-down approach. Public-private partnerships against the backdrop of neoliberal framework conditions are becoming increasingly important for the realisation of renovation measures.

In this context, however, the fundamental question concerning social equilibrium arises. It appears appropriate to speak of a bottom-up course of action when, for example, citizens' action groups successfully press the necessity for redevelopment and renewal measures and are also actively involved in their planning and implementation. As a result of redevelopment measures, a displacement effect on traditional resident and user groups can be observed among other things because historic-city and inner-city renewal processes in many cases also – intentionally or unintentionally – go hand in hand with gentrification tendencies. Particularly in the colonial cities of Latin America, one significant problem presents itself: musealisation, serving at best the tourism industry. The everyday necessities of the local population remain rather neglected in this case.

Developing an adequate model of inner-city renewal is still quite difficult, as it should ideally benefit the whole population and succeed in achieving both the functional objective of urban construction and the mixing of functions, as well as improving the quality of residence in the city centre (in particular in the public space) while simultaneously striving to retain the genius loci. Only a form of inner-city renewal which manages to strike a balance between economic, social and ecological interests could meet the requirements of sustainable urban development (cf. Coy and Zirkl 2001).

8.2 São Paulo: Framework Conditions of City Development and Intra-urban Transformation

The agglomeration of São Paulo with approximately 19.7 million inhabitants is currently the largest megacity in Latin America after Mexico City. The dynamic economic development of the greater São Paulo region since the middle of the twentieth century has attracted immigrants from all regions, in particular, however, from the Brazilian Northeast. The population of the metropolitan region almost quadrupled from less than 5 million in 1960 to an approximate 18 million in 2000 and to an estimated 19.7 million by the year 2006. The most rapid growth took place in the 1960s and 1970s – the crest of the Brazilian "economic miracle" – in order to bottom out again, from the beginning of the 1980s on (cf. Coy 2001, 2005 and Kohlhepp 1997 for a more detailed representation of the development of São Paulo). Disproportionately affected by the population growth were the municipalities situated beyond the core city of the metropolitan area. The population development in São Paulo itself differs significantly from the growth patterns of the suburban municipalities of the metropolitan region, which have registered a consistent population growth. While in 1960 77% of the agglomeration's total population still lived in the city centre, this share decreased to 59% in 2000 (which represents scarcely 10 million inhabitants). The population of the core city almost tripled between 1960 and 2000; however, clear differences between the demographic development of the centre and the periphery can be observed. Especially the quarters near the inner city have suffered

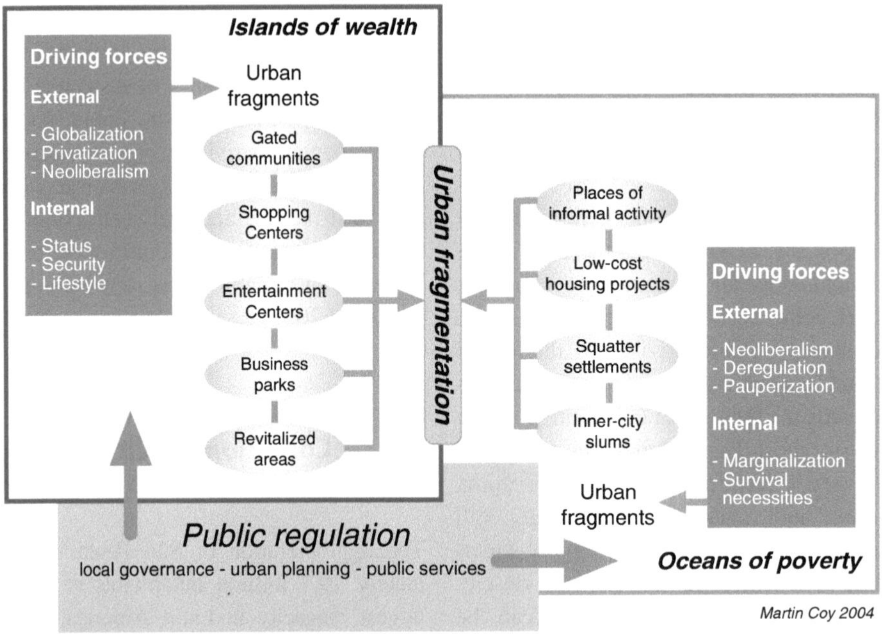

Fig. 8.2 Fragmentation of Brazilian cities

considerable population losses. Thus, for instance, the residential population in the actual city centre decreased by a third in the relatively short period from 1980 to 2000. Likewise, the number of inhabitants in the better situated accommodations near the city centre in the west as well as in the simpler lodging quarters east of the centre decreased clearly. Increases, however, were registered in the quarters located at a greater distance from the centre in the northeast, east, southwest and south (cf. Meyer et al. 2004).

A result of this generally highly dynamic but geographically much differentiated growth process is a metropolitan region which is deeply fragmented in terms of its socio-economic and spatial development dynamics as well as regarding its structures and everyday living conditions (cf. Santos 1990; Villaça 1998 as well as different contributions in Carlos and de Oliveira 2004) (cf. Fig. 8.2). The background for these developments is formed by different novel types of socio-economic and spatial fragmentation which include citadel formation and ghettoisation. They are influenced by processes of privatisation, deregulation and flexibilisation, which are to be highlighted as basic characteristics of the Latin American and Brazilian urban developments of the last years (cf. Coy 2002 as well as Coy and Pöhler 2002). At the same time, altered constellations of actors (e.g. due to the increased significance of transnational enterprises), new political framework conditions (e.g. neoliberalism) as well as diverging value standards and lifestyles under the influence of globalisation are contributing massively to disintegration, disorganisation and destabilisation processes (cf. on the theoretical bases "fragmenting development", Scholz 2002).

During the last few years, against the backdrop of the metropolitan growth patterns and the effects of urban fragmentation, a change in inner-city development has been observed in São Paulo, characterised by mutually induced processes of an overload of traffic capacity, edificial degradation, displacement of the middle-class resident population, a loss of location image and the subsequent outflow of service providers as well as the decline of the formal city centre trade. The inner city of São Paulo is formed by the eight districts of the *Subprefeitura* Sé, in which approximately 370,000 people live (in 1980 the number was about 530,000) and 440,000 work – above all in retailing, public administration, in the finance sector (primarily in the district Sé), but also in the textile industry (mostly in the district Bom Retiro), which today is dominated particularly by Korean immigrants, who employ mainly Bolivians without legal documents in their sweatshops.

The difference between day and night populations in the city centre is extreme: at night completely deserted streets show during the day an hourly frequency of 6,000 passersby and more. The city centre of São Paulo is divided into two areas partitioned by the Vale do Anhangabaú: the so-called old centre (*Centro Velho*) and the new centre (*Centro Novo*). Since the first half of the twentieth century, above all the *Centro Velho* has developed, next to the traditional streets of the retail trade, into the centre of the finance sector with headquarters and branch offices of domestic and foreign banks as well as the stock exchanges. The *Centro Novo* is, in addition to numerous offices, the location of in particular the retail trade, hotels and cultural establishments (*Teatro Municipal*, etc.).

In the past, many of the wide boulevards here were considered high-quality residential addresses, and some still are even today. Characteristic of both central areas are the high building density and the enormous tendency to vertical, that is, multistoreyed buildings that commenced already in the 1920s with the construction of the first high rises, following the North American example (cf. Somekh 1997; Souza 1994). North of the centre, the areas of the large railway stations are located, which were of importance for the trans-regional interconnectedness of the city. Likewise, numerous residential districts in the vicinity of the centre showed traces of ethnic and social segregation early on, and have retained them until today, such as the predominance of specific immigrant groups (e.g. Brás and Bela Vista as quarters dominated by the population of an Italian background and Liberdade as one influenced by Japanese-Brazilians).

Since the 1950s and 1960s, based on the increasing compression of office and retail trade functions in the spatially limited centre area, the enormous transportation load, both public and private, and the lack of investment into available resources, a shift concerning the traditional functions of the centre, could be observed (cf. Fig. 8.3). The city was modified to meet the growing demand of an increasing traffic load through the extension of road axes and above-level streets. This went hand in hand with an associated deterioration of local community conditions as well as a shortage of space reserves for the booming tertiary sector. Thus, for several decades, an almost cyclical realignment concerning residential functions as well as high-ranking service locations took place from within the core city towards an area southwest of the city centre. Already relatively early, high-value residential areas emerged here, for example, so-called *Jardins* (Gardens).

At the latest since the 1970s, a trend has developed in regard to high-ranking residential function which is strongly determined by real estate firms: there has been increasing migration into isolated gated communities equipped with all the necessary infrastructures. These "ghettos of the rich", emerging at the cities' outskirts, are geared to meeting the desire of many city dwellers to achieve their personal happiness in rural harmony and in security without, however, having to do without the advantages and amenities of the city. Nonetheless, favourable traffic connections, above all, are decisive for the success of the new settlements. Consequently, through the planned combination of residential function, work places (with an emphasis on the service sector) as well as shopping and leisure facilities, some particularly large projects are emerging that are developing into edge city-like settlements.

The certainly best-known and most impressive example of this is the privately planned and realised settlement complex *Alphaville* in the north-western part of the metropolitan region of São Paulo, with approximately 34,000 inhabitants and an estimated day population of more than 100,000 persons (cf. to this Caldeira 1996, p. 311 et seqq. as well as Coy and Pöhler 2002). A new type of urban gated community has emerged in São Paulo in the last couple of years with the so-called *condomínios horizontais* (cf. Fig. 8.3). These are smaller, highly secured and mostly luxuriously equipped housing complexes, erected by real estate firms in quarters

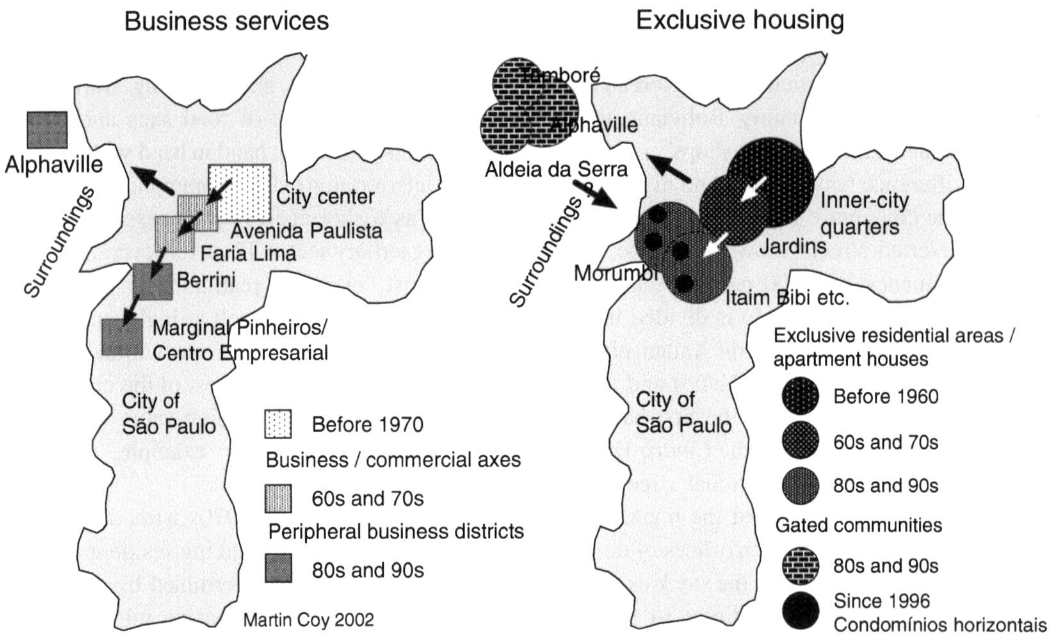

Fig. 8.3 A model of the temporal-spatial shift of business services and exclusive housing in São Paulo

near the inner city after the demolition of the previously existing structures. The *condomínios horizontais* combine the housing comfort of the single-family house with the advantage of inner-city vicinity and therefore also cater to the trend observable among residents of urban edge *condomínios* to return into the core city, due to the long commuting time between home and work place.

At the latest since the 1980s, the inner city as a high-ranking business location ultimately has had to make room for new office cities. This process already began earlier with the transformation of the Avenida Paulista, formerly marked by the splendour villas of the coffee barons, into a high-rise building ravine that today fulfils functions of the central axis of the office and finance sectors; however, in its immediate neighbourhood it simultaneously houses entertainment services such as movie theatres. The repositioning process of the service sector continued in the 1970s, 1980s and 1990s in a south-western direction up to the boundaries of the core city. In these areas of the city, numerous office high rises evolved, equipped with all the modern communication infrastructures (Meyer et al. 2004). Up to this day, it is here where the highest yields of the Paulistano real estate market are obtained.

Along these lines in the last years, in particular along the Avenida Eng. Luis Carlo Berrini and the Avenida Marginal Pinheiros – controlled by the private real estate sector and tolerated by the public institutions – one of the largest and most modern office cities of Latin America has developed. It presents itself as an enclave-like, globally connected "control headquarters", which, until now, has insignificant urban qualities, despite its function as an economic centre (cf. Fix 2007). The public space, for example, no longer plays a significant social or communicative role in these postmodern mono-functional office citadels. The "social costs" of the expansion of these office cities are consequently often high. Thus, numerous lower-stratum quarters with innumerable socially disadvantaged households in the area of the Avenida Berrini had to uncompromisingly yield to profitable office towers (cf. to this in more detail Frúgoli Jr 2000 as well as Fix 2001).

The concept of the better place residential areas and the high-ranking service functions also keeps pace with modifications of the consumption habits and locations that have influenced the retail trade in the "old" centre substantially. São Paulo – as the first Brazilian and Latin American metropolis – has since the mid-1960s adopted the North American model of the shopping centre, which very early on, especially in the south-western expansion zones of the urban development, fulfilled functions of centre development. The first shopping centre in the city emerged in 1966, the *Iguatemi* at the Av. Brig. Faria Lima, which until this day is considered one of the highest-quality shopping precincts in the city. Towards the end of the 1990s, 50 shopping centres existed alone in the municipality of São Paulo, the largest each accommodating over 400 businesses.

In particular in the 1980s and 1990s, numerous new shopping malls developed, even in areas more distant from the city centre. With the exception of the centres that offer high-end goods and services, these also increasingly included those which were based on a clientele with a lower purchasing power in their catchment areas. Apart from the pure business function, the shopping centres as entities produced and controlled by private capital today increasingly are taking on communication and leisure functions in the form of, for example, cinemas, restaurants and open plazas that previously were mainly located in the city centre and in public spaces. This too can be seen as indication of an increasing tendency towards social fragmentation of urban society and urban space (cf. Frúgoli Jr 1995).

The new city fragments and enclave-like centre structures have in common that they are usually owned and developed by single real estate companies who are mainly interested in capital utilisation and for whom the public hand merely creates the conditions for development. In the competition with these new centre structures, the inner city is losing more and more ground. The consequences for the inner cities are high vacancy rates as well as a degradation of public space. Thus, the increasing occupation of public spaces, in particular by the informal street trade, both in the *Centro Velho* and in the *Centro Novo*, is generally considered an indication of change in the inner city. Larger groups of street vendors can be found frequently in public spaces on some traditional shopping streets.

According to Maricato (1996, pp. 49/50), the proportion of the working population employed in the informal sector roughly quadrupled in São Paulo between 1982 and 1992, which is to be rated as an indicator for the deteriorating socio-economic situation in the city. At the same time, the informal street trade for years has been the focus of political and social conflicts around the development of the inner city. Since the 1990s, attempts have been made to contain the street trade through various police operations of varying intensity. Simultaneously, many street vendors became victims of illegal collection activities of so-called "protection funds", thus being drawn into one of the largest corruption scandals in the history of the São Paulo city administration (cf. Frúgoli Jr 1999).

A further aspect that plays a crucial role in the discussion concerning the degradation of the inner city is the increase of prostitution, violence and crime (theft, etc.), whereby many parts of the inner city have become danger zones in the perception of the city population. These developments, among other things, have substantially contributed to the prosperous population groups moving permanently away from the highly condensed inner-city areas of São Paulo in the last years. Thus, public space is affected by spatial and socio-economic fragmentation, particularly in the inner-city areas. Public space is commonly considered a place of daily encounter and communication, associated with the conflicts resulting from a diverse, heterogeneous society that exhibits variety and cultural diversity.

Here it is possible that citizens with different interests and expectations interact with one another, leading to sometimes conflicting discussions; such encounters can be new experiences and serve to expand participants' horizons and stimulate new ideas. Public space is thus inseparably connected with the social life of a society (Borges 2005, p. 20 f.). However, everyday life in Brazilian public plazas, in parks and other public spaces looks different at times. Frúgoli Jr (1995), p. 37 et seqq.) describes the use of public spaces close to the inner city in the case of São Paulo and makes clear that it is first of all members of lower-income groups and other disadvantaged segments of society who use these open spaces. Therefore, in the public perception – that is, that of the middle and upper classes – these open spaces are considered dangerous and derelict and are therefore to be avoided.

A possible strategy for the members of the middle and upper classes to escape the alleged danger in public spaces is to retreat into private areas. It is common for these private spaces that using and gaining access to them requires substantial financial means, and they are therefore only accessible for the high-income population. The availability of these private spaces to a segment of the population makes this group at least to some extent independent of public spaces. As a result, privileged city dwellers use the public space more and more rarely; due to the availability of privatised areas, it has become secondary and even less attractive and is consequently avoided (Frúgoli Jr 1995, pp. 90, 105). Public space in the form of roads and plazas is serving traffic and transit to an increasing extent, and long-term stay becomes ever rarer (Borges 2005, p. 30). In contrast to public space, a limited spectrum of purposes which is regulated by private utilisation rules characterises private spaces (Frúgoli Jr 1995, p. 102 f.).

The growing importance of private spaces with a simultaneous loss of the importance of public space entails two consequences. The growing utilisation of private space is on the one hand part of a process of self-segregation that is characterised by a decreasing interest in the city itself and instead serves the isolation and protection of certain privileges. Sennett considers this a "celebration of the ghettos" and sees a tendency that citizens living in these conditions are giving up the chance to extend their individual impressions and experiences, with the consequence that they lose the possibility to question their once-accepted circumstances of life and to tolerate differences (cf. Souza 2000, p. 207). On the other hand, public space increasingly is losing its meaning for the middle and upper classes, so that their retreat is leading to a social homogenisation of the user groups of public spaces; such groups subsequently consist of large parts of low-income residents (cf. Fig. 8.4).

Overall, the problems and processes addressed here show that the functions as well as the dynamics between the actors in the inner city have fundamentally changed during the last decades. Functions that were formerly fulfilled by the inner cities and their public spaces are now increasingly being taken over by the newly developing city fragments. This process is intensifying to this day, so that for those who can afford it, the "artificial worlds" of the shopping centres and the housing and leisure ghettos now embody the places for the realisation of their consumption and representation needs as well as the new social contact spaces (cf. also Frúgoli Jr 1995). The socially weaker and the poor, for whom the inner city is still an important space of action and survival, are remaining behind. The group-specific everyday action spaces of the privileged, in contrast, are increasingly concentrated in enclaves with controlled access (residential ghettos, shopping centres and business parks). Consequently the dividing lines between public space and that controlled by private capital are increasingly becoming insurmountable. Fragmentation is becoming much more visible, also on a small scale, so that the Brazilian city today is more than ever in accordance with the global pattern of "islands of wealth in oceans of poverty" (cf. Scholz 2002).

8.3 Actors and Conflicts of Interest in the Sphere of Inner-City Development

To be able to understand the existing open and virulent conflicts in the inner city of São Paulo, it is reasonable to compare the actors and their

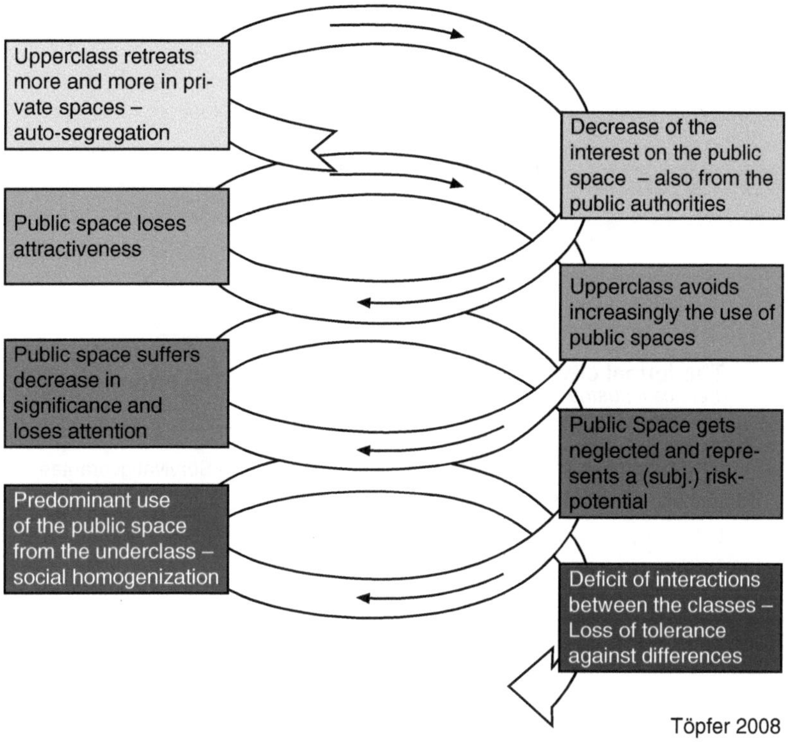

Fig. 8.4 Spiral of degradation of the public space and possible causes and consequences

different perceptions of existing problems (cf. Fig. 8.5). Thereby, actors who shape the "formal city" can be summarised in a first group. Their perception of the inner city is above all determined by the criteria of image and quality of location and environment. In this context, a special significance can be attributed to the nongovernmental organisation *Viva o Centro*, which was established in 1991 under substantial participation of the Brazilian branch of the Bank Boston, located in the *Centro Velho*. It had a crucial part in counteracting the loss of quality and image of the inner city in its role as enterprise location (cf. as a critical analysis of the activities of *Viva o Centro* Oliveira 1999; Frúgoli Jr 2000, 2001 as well as Moser 2006).

Besides the aforementioned bank, numerous other firms from the finance sector, as well as different interest groups and professional retail organisations represented in the inner city, participate in the relatively unconventionally assembled nongovernmental organisation. On this basis, a professionally operating organisation has developed that represents the concept of inner-city revitalisation with planning competence and scientific claim as well as with high-publicity marketing. At the same time, from the start, it has focussed not exclusively on lobbying and the application of political pressure but also on the development of its own concepts as well as their implementation. Even though the NGO has never lost sight of its own economic interests, it has nevertheless succeeded in putting the inner city on the local political agenda and in stimulating public-private partnerships in a wide range of domains. Thus, in the past years different large events were staged during which many different problems of the inner city as well as adequate planning projects were discussed. Furthermore, for many years *Viva o Centro* has skilfully pursued lobby work, with which the city administration was successfully forced into the formulation and at least partial realisation of a specific city centre policy.

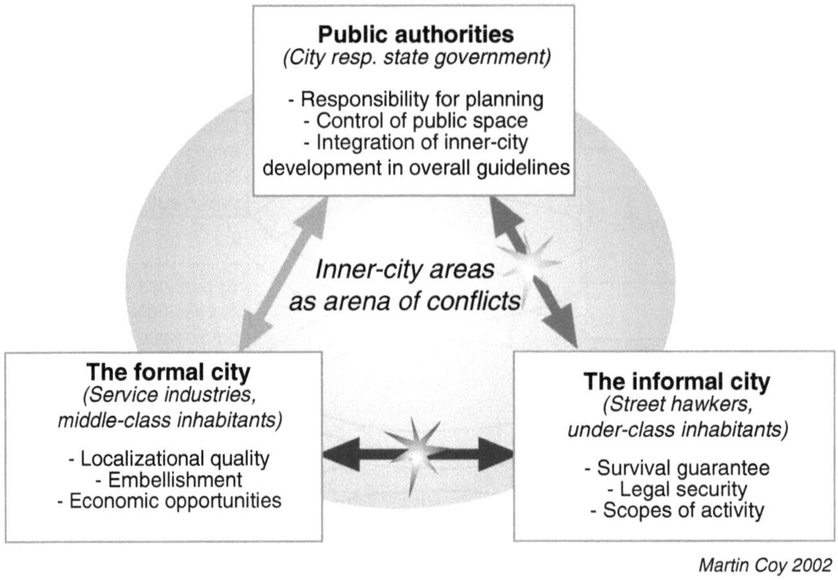

Fig. 8.5 Actors in the inner city of São Paulo and their interests

Viva o Centro meanwhile, in a special programme under the name *Ações Locais*, supports more than 100 private initiatives of businesses and residents in many different parts of the inner city (cf. Moser 2006), whose aim is not only to point out shortcomings in the closer surroundings but also to implement rehabilitation measures in self-responsibility and self-regulation (a rather successful example in this context is the reorganisation and revitalization of the *Grandes Galerias* on Largo do Paissandu.) Overall, nonetheless, the vested interests of the economic circles involved are in the foreground. From their perspective, the degradation and misalignment tendencies described above represent a danger to the image of the location in the long term, which will also affect the firms that have remained in the inner city.

The organisation is still an influential and highly successful actor that in the meantime has found imitators, also in the newer centres. In the mid-1990s – under the leadership of a former prefect and supported by a major bank (Banco Itaú) – the Associação Paulista Viva evolved, which has taken upon itself the task of maintaining the location quality and image of the Avenida Paulista, depicted as a "symbol of the city", through public relations, political pressure and community actions (e.g. in the sense of "beautification" as well as through a reduction of the traffic load and an improvement of the "security situation" through the eviction of the street vendors). Even in the area of the new service centre on Avenida Berrini, an NGO (the *Associação de Promoção Habitacional*) has been active for several years. It is comprised of local enterprises (first and foremost the real estate firm Bratke-Collet, which has built a majority of the high-rise office blocks of the new service centre). Its objective is to, in cooperation with the public sector, "resettle" the lower-stratum population that has remained in this part of the city through strategic measures and to in this way lastingly protect its own "capital usage interests" (cf. Frúgoli Jr 2001).

A second, growing group of actors – low-income residents, homeless and street vendors, among others – is part of the informal sector. Specific vulnerabilities and risks influence this group's perception of the inner city, which is viewed as a survival space. Their principal interest is to create and maintain room for manoeuvre. During the last years, parallel to the retreat of the better off, low-income segments have increasingly moved into the inner city and into quarters close to it. Degraded or vacated residential and office buildings have been occupied without permission, and property or renting conditions remain unsettled, leading to an altogether extremely precarious housing situation which is typical for the informal living situation, referred to as *cortiço* in São Paulo (cf. Kowarick and Ant 1994).

According to recent information, approx. 600,000 persons in the core city of São Paulo – with rising tendency – live in *cortiços*, many of them in quarters close to the inner city. The propagation of this development in the inner city, problematic from a social point of view, is promoted by comparatively high vacancy rates (in some city areas up to 30% of the housing stock) (cf. Sampaio and Pereira 2003). However, the living situation of the (according to official estimates) approximately 14,000 homeless in the São Paulo inner city presents itself as still far more precarious than that of the *cortiço* residents. After all, the number of the informal street vendors has clearly increased during the last years in the inner city of São Paulo. According to the *Sindicato dos Trabalhadores da Economia Informal*, by the end of the 1990s, approximately 16,000 ambulatory traders were purportedly active in the inner city of São Paulo, from whom at most a third were recorded by the registration attempts of the city administration. Akin to the actors of the "formal city", those of the informal city have also organised themselves in order to be able to represent their own interests more effectively.

Already since the end of the 1980s an organisation of the *cortiço* residents has existed, *Unificação das Lutas dos Cortiços* (ULC), which acts on their behalf against evictions by the police and for the supply of dignified living space. At the end of the 1990s, other organisations emerged from this organisation, in the sense of political spin-offs, like the *Fórum dos Cortiços*, the *Movimento de Moradia do Centro* (MMC) and the *Movimento dos Sem-Teto do Centro* (MSTC), which all engage in the "entitlement to inhabitation" in the inner city (cf. Teixeira et al. 2005, p. 5 et seqq.). Particularly in cases of squatting (i.e. house occupations) and the frequently occurring subsequent evictions by the police, these inhabitant organisations publicly appear as supporters of the affected persons.

Then, at last, in 1999 – as an "antipole" so to speak to *Viva o Centro* – the *Fórum Centro Vivo* was formed as an "umbrella organisation" and as the representation of the interests of the disadvantaged actors in the Paulistano inner city. It offers an organisational basis for the collaboration between the aforementioned residents and labour unions, culturally creative as well as basis-oriented research establishments and other social movements for the purpose of a "democratic inner-city development". Their principal demands aim at the entitlement to inhabitation, the entitlement to work and the observance of civil rights for all user groups in the Paulistano inner city. Finally, a third group of actors is to be allotted to the public institutions. Its actions are aimed at process control or at least problem control. At least four stakeholder levels are to be named in connection with this: first, a national level to which, for example, IPHAN, the national office for conservation of monuments and historical preservation, is to be counted. The Federal State of São Paulo as the second level is among other things involved in the residential construction authority CDHU (*Companhia de Desenvolvimento Habitacional e Urbano*) in the production and regularisation of housing lots in the Paulistano inner city; it is jointly responsible with the *Polícia Militar* for the increasingly more important security questions.

The numerous municipal authorities represent a third level. On the one hand they have to manage all important issues of everyday life in the inner city (e.g. traffic, living, provision and disposal, security and trade supervision and culture); on the other hand, a coordinating function is taken over by the city planning department SEMPLA (*Secretaria Municipal de Planejamento Urbano*), whose major task in recent years has consisted of compiling a new master plan for the development of São Paulo for the period of 2002–2012, in accordance with the rules of the national city legislation (SEMPLA 2004). Finally, a municipal project developer,

EMURB (*Empresa Municipal de Urbanização*), has been appointed the responsibility for the concrete execution of projects of city development and city reconstruction.

Besides these municipal institutions which are responsible for the entire municipal level, a few years ago, as a fourth level in the course of a municipal administration reform with the aim of decentralisation, so-called *Subprefeituras* were created which should provide better public relations with the city administration in 31 municipal districts. In this framework, the *Subprefeitura* Sé is responsible for the concerns of the inner city. From this rather cursory compilation of the numerous public institutions that are involved in one way or another in questions of inner-city development, it can already be determined that the result will be multiple horizontal (e.g. between different municipal offices) as well as vertical (e.g. between state and municipal authorities) coordination and voting requirements. If the respective political governance is comprised of members of different parties, the latter can lead to serious conflicts and even to a regulatory impasse.

From the interaction of these different groups of actors, numerous fields of conflict arise that are reflected in at least three conflict dimensions. A first conflict dimension arises through the debate on the everyday problems of the inner city. Thereby, the form of pursuit of conflict reveals itself in the respective actions motivated by the special interests of the various groups of actors in the subsequent public debate (e.g. through the local press) and at last in the "stout" intervention of the police or other authorities. Typical examples of such everyday conflicts can be found in the environment of the street trade (e.g. expulsion), squatting (e.g. evictions) or the degradation of the public space.

A second conflict dimension appears in highly contradictory discourses lead by the different groups of actors about the inner city and its functions (cf. to this more detailed Frúgoli Jr 2000, 2001). In the inner-city discourse of the actors who are to be apportioned to the "formal city", aspects such as location quality, quality of stay, security, cleanliness or the situation of the (individual) traffic play a central role. Background for this is a conception of the city centre and its functions, which is derived on the one hand from its specific local and historic, socio-economic and cultural importance and on the other hand is based on the general model of the European (partly also North American) city as well as on a general "cosmopolitan city discourse".

In contrast to this, the inner-city discourse of the representatives of the "informal city" is much more determined by conceptions of a "democratic city", from principles of the satisfaction of basic needs, participation, multiculturalism and above all a guarantee of survival. In this sense, questions of the safeguarding of living space, the better provision with basis infrastructure, the promotion of small trade and the prevention of displacement processes take precedence. The inner-city discourse of the representatives of the public sector sways in certain respects between both represented poles, depending upon political alignment and the necessity to "serve" the interests of the respective political clientele.

A third conflict dimension finally arises in the discussion around the objectives of renewal strategies, which can, according to the predominant discourses, be very differently aligned. It concerns essentially the question for whom inner-city renewal should be pursued. While organisations like *Viva o Centro* assign priority to a beautification of the city centre, an economic revitalisation for the purpose of competitive power versus the newly emerged office cities and business centres as well as cultural and tourist upgrading; while these organisations strive implicitly for gentrification and thereby gratify the notions of the economic circles and the more affluent city dwellers, the groups represented in the *Fórum Centro Vivo*, on the other hand, advocate renewal strategies that serve the social and economic stabilisation of the vulnerable inner-city groups (street vendors, squatters, the homeless, etc.). These different strategy discourses mark the corner points by which the specific renewal programmes and measures are determined in the respective local political and economic situation and the respective interest relations.

8.4 Inner-City Renewal: Local Politics in a Conflict of Interests

Up to this day the normally limited scope of action at the local level can be emphasised as a common characteristic of city politics in Latin America, which is among other things caused by the desolate financial situation of the municipal households. Thus, under the omens of neoliberal politics, as it has been practised by many Latin American countries in the last years, the framework conditions for city politics have generally worsened. Nevertheless, the political concepts and instruments on the municipal level for resolving the issues in the multilayered urban problem areas are changing in many places.

Depending on political framework conditions, city planning, infrastructure and renovation measures as well as residential construction, politics are either primarily regulatory, that is, programmes and directives are driven in a "top-down" manner, or they give priority to locally adapted attempts towards implementation with stronger participative components, for example, in the context of general decentralisation attempts (cf. Coy 2003). The recent legal efforts in Brazil towards a reorientation of city policy and planning in the sense of a city reform essentially correspond with the attempt to call for and to realise the "right to the city" for all social groups of the urban society. The articles of the Brazilian constitution of 1988 dedicated to city policy (Articles 182 and 183) are based on this principle as well as on the *Estatuto da Cidade*, passed in 2001 as an executive law, whereby the latter grants an extensive scope of action as well as a variety of innovative planning-based steering instruments to the cities.

After a long phase of undifferentiated modernisation, politicians and planners in many Brazilian cities have in recent years recognised the problematic nature of the inner-city blights and have initiated renovation projects in historic cities and the revitalisation and renewal of inner-city areas. A common aim of redevelopment programmes in historic old towns and inner-city renewal projects is to preserve the local historical heritage (building substance, public spaces, etc.) on the one hand and to increase the urban attractivity of the city centre areas on the other while simultaneously creating the infrastructural framework conditions for the improvement of inner-city locations (i.e. trade function as well as residential function).

In many cases, however, the revitalisation of the city centre is connected to a risk of socially weaker groups of residents and other actors (in particular members of the informal sector) being displaced from the inner-city areas, initiating a process of gentrification – either deliberately or unconsciously. Likewise, it is to be observed that, as a consequence of some renovation projects, instead of supporting an organic and vital inner-city life, support is given to a rather sterile "musealisation", especially of the historic-city areas, from which at most the tourist industry can benefit (in Brazil, a cited example for this tendency is the renewal of the *Pelourinho* in Salvador, cf. Augel 1998).

In the largest city of Brazil, in view of the degradation tendencies described above, enhancing the attractiveness of the city and the revitalisation of the inner city attained an increasing importance in the 1990s and continue to be of great interest to this day (cf. Simões Jr. 1994, p. 37 et seqq.). At the same time, based on the prevailing building substance, classic historic-city rehabilitation is of a lesser concern, and the fundamental question arises, which parts of the Paulistano inner city are identity defining and in fact worthy of preservation. However, São Paulo's phase of ascent-and-blossom of the early twentieth century manifests itself in places and buildings, in the early high rises and indeed in the whole city layout. In view of the described decay and the massive conflicts, the concern is essentially about the recovery of an urban quality that the centre undoubtedly once possessed as well as the determination of the functions that can help accomplish this quality. Thus, all city governments of the last years have declared the revitalisation of the Paulistano inner city a field of prioritised activity that serves various superordinate objectives and must be reflected in different types of measures within municipal politics (cf. Fig. 8.6).

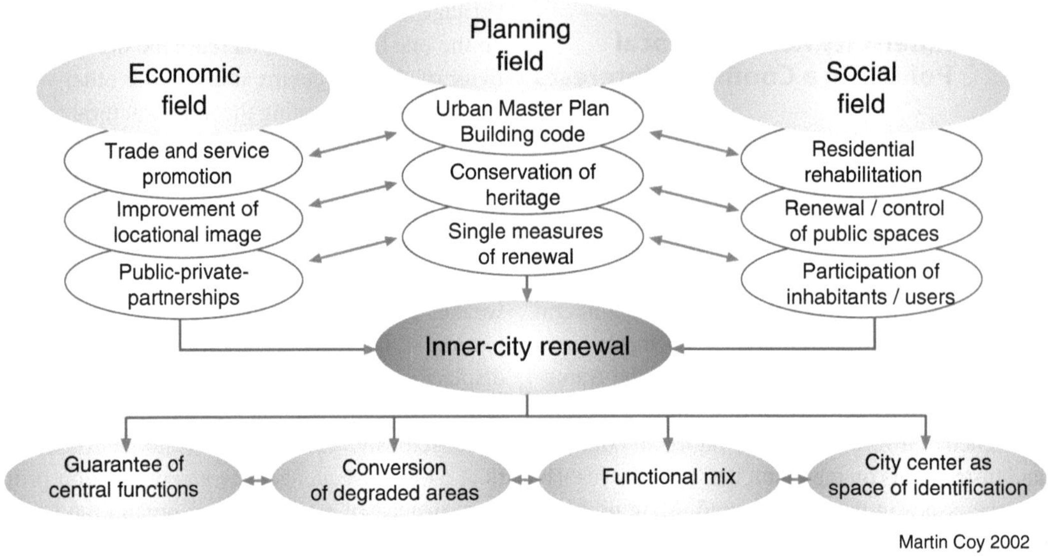

Fig. 8.6 Fields of activity and objectives of inner-city renewal in São Paulo

In the municipal election campaign in 2008, however, the subject "city centre" played no dominant role among the candidates vying for office.

From the end of the 1980s on, measures for traffic abatement in the inner city and for the improvement of the quality of public space were at the core of a policy of inner-city renewal in São Paulo. In this way, numerous streets of the *Centro Velho* and the *Centro Novo* were converted into pedestrian precincts whose length meanwhile amounts to approximately 10 km. A large-scale project of special importance was the successful transformation of the Vale do Anhangabaú from a noisy traffic artery to a public space with high quality of stay. In addition, there is the renovation and repurposing of important buildings: for example, the restoration of one of the oldest multistoreyed buildings of the city, the *Edifício Martinelli*, which today accommodates different departments of the city administration; the renewal of one of the central railway stations (*Estação Júlio Prestes*) and its partial transformation into an enormous concert hall (*Sala São Paulo*); the preservation of the former administration building of the electricity company *Light* and its conversion into a shopping centre; the adaptation of the former central post office, turning it into an arts centre, to date not yet opened; the transformation of the railway station *Estação da Luz* into a museum (*Museu da Língua Portuguesa*); as well as many other projects.

Overall however, the renewal projects realised so far mostly fall into the category of more or less successful "islands of restauration" in a still degraded environment, and the principal question therefore should be to whom this concept of inner-city renewal should be useful. This depends essentially upon who seizes the initiative and who prevails. The city administration has striven for some years, with varying success, to coordinate the numerous individual projects under the umbrella of a master plan and to integrate them into a higher framework of the city development (to view the presently valid *Plano Diretor Estratégico*, cf. SEMPLA 2004). Likewise, the framework conditions for the persistent renewal of the centre are to be created in special programmes (so-called *Operações Urbanas*) with adapted building codes and incentive systems. Particularly the firms located in the centre have large interests in these revitalisation measures. This is also recognisable in the fact that in the context of numerous public-private partnerships, the private initiative is attaining an increasing importance for the realisation of the projects (cf. *Ações Locais*).

Which value has inner-city renewal had for the agendas of the city governments of the last years? The respective priorities and concepts are closely connected with the framework conditions of urban governance, and, according to the political colour of the city governments, different external influences are to be considered dominant. While under Mayor Luiza Erundina, belonging to the political left, at the beginning of the 1990s, the first inner-city renewal measures (e.g. in the context of the *Operação Urbana Anhangabaú*) were undertaken, quite different objectives could be perceived under the successive city governments of the prefects Paulo Maluf and Celso Pitta, both belonging to the conservative camp. While inner-city renewal was not a priority for prefect Maluf, who subscribed to a modernisation of the city "at all costs", in the second half of the 1990s, a city centre policy committed to upgrading – claimed also by *Viva o Centro* – was pursued: execution of individual renovation projects and, above all, the combat against the informal sector were the key points of the inner-city politics of prefect Pitta.

At the beginning of the new millennium, with the assumption of office by Marta Suplicy as Mayor of São Paulo, nominated by the leftwing labour party PT (*Partido dos Trabalhadores*), a clear strategy turn took place. Thus, under pressure from supporters' groups that were organised in the *Fórum Centro Vivo*, among others, the social function of the inner city was moved into the foreground, whereupon a holistic new strategy for the inner city which went beyond upgrading intentions and object redevelopment was to be implemented. In this context, the field of strategic city planning is to be named first of all. The new master plan designed by the PT government for São Paulo, *Plano Diretor Estratégico* (SEMPLA 2004), sees itself as being committed explicitly to the social function of property and the principle of the right to the city for all residents. For the implementation of these principles, new steering tools, which are provided by the federal city law, are deliberately installed: the possibility of expropriation, construction obligations, purchase options, building law transmissions, defining of redevelopment areas with special instruments, etc. Basic principles of the master plan are decentralisation and participation. For the 31 subprefectures of the municipality of São Paulo – and also for the inner city – small-scale adapted district zoning maps and district development plans were compiled in cooperation with the different stakeholders, in which, besides new types of general zoning, specific developmentally assisted areas and the scope of action are defined by special programmes of renewal.

Three new priorities can be named here: the transformation of former railway and industrial areas, the strengthening of centre functions and public space as well as, above all, the definition of social priority zones in property regulation and the housing renovation (so-called *Zonas Especiais de Interesse Social*). For the implementation of the perspectives given in the *Plano Diretor*, different special programmes were mentioned. In this context the *Ação Centro* programme is undoubtedly the principal item of inner-city renewal (cf. for a critical evaluation of this programme Teixeira et al. 2005). It consists of 130 subprojects, and its total costs of 170 million US$ are financed by a loan from the Inter-American Development Bank. The programme's objectives represent the ambitious to provide proof that inner-city renewal, if it is conceived mindful of social balance, must not inevitably lead to displacement.

Three of the programme's fields of action shall be mentioned as examples. Revitalisation projects should be guided by objectives of historic-site preservation, but at the same time they should place the functional aspect in the foreground and should likewise be adapted to the surrounding environment. A positive example of a revitalisation project is the renovation of the Paulistano market hall *Mercado Municipal*. Principal aspects of the programme are on the one hand measures which support social inclusion, above all encompassing those active in the informal sector, and on the other hand measures linked to the area of social housing construction. The target group of the sub-programme "Living in the Centre" (*Morar no Centro*) are low-income members of society; the measures are aimed with priority at *Favela* and *Cortiço* renovation. A specific example is the renovation of the 26-storey, strongly degraded and in part irregularly occupied high-rise *Edifício São Vito*, which was initiated in 2004. The project is based on the cooperation of different institutions with the objectives of renewing approximately 600 housing units for nearly 2,000 persons, of providing intermediate housing and of guaranteeing that the original residents are able to return after the renovation is finished.

However, just as in many other places, political "life cycles" are extremely short in São Paulo, so that with the accession of a liberal city government under prefect José Serra (later replaced by Gilberto Kassab) at the beginning of 2005, a further trend reversal was initiated which, concerning inner-city renewal, indicated a return to the paradigm of upgrading. In this context, even today questions concerning the future of the *Ação Centro* programme arise. The goal of a strategic reorientation has already been clearly formulated: the target group for "living in the centre" should no longer be the lower class, but rather the middle class. This affected, for example, the continuation of the renovation of the *Edifício São Vito*. After a certain period of vacancy without any noticeable ongoing renewal measures, the municipal government decided to demolish the building to make room for cultural and related facilities.

8.5 Scenarios of Inner-City Development in São Paulo

In conclusion, the processes of change and possible development perspectives of the Paulistano inner city in the context of current conflict constellations and fragmentation tendencies are shown in a model of scenarios (cf. Fig. 8.7). Scenario I corresponds to what is to be observed in many inner cities of Latin America today: an almost cyclical sequence of displacement processes and translocations, accompanied by an increasing degradation. While the well positioned abandon the inner city as a residential location, marginalised population groups and informal activities expand in the degraded areas of the city centre. Companies as well as cultural and leisure facilities shift their locations into the new island-like city fragments situated in the peripheries, which increasingly adopt centre functions. Therefore, in the final analysis, due to these centrifugal processes, for many cities the danger of a "loss of the centre" exists. Thus, there is an increasing pressure for action with respect to the future design of inner-city areas. Thus, there is an increasing pressure for action with respect to the sustainable design of inner-city areas for the future.

Scenario II describes what is to be diagnosed today in São Paulo as well as in many other places, as a concomitant phenomenon of inner-city renewal programmes: so far measures of rehabilitation are limited all too often to the "beautification" of the city centre and therefore remain incomplete. Thus, "rehabilitated islands" emerge in degraded environments, in the sense of a social "city hygiene" which can in one way achieve impressive results through the conversion of building complexes worthy of preservation. Too often, however, their effects on social and socio-spatial integration remain doubtful as these measures are usually implemented at the expense of the socially weaker members of society. This can be seen as a consequence of the growing importance of private-sector initiatives, which in the last years have increasingly begun to fill the vacuum in municipal politics, caused in part by neoliberal framework conditions. As a result, the danger arises that the fragmentation of the inner cities deepens even further.

Scenario III even today still represents a vision of the future, for São Paulo as well as for most Latin American cities. Nevertheless, from our point of view, it fulfils an important function as an ideal model. In view of the inner-city fragmentation processes, a functional *and* socially well-balanced process of renewal is of special urgency. Such a renewal strategy, founded on the basic substance of the inner-city structures, requires the participation of all actors with their often strongly diverging interests. First attempts in this direction were initiated in São Paulo under the administration of Mayor Marta Suplicy. Nevertheless, such a renewal strategy requires above all staying in power, which in reality is missing nearly everywhere due to discontinuities in local politics. This means that

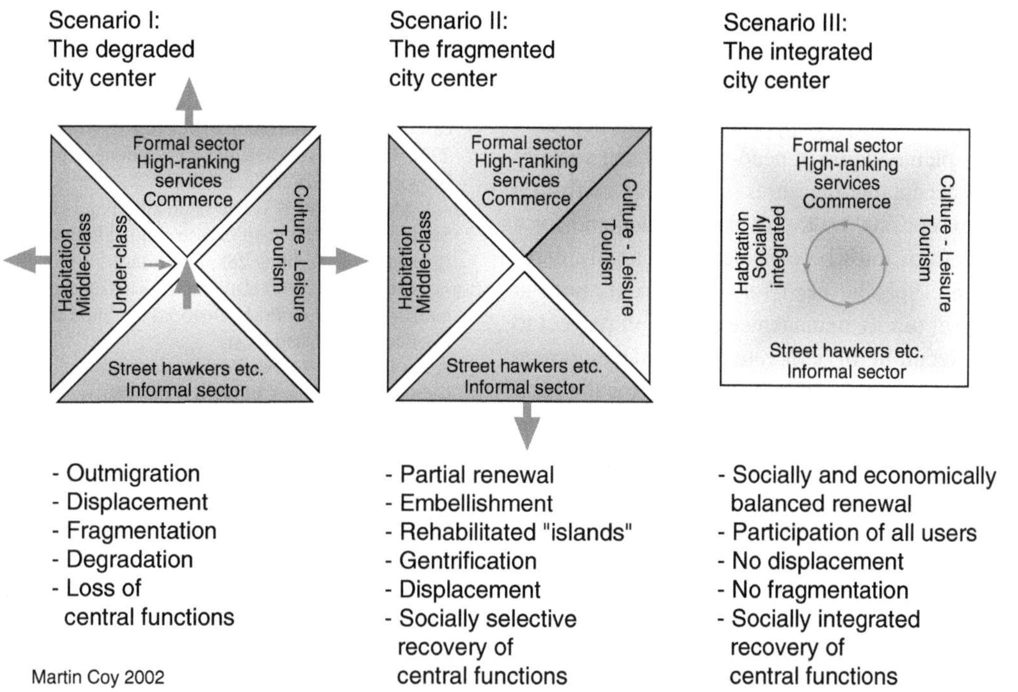

Fig. 8.7 Inner-city development in Brazil: a scenario model

only an economically viable and socially compatible inner-city renewal, sensitive to historic architectural parameters, can be considered an integrative component of the total concept of a sustainable city development, to which more and more Latin American cities profess.

The current situation in Brazilian cities in general and in the megacity São Paulo in particular reflects a disparate society which is not able to lastingly resolve social conflicts and to improve the living situation of the majority of the population, in spite of successes in macroeconomic development and stabilisation in the last decades. Neoliberalism and the integration of the country into a globalised world economy also have had contradictory results and have affected the city in the form of increasing socioeconomic and spatial fragmentation. Thus, the social conflict field "city", as it was presented in the case study on São Paulo, is influenced by a number of most contradictory processes, which lead to ever stronger social and economic divergence. The social groups, however, are affected in different ways in this context. For the rich, money has the function of a "buffer", and it helps them adapt, even to unclear situations (cf., e.g. the gated-community phenomenon). The most vulnerable are, as always, the poorer social classes, which constitute an ever-larger part of the city's residents.

These superordinate conditions also have an effect on whether individual cities can use the new manoeuvring room, which has been created by the re-democratisation and decentralisation processes of the last years in Brazil. Thus, efforts have existed since the end of the 1980s towards a new formulation of city politics and a reorientation of city planning in the sense of a "city reform" that correspond essentially to the attempt to demand and to guarantee the "right to the city" for all social groups of urban society. In this context, meanwhile, a number of cities have demonstrated through innovative best practices that, even under unfavourable framework conditions, possibilities for updating of alternative city politics indeed exist at the local level (Hall et al. 2000) (one example for participative budgetary household planning is Porto Alegre, which by now has found imitators beyond the country boundaries; cf. detailed Coy and Zirkl 2001 as well as Coy 2003).

However, it must be stated in retrospect that numerous obstacles opposed – and still oppose – the implementation of need-oriented and socially more acceptable measures, so that, apart from a few exceptions, up to this day the demand for a city reform must be considered as largely unredeemed. Specific need for action exists both at the level of city management and with respect to the execution of concrete city redevelopment measures. Apart from the measures for the revitalisation of city centres, as described in the case of São Paulo, these are in particular the improvement of the housing situation, solutions for the traffic problems as well as improvement of the management of the communal environment. Many meaningful approaches have failed because of the disastrous financial situation of most cities, or also because of still existing nepotistical political structures.

Does a persistent inner-city renewal generally even have a chance? This depends first of all on the superordinate framework conditions under which it takes place: the elbow room for local politics, the specific structural characteristics and conflicts of interest to be found in the inner cities as well as predominant superordinate spatial fragmentation processes. However – shown by the example of São Paulo – a set of specific factors influences the chances and limitations of inner-city renewal: the change of strategy based on local political discontinuities and frequently connected with personnel changes in the administration, the associated different paradigms of renewal and also most notably the different standpoints of the involved groups of actors, resulting in substantial conflicts of interest.

Undoubtedly, inner cities are never static, but rather subjected to ongoing structural transformation and functional change. If one wants to steer their development proactively, however, one basic question must first of all be answered: "To whom does the inner city belong?" Answers to this question are determined, as the example São Paulo shows, now as ever by the unequal odds of interest implementation by the various groups of actors in the inner city.

References

Augel J (1998) Sustentabilidad de la revitalización. El casco antiguo de Salvador/Bahia, Brasil. Working paper 294, Universität Bielefeld, Forschungsschwerpunkt Entwicklungssoziologie, Bielefeld

Bähr J, Mertins G (1995) Die lateinamerikanische Groß-Stadt. Verstädterungsprozesse und Stadtstrukturen. Erträge der Forschung 288, Darmstadt

Borges D Garreto (2005) Usos e territórios do espaço livre público. O caso da Praça Deodoro em São Luís – MA, Recife, unpublished thesis

Caldeira TPR (1996) Fortified enclaves: the new urban segregation. Public Culture 8:303–328

Carlos AFA, de Oliveira AU (eds) (2004) Geografias de São Paulo vol 1: Representação e Crise da Metrópole, vol 2: A Metrópole do Século XXI, São Paulo

Coy M (2001) São Paulo. Entwicklungstrends einer brasilianischen Megastadt. Geographica Helvetica 56(4):274–288

Coy M (2002) Jüngere Tendenzen der Verstädterung in Lateinamerika. In: Bodemer K, Nolte D, Sangmeister H (eds) Lateinamerika Jahrbuch 2002. Frankfurt am Main, pp 9–42

Coy M (2003) Nachhaltige Stadtentwicklung in Brasilien. In: Struck E (ed) Ökologische und sozioökonomische Probleme in Lateinamerika. Passauer Kontaktstudium Erdkunde 7, Passau, pp 71–86

Coy M (2005) Zwischen Fragmentierung und Erneuerung. Das Beispiel der Megastadt São Paulo. Staden-Jahrbuch 52:161–192

Coy M, Pöhler M (2002) Condomínios fechados und die Fragmentierung der brasilianischen Stadt. Typen - Akteure - Folgewirkungen. Geographica Helvetica 57(4):264–277

Coy M, Zirkl F (2001) Handlungsfelder und Lösungsansätze nachhaltiger Stadtentwicklung in der Dritten Welt. Beispiele aus Brasilien. Petermanns Geographica Mitteilungen 145(5):74–83

Fix M (2001) Parceiros da exclusão: Duas histórias da construção de uma "nova cidade" em São Paulo: Faria Lima e Àgua Espraiada. São Paulo

Fix M (2007) São Paulo – cidade global. Fundamentos financeiros de uma miragem, São Paulo

Frúgoli H Jr (1995) São Paulo: espaços públicos e interação social. São Paulo

Frúgoli H Jr (1999) A questão dos camelôs no contexto da revitalização do centro da metrópole de São Paulo. In: Souza MAA de et al (eds) Metrópole e globalização. Conhecendo a cidade de São Paulo. São Paulo, pp 151–165

Frúgoli H Jr (2000) Centralidade em São Paulo. Trajetórias, conflitos e negociações na metrópole. São Paulo

Frúgoli H Jr (2001) A questão da centralidade em São Paulo: O papel das associações de caráter empresarial. Rev Sociol Política 16:51–66

Hall P, Pfeiffer U (2000) Urban 21. Der Expertenbericht zur Zukunft der Städte Stuttgart, München

Kohlhepp G (1997) São Paulo: größter industrieller Ballungsraum Lateinamerikas. Der Bürger im Staat 47(2):137–143

Kowarick L, Ant C (1994) Cem anos de promiscuidade: o cortiço na cidade de São Paulo. In: Kowarick L (ed) As lutas sociais e a cidade. São Paulo, pp 73–93

Lichtenberger E (1990) Stadtverfall und Stadterneuerung. Beiträge zur Stadt- und Regionalforschung 10. Wien

Maricato E (1996) Metrópole na periferia do capitalismo: illegalidade, desigualidade e violência. São Paulo

Meyer RMP, Grostein MD, Biderman C (2004) São Paulo Metrópole. São Paulo

Moser M (2006) Innenstadterneuerung in São Paulo. Innsbruck, unpublished thesis

Oliveira MAAJ de (1999) Associação Viva o Centro: Estratégias, Propostas e Compromissos para São Paulo, Univ. Mackenzie, Tese de Mestrado. São Paulo

Porta P (org) (2004) História da cidade de São Paulo. vol 3: A cidade na primeira metade do século XX. São Paulo

Sampaio MRA, Pereira PCX (2003) Habitação em São Paulo. Estudos Avançados 17(48):167–183

Santos M (1990) Metrópole corporativa fragmentada. O caso de São Paulo, São Paulo

Scholz F (2002) Globalisierung und Fragmentierung. Eine Welt in "Bruchstücken" In: Ehlers E, Leser H (eds) Geographie heute – für die Welt von morgen. Gotha, pp 121–127

SEMPLA (org) (2004) Plano Diretor Estratégico do Município de São Paulo 2002 – 2012, Secretaria Municipal de Planejamento Urbano. São Paulo

Simões Jr JG (1994) Revitalização de Centros Urbanos. Publicações Pólis19. São Paulo

Somekh N (1997) A cidade vertical e o urbanismo modernizador. São Paulo 1920–1939. São Paulo

Souza MAA de (1994) A identidade da metrópole. A verticalização em São Paulo. São Paulo

Souza M Lopes de (2000) O desafio metropolitano: um estudo sobre a problemática sócio-espacial nas metrópoles brasileiras. Rio de Janeiro

Teixeira A, Chaves C, Comaru F De Assis, Cymbalista R, Sutti W (2005) Conflitos em torno do direito à moradia na região central de São Paulo. – MAPAS (Monitoramento Ativo da Participação da Sociedade) – Relatório de Projeto. Rio de Janeiro

Villaça F (1998) Espaço intra-urbano no Brasil. São Paulo

9 Agile Firm Organisation and Upgrading in the Greater Pearl River Delta

W.-H. Liu, F. Bickenbach, W. Fu, P. Hartmann, S. Meyer, S. Ohm, S. Hennemann, D. Schiller, Frauke Kraas, I. Liefner, J. Revilla-Diez, and R. Soltwedel

Abstract

This chapter analyses an original firm-level dataset collected by our own Hong Kong (HK) Company Survey 2007 to investigate agility patterns applied by HK companies in the Pearl River Delta (PRD) in organising relations to their producers in the PRD and their customers. It also investigates the role of informality proxied by personal relationships for companies in the PRD in governing their production as well as their innovation activities, which have gained greatly in importance in the PRD in the new century. Our findings suggest that different agility patterns can be identified that are applied by companies to organise their relations to their producers and customers. Moreover, the findings show that companies generally tend to vertically integrate their production or innovation activities rather than carrying out such activities under a cooperative governance structure. With regard to cooperative governance structures, there is evidence of a complementary relationship between the importance of informality proxied by personal relationships and the importance of cooperative governance structures both for production and for innovation.

Keywords

Agility • Informality • Personal relationships • Governance • Innovation • China • Pearl River Delta

W.-H. Liu (✉) • F. Bickenbach • R. Soltwedel
Kiel Institute for the World Economy,
Kiel, Germany
e-mail: wan-hsin.liu@ifw-kiel.de

W. Fu • D. Schiller • J. Revilla-Diez
Institute of Economic and Cultural Geography,
Leibniz University Hanover, Hanover,
Germany

P. Hartmann • F. Kraas
Institute of Geography, University of Cologne,
Cologne, Germany

S. Meyer
Joanneum Research Forschungsgesellschaft mbH,
Graz and Vienna, Austria

S. Ohm • S. Hennemann • I. Liefner
Department of Geography, Justus Liebig University
Giessen, Giessen, Germany

F. Kraas et al. (eds.), *Megacities: Our Global Urban Future*, International Year of Planet Earth,
DOI 10.1007/978-90-481-3417-5_9,

9.1 Introduction

Since the beginning of the open-door policy in 1979, China has recorded very high economic growth rates and has developed into one of the three largest economies in the world (World Bank 2009). The Pearl River Delta (PRD) in Guangdong province was the pioneer region of the Chinese economic reforms. In recent decades, it has been one of the regions in China with the highest growth rates and became a well-known "factory of the world" for especially low-priced products in the 1990s (BrandHK 2005). The huge and early investments of the overseas Chinese, especially from nearby Hong Kong (HK), which facilitated the emergence of dynamic private companies on site, were one of the most substantial drivers of its rapid development (e.g. Sit and Yang 1997). With continuously strengthening interactions between economic agents in HK and in the PRD, a mega-urban area known as the "Greater Pearl River Delta (GPRD)" consisting of HK and the PRD emerged.

Companies operating in the PRD, irrespective of foreign or domestic ownership, have mainly carried out lower-end tasks of the global production value chain. Constrained by their limited innovation experiences and the scarcity of financial and human capital, companies tend to focus on searching for ways and means to enhance their production efficiency to maintain their price competitiveness rather than carrying out costly innovation projects. One possible way of increasing competitiveness is to apply agile firm organisation, an organisational innovation aiming at achieving a sustained competitive advantage by relying more on informal business practices to increase firms' flexibility so that they are more capable of seizing opportunities and coping with threats and uncertainties in volatile markets. Informal business practices may be carried out in several different business operations such as organising the labour force, extending and maintaining producer and customer relationships, governing production and innovation activities and financing. This study aims to provide some empirical evidence on agility patterns applied by companies in the PRD in organising their relations to their customers and producers.[1] In addition, given the growing need for upgrading as well as innovation incentives among companies in the PRD as observed by Lu and Wei (2007) and Lin (2004), it is also of considerable interest to see whether the concept of agile firm organisation, which is assumed to be especially relevant for companies producing labour-intensive low-cost products in an institutional environment characterised by deficient formal institutions, is still applicable when companies engage more intensively in innovation activities.[2] Thus, this chapter also aims to provide some empirical evidence on the features of agility patterns applied by companies in the PRD in governing their innovation activities. It compares the empirical findings on the role of informality (a core element of the agility concept) in governing companies' innovation activities with its role in governing their production activities. The analysis is mainly based on an original firm-level dataset collected from our own Hong Kong Company Survey in 2007.

The remainder of the study is organised as follows. In Sect. 9.2, previous literature on informality derived from an institutional economics perspective is summarised. In Sect. 9.3, the research design concerning the institutional and economic background of the research region (GPRD) and the technical information collected in the HK Company Survey 2007 is summarised. In Sect. 9.4, core empirical findings based on the analysis of secondary statistics and the analysis of the survey dataset are presented. Section 9.5 concludes the study.

9.2 Previous Literature

The concept of agile firm organisation complements the flexibility concept by focusing on informality in business practices as a means to achieve

[1] Producers and suppliers are used as equivalent terms in this chapter.

[2] One may argue, for example, that companies carrying out R&D may benefit from stable relations with research partners, for example, universities, and that these stable relations may need to be based on formal and more rigid contractual agreements.

flexibility. Informality here is defined from an institutional economics perspective. Institutions are the formal and informal rules that constrain human economic behaviour (North 1990). According to North, informal institutions can be understood as self-enforcing constraints (e.g. traditions, conventions, norms and customs) that evolve over time, whereas formal institutions (e.g. constitutions, laws, property rights) are intentionally created and enforced by legitimate power. The main role of institutions is to reduce transaction costs by introducing regularity and stability into human interaction. This common distinction between informal and formal institutions can easily be operationalised for empirical analysis. Contract theory reminds us that not all contracts or formal institutions are necessarily always and entirely fixed in a written form. Hart and Moore (1999, 2004) have discussed incomplete contracts and orally based but personally embedded agreements which offer flexibility for informal adaptation over time.

Exchange hazards and the resulting transaction costs are particularly high if transactions are characterised by a combination of environmental uncertainty and asset specificity or behavioural uncertainty (for the following, see Bickenbach and Liu 2010). Transaction costs economics argues that market governance will generally be the transaction cost-efficient governance mode if product quality or individual performance can be measured easily *and* the transaction involves low levels of asset specificity. Transactions that involve products or tasks whose quality is hard to measure and/or substantial investments in specific assets pose greater exchange hazards to the transacting parties. More complex contractual arrangements (hybrid forms of governance) or hierarchical governance will generally be needed to effectively control these hazards.[3] For transactions for which the hazards related to measurement problems or specific assets are particularly great, internal (within-firm) organisation (hierarchical governance) will tend to be the most efficient form of governance. For intermediate levels of measurement, problems and asset specificity hybrid governance, that is, long-term cooperative agreements between legally autonomous partners,[4] may often be the most efficient form of governance. However, even for this intermediate case, hybrid forms of governance will not be optimal if environmental uncertainty is too high (Williamson 1985: 80, 1996: 116–117). As the degree of uncertainty increases, the transactions costs of hybrid governance increase particularly markedly, and one of the polar extremes, hierarchy or market, will generally become the transaction cost-minimising governance structure (Williamson 1985: 80). While frequent and logical changes in the economic environment to which the terms of exchange have to be adjusted (high environmental uncertainty) may increase transaction costs under all forms of governance, the hybrid mode is arguably the most affected. Under hybrid governance, adjustments to environmental changes "cannot be made unilaterally (as with market governance) or by fiat (as with hierarchy) but require mutual consent. Consent, however, takes time" (Williamson 1996: 116).

The comparative efficiency of the alternative modes of governance does not only depend on the characteristics of the transactions but also on the institutional environment. It depends in particular on the quality of formal and informal contracting institutions. While the quality of contract law and of contracting institution more generally affects the transaction costs under all three modes of governance (hierarchy, hybrid and market), its effects are arguably greatest for hybrid governance structures (Williamson 1991, 1996: 115–116). The argument that parties will choose complex forms of contractual arrangements (hybrids) for intermediate levels of asset specificity or behavioural uncertainty depends on the assumption that sufficiently effective contracting institutions exist to support these arrangements.

[3] For details on the argument with respect to asset specificity, see Williamson (1985, 1996); with respect to behavioural uncertainty, see Alchian and Demsetz (1972) for the case of moral hazard and Barzel (1982) for the case of adverse selection.

[4] Cooperations may be "equity based", in which case cooperation partners are linked by equity relations, such as partial ownership arrangements or joint ventures and/or "non-equity" or "agreement based", in which case cooperation partners are not linked by certain type of equity relationships but by long-term contractual agreements only.

Problems associated with weak formal contracting institutions may, however, be alleviated by employing informal relation- or reputation-based institutions of contract enforcement. These informal institutions can attenuate incentives to behave opportunistically and thereby reduce the transaction costs of interfirm as well as intra-firm exchange. The benefits of effective interfirm reputation and relation-based enforcement mechanisms are likely to be greatest for hybrid governance, which is associated with the greatest hazards of opportunism in interfirm trading (and relies on more complex contracting than the other modes of governance). Hybrid contracting will therefore be more prevalent in environments where interfirm reputation mechanisms are working well, ceteris paribus (Williamson 1996: 116).[5] In environments with particularly weak legal contract enforcement institutions, hybrid governance may only be viable in the presence of effectively functioning informal (relation or reputation based) contract enforcement institutions. This last argument suggests that there may be a more general complementary relationship between (complex) formal contracts and informal institutions (personal relationships) in the governance of exchange – in particular, in institutional environments with weak formal contract enforcement institutions.

9.3 Research Design

9.3.1 Background of the Research Region

The GPRD is especially suitable for investigating the relative importance of agile firm organisation for different business operations and the role of agility for sustaining companies' competitiveness either with or without an innovation focus. Firstly, the PRD in Guangdong was a pioneer region of the Chinese economic reforms. Ever since the beginning of reforms, the business and investment environment in Guangdong has been more business friendly than in many other Chinese provinces. A recent World Bank report (World Bank 2008) compared the business and investment environments of the 30 major cities in China, including the capital city of each province (with the exception of Tibet) and four centrally administered municipalities. Four indicators of a business-friendly environment were compared: the ease with which a business can be started, property registered, credit acquisition and contract enforcement. Guangzhou, the capital of Guangdong, ranked among the top three cities for all four indicators, and it ranked first for contract enforcement. As there may be substantial variation in the quality of institutions between different cities in the same province, this finding cannot be generalised to all municipalities within Guangdong province. Nevertheless, it seems reasonable to surmise that business environment is more business friendly in Guangdong than in most other Chinese provinces. Given such a business-friendly investment environment, the business operations of local companies are likely to be less (politically or legally) constrained with regard to their decisions concerning the use of organisational modes or business tools.

Secondly, HK was returned to China in 1997 under a scheme that preserves market-liberal institutions and governance in HK against the more rigid mainland economy. The integration of HK has been based on informal governance. Formal agreements on economic integration (e.g. the Closer Economic Partnership Agreement, CEPA) came into effect in 2003 (Yang 2004). Therefore, cross-border economic relations have to function within two different institutional frameworks. The interplay of the two systems affects numerous companies that are running businesses in both parts of the GPRD (e.g. headquarters and business services in Hong Kong and production in Guangdong Province). Thus, it provides several opportunities for integrating informal practices into usual business operations.

Thirdly, regional economic restructuring is evident in the region under investigation: HK's economy has undergone rapid tertiarisation. The core of the PRD, in particular, the cities of Shenzhen, Guangzhou and Dongguan, is currently undergoing a transition from an economy based on low-tech manufacturing of products relying on strategic

[5] Similarly, if intra-firm reputation effects improve, opportunism within the firm will be reduced, and the costs of hierarchical governance will fall (Williamson 1996: 116).

management from HK towards an economy focusing technological innovation and managing complex business activities itself (Lu and Wei 2007: 234). Thus, the core cities of PRD are starting to compete with HK. This restructuring pushes low-tech manufacturing activities to relocate to less developed cities in the PRD where the labour and land costs are comparatively low. As a consequence of this economic restructuring, the spatial differentiation of the region is becoming more pronounced. HK leads with respect to high value-added activities, but the most developed cities in the Delta, for example, Shenzhen and Guangzhou, are developing similar features. Smaller cities and locations that are more distant from HK still focus on low-tech and labour-intensive production. As the latter manufacturing firms are functionally integrated with the core of the PRD and HK, their locations at a distance from HK are still viewed as the hinterland of the GPRD. This economic restructuring makes the GPRD a mega-urban region responsible for a more comprehensive part of the global value chains of electronics products over time, and these more comprehensive economic activities serve as a suitable base to investigate the application of agile firm organisation by firms with different competences and different business focuses including both production and innovation.

9.3.2 Research Methods

In order to achieve the research aims of this chapter, an original firm-level dataset collected from our own "HK Company Survey 2007" is analysed. This survey aimed to collect first-hand firm-level data concerning companies' use of informal practices in their different business operations in the PRD. The survey focused on electronics companies only. The specification of a narrow sectoral scope was necessary to produce relevant and detailed results because firm organisation is expected to differ considerably between sectors. It would be difficult to distinguish between sectoral and firm-level effects from a small sample of firms if too many sectors were included. Our choice of the electronics industry is supported by secondary statistics, which suggest that the electronics industry is the most important sector in the PRD in general and that several municipalities are dominated by this industry (Shenzhen, Dongguan, certain districts in Guangzhou). Other sources such as trade statistics from the UN COMTRADE database[6] further substantiate the importance of companies from HK and the Pearl River Delta in the global electronics industry.[7]

For the HK Company Survey 2007, a standardised questionnaire was used. The questionnaire consisted of five sections: (1) company characteristics, (2) industry and market conditions, (3) business strategies (4) concluding questions concerning guanxi (personal relationships) and (5) fact sheet about company background information. In order to obtain high-quality first-hand data such as sensitive information on business strategy, face-to-face interviews with senior executives were conducted. Each interview took about 60 minutes. We focused on the small- and medium-sized enterprises (SMEs)[8] which make up more than 98% of all Hong Kong companies (HKSCC 2008). Also, the operations of SMEs are easier to monitor via a standardised survey.[9]

For the identification and selection of companies, we cooperated with the Research Division of the HK Trade Development Council (HKTDC). HKTDC provided a database of 4572 HK SMEs (September 2007) engaged in the electronics sector and with operations in the PRD. A random sample of 3000 HK SMEs was the basis for the Social Science Research Centre (SSRC) at Hong Kong University to implement the survey. After contacting the firms to collect the senior executives' contact details, SSRC mailed the questionnaire to the firms and made appointments with the executives for a face-to-face interview. In total, senior executives from 104 HK electronics

[6] See http://comtrade.un.org/ for more information.

[7] According to Enright et al. (2005), industrial output in the GPRD is mainly from light manufacturing (e.g. toys, textiles and garments) and the electrical and electronics industry (e.g. consumer electronics, watches). The few heavy industries (e.g. chemicals or plastics) mainly serve as suppliers for these industries.

[8] According to the HK statistical standard, firms with 100 employees or less are defined as SMEs in HK (HKSCC 2008).

[9] Guided interviews with 15 large HK electronics companies were additionally carried out to cover their more complex operations.

companies were interviewed from August till December 2007. Two of the 104 responding companies were excluded from the following analysis because of the disparity between their real size and the HK definition of HK SME.

In addition to the survey, two further data sources were used to obtain supplementary data and background information. First, statistical yearbooks were used as sources to obtain secondary statistics on macroeconomic variables of industrial development in the GPRD.[10] Second, earlier surveys on the interaction between HK and mainland companies (e.g. "Made in PRD" studies by the Federation of Hong Kong Industries (FHKI 2003, 2007) and a study by Michael Enright et al. (2005) "Regional Powerhouse: The Greater Pearl River Delta and the Rise of China") provided a useful general overview of the division of labour between HK and the PRD. However, information on the actual organisation of business functions and evidence for the prevalence of different types of contractual relationships were not available from these studies, making it necessary to collect primary data by carrying out our own company surveys and interviews.

9.4 Findings

In this section, we analyse the original dataset collected from our own HK Company Survey 2007 to provide some empirical evidence of the agility patterns applied by the HK electronics SMEs operating in the PRD in organising their relations to their customers and producers on the one hand and on the integration of informal institutions proxied by personal relationships into their governance structures for production and innovation activities on the other hand. This enables us to investigate whether informality, that is, personal relationships, plays different roles for companies in governing their production and innovation activities.

[10]The "Guangdong Statistical Yearbooks" were the main sources in this regard. In addition, the "China Statistical Yearbooks on Science and Technology" were used to obtain information about the innovation tendency of innovators in the PRD in Guangdong compared to the other Chinese provinces over the past decade.

The innovation statistics presented in Box 9.1 show that over time innovation has substantially gained in importance for China in general and for Guangdong with the PRD as its core economic region in particular.

Box 9.1 Selected innovation indicators in China

Innovation inputs: Although the rapid development of the Chinese economy over the past decades was largely attributable to the rapid development of labour-intensive industries and the export of low value-added products to the world market, the Chinese government has now recognised the crucial role of innovation for sustaining a long-term stable economic growth in China. Against this background, several innovation-friendly policies were applied. Preferential treatments provided by such policies, together with the increasing competition on global markets, enhanced Chinese companies' willingness to invest in innovation activities. R&D expenditures in China have increased substantially from 35 billions Yuan in 1995 to 370 billions Yuan (about 49 billions USD) in 2007, resulting in a strong rise of R&D-to-GDP ratio from 0.57% in 1995 to 1.5% in 2007 (NBSC(a) 2001–2008). Most of the R&D expenditures were used for experimental development activities, which aim to transform the existing knowledge into innovative products and processes for practical use. Correspondingly, most of the R&D expenditures were spent by companies rather than by academic institutions.

At the provincial level, data show that R&D expenditures in different provinces in China have increased at grossly different speeds. Beijing dominated with respect to the absolute amount of the R&D expenditure from 2000 to 2007. However, the growth rate of the R&D expenditure in Beijing over the same period was lower than the average growth rate of the R&D expenditure of the

(continued)

Box 9.1 (continued)
nine most innovative provinces (23.4%) by more than five percentage points. The growth rate of the R&D expenditures of our focus province – Guangdong which realised the second highest amount of R&D expenditure before 2004 and the third highest amount of R&D expenditure from 2004 to 2007 was also lower than the average rate by about 2 percentage points. Guangdong also featured a continuously increasing number of R&D personnel over time. The growth rate of the number of R&D personnel deployed was so strong that Guangdong became to dominate the other eight most innovative provinces in China in this aspect in 2007 (NBSC(a) 2001–2008).

Innovation outputs: The number of patent applications in Guangdong is also comparatively large. Between 2000 and 2007 the number of patent applications by innovators in Guangdong increased substantially and in 2007 innovators in Guangdong applied for more patents than those in the other eight most innovative provinces in China, including Beijing and Jiangsu which featured even higher innovation inputs than Guangdong. Such a dominant role of Guangdong can also be identified when looking at invention patent applications instead of total patent applications. Starting from a relatively small number of invention patent applications in 2000, Guangdong became to dominate the other provinces in this respect since 2004. Focusing on companies' patenting behaviour, the number – and also the growth rate – of invention patent applications applied by "companies" in Guangdong was even much higher than that in the other provinces. The dominant role of firms in Guangdong remains after taking effects of population size into account (Fig. 9.1). These results suggest that innovators especially companies as innovators in Guangdong seem to be more efficient innovators than those in the other provinces.

9.4.1 Agility Patterns in Networks with Producers and Customers

This section provides empirical evidence on the governance structures applied between companies in the GPRD and their producers and between the companies and their customers within the value chain. In our survey, hierarchies, equity-based cooperation, agreement-based cooperation and markets were distinguished (cf. Sect. 9.2). As argued in Sect. 9.2 and further discussed below, equity-based and agreement-based cooperations may be expected to be related to more informal modes of interaction and enforcement as they involve more complex arrangements that are more difficult to be enforced by formal enforcement mechanisms, particularly in an environment with weak formal institutions.

Table 9.1a summarises the responses of companies with regard to the importance of these four different governance structures for organising relations to their producers (their production activities) in the PRD. It shows that for more than half of the responding companies, placing production orders to producers (53%) and/or hierarchical production (52%) is very important or important governance structure in organising production. In contrast, equity-based cooperations and agreement-based cooperations are very important or important governance structures for production for only for 17.6 and 21.6% of companies, respectively. A large majority of companies never engage in such cooperative governance modes of production (about 76.5 and 65.7%, respectively).[11] This suggests that HK SMEs tend to avoid cooperative (hybrid) governance structures and to prefer either market or hierarchical forms of governance to organise production in the PRD. The high risks of opportunistic behaviour and the high transaction costs of hybrid

[11] Pairwise two-tailed Wilcoxon Signed-Rank Tests (WSRTs) confirm that cooperative forms of production are of statistically significantly lower importance than the other two alternatives. No significant difference in importance is found between the two cooperative forms of production.

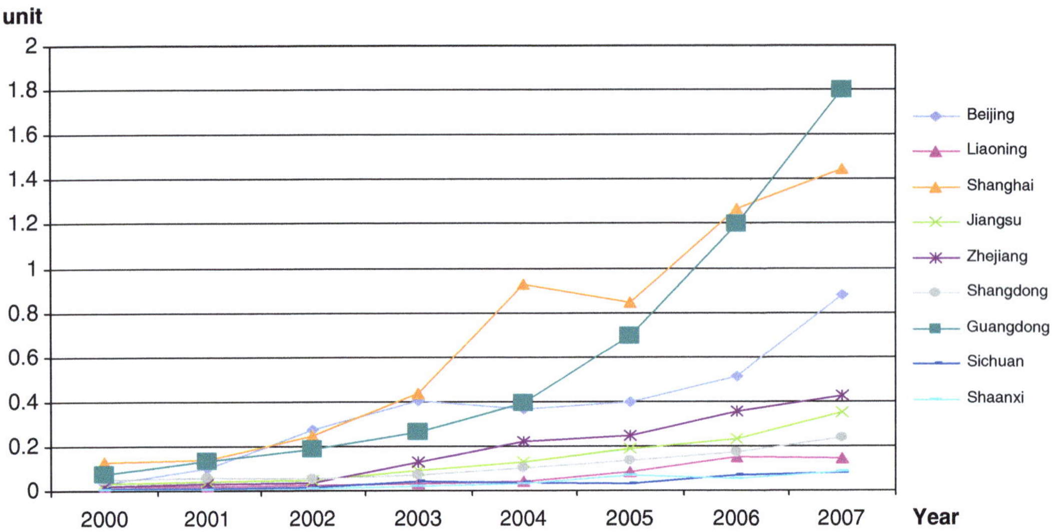

Own presentation based on data from China Statistical Yearbook on Science and Technology (NBSC(a) 2001–2008) and China Statistical Yearbook (NBSC(b) 2001–2008)

Fig. 9.1 Number of invention patent applications by companies per 10,000 persons in the nine most innovative provinces

Table 9.1 Importance[a] of governance structures for organising supplier (9.1a) and customer relationships (9.1b) (company share in %)

	1	2	3	4	5	Never applied	Total
Table 9.1a – Governance structures for organising supplier relationships (production activities) in the PRD							
Own production site (hierarchical production)	39.2	12.8	6.9	1.0	2.0	38.2	100 (n=102)
Equity-based cooperation	9.8	7.8	4.9	1.0	0.0	76.5	100 (n=102)
Agreement (non-equity) -based cooperation	15.7	5.9	8.8	2.9	1.0	65.7	100 (n=102)
Placing orders	26.5	26.5	17.7	2.9	3.9	22.6	100 (n=102)
Table 9.1b – Governance structures for organising customer relationships							
Customers are 100% owned subsi. or parent company	2.0	2.0	5.9	2.9	2.9	84.3	100 (n=102)
Equity-based cooperation	2.9	4.9	2.0	2.0	0.0	88.2	100 (n=102)
Agreement (non-equity) -based cooperation	15.7	11.8	13.7	5.9	2.9	50.0	100 (n=102)
Customers just place orders	60.8	20.6	9.8	4.9	0.0	3.9	100 (n=102)

Source: Own HK Survey 2007

[a]A 5-level Likert scale was used, with "1" to "5" referring to very important, important, of normal importance, of little importance and not important, respectively

governance structures in an environment characterised by high levels of dynamism in technological and market conditions (environmental uncertainty) and weak legal and formal market institutions may at least form part of the explanation of these governance choices. However, if a time dimension is taken into account, our survey data suggest that an improving formal institutional environment in the PRD may encourage more HK companies to opt for cooperative relationships or even market relationships with their PRD producers (Meyer 2009).

With regard to government structures used to organise companies' relationships to their customers,[12] Table 9.1b shows that hierarchical customer relationships are generally of little relevance for the companies surveyed. About 84.3% of responding companies never engage in such hierarchical relationships with their customers. In contrast, only 3.9% of companies never have pure market relationships with their customers. Moreover, about 81.4% of them evaluated pure market relationships with their customers as very important or important. Among the other two governance structures, agreement-based cooperations seem to be more important and relevant for the HK companies than equity-based cooperations.

A further comparison of HK companies' relations to customers and producers reveals that the selection of governance structures (especially hierarchy vs. market) depends strongly on the location of partners, which is strongly linked to a particular institutional environment. Relationships to international customers from countries with highly developed legal and formal business institutions are mainly market oriented, whereas relationships to producers located in the PRD are still mostly organised in hierarchies, despite recent trends towards cooperative and market-based forms of governance (Meyer 2009).

This may be interpreted as support for the theoretically motivated hypothesis that the institutional environment affects the choice of governance structures and how it does so. However, there may be other determinants, such as general business practice, that may be partly responsible for companies' choice of governance structures. The survey evidence shows that in addition to the location of business partners, relationship-specific characteristics are significantly associated with certain governance structures. Companies with a higher dependency on individual customers or producers are more likely to chose hierarchical governance structures (in-house production), which reduce the hazards associated with the dependence on a single customer of supplier (Meyer 2009).

Our survey results also suggest that governance structures based on long-term cooperative business relationships reduce the time required by companies to negotiate supply contracts with their producers – relative to market procurement. Cooperative production agreements thus increase companies' ability to respond flexibly to changing market conditions and customer demands.[13] This result is in apparent conflict with the view that hybrid (cooperative) governance is generally comparatively slow to adjust to environmental changes because adaptations of contract requires mutual consent (Williamson 1996). The conflict may be resolved by recognising that market transactions are exacerbated by the thinness of markets and weakness of institutions of (public) information provision and formal contracting institutions in the PRD.

9.4.2 Integration of Informality in Governing Production and Innovation Activities

Personal relationships are an important element of informality, and they are of considerable importance for companies doing business in the PRD (Bickenbach and Liu 2010). Going beyond mere description of companies' relations with their producers and suppliers, in this section, we investigate how informality – proxied by the importance of personal relationships – is related to companies' governance choices in organising their production activities as well as their – increasingly important – innovation activities.

[12] Customers are not geographically restricted to the PRD only.

[13] See Meyer (2009) for more information.

Table 9.2 Importance[a] of governance structures for innovation (company share in %)

	1	2	3	4	5	Never applied	Total
Own R&D and innovation activities	24.3	24.3	30.0	10.0	2.9	8.6	100 (n=70)
Cooperation with partners	15.7	25.7	27.1	4.3	0.0	27.1	100 (n=70)
Acquisition of licenses or innovations	5.7	22.9	21.4	0.0	2.9	47.1	100 (n=70)
Reverse engineering	17.1	30.0	15.7	12.9	0.0	24.3	100 (n=70)

Source: Own HK Survey 2007
[a]Same as in Table 9.1

The discussion in Sect. 9.2 suggests that, in the case of weak legal environments, there may be a complementary relationship between the complexity of formal contracts and personal relationships in the governance of production- or innovation-related business transactions. More specifically, the discussion suggests that personal relationships should be of greater importance for hybrid governance structures (equity-based or agreement-based cooperation) than for simple market transaction or vertically integrated operations. We may therefore expect that companies that choose cooperative (hybrid) governance structures to organise their production and innovation activities tend to consider personal relationships more important than companies choosing hierarchical or market governance. And companies that do not rely on personal relations may be expected to avoid using hybrid forms of governance.

The findings concerning companies' choices of governance structures for production are presented in Table 9.1a. The corresponding findings regarding the governance structures for innovation are summarised in Table 9.2. In the case of innovation, a slightly different set of governance structures was specified in the questionnaire: own R&D and innovation, cooperation with (external) partners, acquisition of licences and/or innovations, and reverse engineering.[14] As shown in Table 9.2, about 48.6% (47.1%) of the 70 "innovating" companies consider in-house R&D and innovation (reverse engineering) as a very important or important governance structure for governing their innovation activities. For a slightly smaller share of companies (41.4%), cooperative governance of innovation is very important or important. In contrast, for only 28.6% of companies are acquisitions of innovations (licences) very important or important for innovation. In contrast to the case of production activities, only a minority of 27.1% of companies never engage in cooperative innovation at all. Pairwise two-tailed WSRTs indicate that there is no significant difference in the importance of cooperative innovation on the one hand and of own innovation and reverse engineering on the other. In contrast, innovation by acquisition is found to be significantly less important than the other three governance modes. Due to high uncertainty and (asset) specificity, acquiring innovations via simple market transactions may often be associated with high transaction costs. At the same time, the lack of technological capabilities and innovation experience of many electronics SMEs, as well as limited resources, may enhance (relative to the case of production) companies' willingness to cooperate with more experienced partners in innovation.

Against the background of a deficient legal contract enforcement system in the PRD (at least compared to HK), personal relationships may be expected to play some role in enforcing cooperation agreements. More specifically, as argued in Sect. 9.2, under weak formal contract enforcement, there may be a complementary relationship between the importance of personal relationships and the use of cooperative (hybrid) governance

[14]Note that only executives from companies which carry out innovation activities (*n*=70) were asked to answer the corresponding question.

Table 9.3 Spearman-rho correlations between "importance of hybrid governance structures" and "importance of personal relationships as criterion for partner decisions (production and innovation)"

	Equity-based production coop.[b]	Agreement-based production coop.[b]	Innovation cooperation[c]
PPersRP[a]	0.249*	−0.069	
IPersRP[a]			0.143

Source: Own HK Survey 2007
*10% significance level in two-tailed tests
[a]PPersRP (IPersRP) refers to the importance of personal relationships as a criterion for companies to decide their production (innovation) partners. The same 5-level Likert scale as in Table 9.1 was used
[b]46 valid responses
[c]51 valid responses

structures. In order to find out whether there is evidence for such a relationship in our survey results, we calculated pairwise Spearman rho correlations between the importance of personal relationships for the choice of production partners and of innovation partners on the one hand[15] and the importance of cooperative (hybrid) governance structures for production and of cooperative governance for innovation on the other hand. Table 9.3 reports the results.

The correlation is statistically significantly different from zero only for the case of equity-based cooperation for production, suggesting that companies that consider equity-based cooperation more important for organising their production in the PRD also consider personal relationships more important in their choice of cooperation partners.

Additional evidence concerning the complementary relation between personal relationships and hybrid governance structures may be gained from analysing the interrelations between the executives' assessments of the importance of different cooperative governance structures and the importance of different reasons for using personal relationships for business in general. Table 9.4 reports the corresponding Spearman rho correlation coefficients and shows that there are several statistically significant positive correlations between the importance of cooperative governance structures for production and innovation on the one hand and the importance of some of the potential reasons for using personal relationships for business on the other hand. Companies that attach a higher level of importance to contract- or agreement-based cooperation for production also tend to consider increasing the "performance goals" – that is, increasing responsiveness to customer needs, increasing companies' operational flexibility and reducing risks and uncertainties – more important as reasons for using personal relationships for business in general. The same is true for companies that attach greater importance to equity-based cooperation for production (in order to reduce risks and uncertainty). For cooperative governance structures in innovation, the situation is quite different. Here, companies that attach a high level of importance to cooperative governance structures also tend to stress the importance of institutional deficiencies – substituting personal relationships for nonexistent governmental laws and regulations and nonexistent formal business rules and procedures – and/or cultural traditions as reasons for using personal relationships.

Overall this seems to suggest that companies applying (or stressing the importance of) hybrid governance structures for production tend to use personal relationships mainly to reduce business risks and to increase their customer responsiveness and flexibility. In contrast, companies applying hybrid governance structures to organise innovation activities may use personal relationships mainly to cope with weak legal institutions and deficient regulatory policies and nonexistent formal business rules and procedures in the field of innovation (or simply to follow cultural traditions). They may use personal relationships, for example, to increase their ability to enforce contractual agreements with (private) innovation partners.

[15]More detailed analysis regarding criteria for partner decisions and the role of personal relationships in this context can be found in Bickenbach and Liu (2010).

Table 9.4 Spearman-rho correlations between "importance of hybrid governance structures" and "importance of reasons for applying personal relationships for business"

	Equity-based production coop.[a]	Agreement-based production coop.[a]	Innovation cooperation[b]
(Follow) culture-determined way for doing business	0.119	0.109	0.265**
Substitute for missing governmental laws	0.095	–0.047	0.237**
Substitute for missing formal business rules	0.142	0.139	0.231*
Respond quickly to customer demands	0.097	0.245**	0.147
Increase operational flexibility	0.056	0.235**	0.192
Reduce risks and uncertainty	0.239**	0.210**	0.164

Source: Own HK Survey 2007
**5%; *10% significance level in two-tailed tests
[a]102 valid responses
[b]70 valid responses

These differences in the motives for using personal relationships just described may have important implications if formal contracting institutions and business rules in the PRD further improve over time. While personal relationships can be expected to remain important for the (hybrid) organisation of production (as they may continue to increase companies' flexibility and responsiveness), they may become less important for the (cooperative) organisation of innovation activities. Personal relationships as substitutes for weak legal institutions and nonexistent formal business rules will become less important as formal institutions become more effective in enforcing agreements on cooperative innovation and limiting opportunistic behaviour of innovation partners.

9.5 Conclusion

The concept of agile firm organisation complements the flexibility concept by focusing on informality in business practices as a means to achieve flexibility. It has been argued that agile firm organisation was of considerable importance to HK companies operating in the PRD over the past decades.

This chapter analyses an original firm-level dataset collected by our own HK Company Survey 2007 and provides selected empirical evidence on the role of informality (agility) for the organisation of customer and producer relations by HK electronics SMEs in the PRD. It also compares the importance of different governance structures and the role of informality and personal relationships for organising companies' production activities and for organising their innovation activities.

A comparison of HK firms' relations to customers and producers reveals that the selection of governance modes (hierarchy vs. market) depends strongly on the location of customers and suppliers, which is strongly linked to a particular institutional environment. The results also suggest that improvements in the formal institutional environment in the PRD have already encouraged a growing number of HK SMEs to opt for cooperative relationships or market relationships with their PRD producers. Currently most companies still prefer hierarchical modes of governance for production (in-house production).

We also find some evidence for a complementary relationship between the importance of informal institutions such as personal relationships on the one hand and the importance of cooperative governance structures on the other. Companies that chose hybrid governance structures to organise their production tend to emphasise increasing their operational flexibility and their ability to respond quickly to customer needs, as well as the reduction of risks and uncertainties, as important reasons for using personal relationships. In contrast, companies that chose hybrid governance structures to organise innovation tend to emphasise the deficiencies in legal institutions and formal business rules as well as cultural factors as important reasons for using personal relationships. Results suggest that informality – and agile firm organisation more broadly – may remain important for companies operating in the PRD even if they increasingly focus on innovation rather than mere cost minimisation strategies for sustaining their long-term competitiveness – at least as long as serious deficiencies in the legal and formal business institutions remain. However, if the deficiencies of the formal institutional environment in the PRD are removed by further reforms, personal relationships and informality, more generally, may lose significance for companies operating in the PRD – at least for those companies following an innovation strategy.

Based on our research so far, at least three possible useful research extensions can be suggested. Firstly, the difference in informality patterns for production and innovation encourages further research on the determinants and organisation of upgrading in the GPRD. Secondly, future research could extend the analysis to additional types of business operations such as business financing to investigate the nature of agile firm organisations as applied by companies in the GPRD more comprehensively. Thirdly, future research could try to investigate whether such agile firm organisations are also prevalent and/or can be easily replicated in other developing economics.

Acknowledgement The authors would like to thank the German Research Foundation (DFG) for its financial support of the cooperative projects – "Informal Dynamics of Agile Firm Organisation in the Greater Pearl River Delta" (Phase 1) and "Regional Agility and Upgrading in Hong Kong and the Pearl River Delta" (*Phase* 2) within the *Priority Program 1233: Megacities – Megachallenge: Informal Dynamics of Global Change.*

References

Alchian A, Demsetz H (1972) Production, information costs and economic organization. Am Econ Rev 62:777–795

Barzel Y (1982) Measurement cost and the organization of markets. J Law Econ 25:27–48

Bickenbach F, Liu W-H (2010) On the role of personal relationships for doing business in the Greater Pearl River Delta, China, Econ J 3(3):281–306

BrandHK (2005) Pearl River Delta – factory of the world. Brand Hong Kong. URL: http://www.brandhk.gov.hk/brandhk/e_pdf/efact12.pdf

Enright M, Scott E, Chang K-M (2005) Regional powerhouse: the greater Pearl River Delta and the rise of China. Wiley, Singapore

FHKI Federation of Hong Kong Industries (2003) Made in PRD—the changing face of Hong Kong manufacturers. FHKI, Hong Kong

FHKI Federation of Hong Kong Industries (2007) Made in PRD—challenges and Opportunities for HK Industry. FHKI, Hong Kong

Hart O, Moore J (1999) Foundations of incomplete contracts. Rev Econ Stud 66:115–138

Hart O, Moore J (2004) Agreeing now to agree later: contracts that rule out but not rule. In: NBER working paper no. 10397. Harvard University

HKSCC, Hong Kong Support and Consultation Centre for SMEs (2008) SMEs in HK. URL: http://www.success.tid.gov.hk/english/lin_sup_org/gov_dep/service_detail_6863.html

Lin GCS (2004) Toward a post-socialist city? Economic tertiarization and urban reformation in the Guangzhou metropolis, China. Eurasian Geogr Econ 45(1):18–44

Lu L, Wei YHD (2007) Domesticating globalisation, new economic spaces and regional polarisation in Guangdong Province, China. Tijdschr Econ Soc Geogr 98(2):225–244

Meyer S (2009) Informal modes of governance in customer producer relations. The Electronics Industry in the Greater Pearl River Delta (China). PhD thesis. University of Hannover

NBSC(a), National Bureau of Statistics China (2001–2008) China statistical yearbook on science and technology. China Statistics Press, Beijing

NBSC(b), National Bureau of Statistics China (2001–2008) China statistical yearbook. China Statistics Press, Beijing

North D (1990) Institutions, institutional change and economic performance. Cambridge University Press, Cambridge

Sit VFS, Yang C (1997) Foreign-investment-induced exo-urbanisation in the Pearl River Delta, China. Urban Stud 34(4):647–678

Williamson OE (1985) The economic institutions of capitalism. The Free Press, New York

Williamson OE (1991) Comparative economic organization: the analysis of discrete structural alternatives. Adm Sci Q 36:269–296

Williamson OE (1996) The mechanisms of governance. Oxford University Press, New York

World Bank (2008) Doing business in China 2008. World Bank Group, Washington, DC. URL: http://www.doingbusiness.org/subnational/exploreeconomies/China.aspx

World Bank (2009) Gross domestic product 2008, PPP (Quick reference tables), World Development Indicators Database. URL: http://siteresources.worldbank.org/DATASTATISTICS/Resources/GDP_PPP.pdf

Yang C (2004) From market-led to institution-based economic integration: the case of Pearl River Delta and Hong Kong. Issues Stud 40:79–118

10 Global Change, National Development Goals, Urbanisation and International Migration in China: African Migrants in Guangzhou and Foshan

Tabea Bork-Hüffer, Birte Rafflenbeul, Frauke Kraas, and Zhigang Li

Abstract

The migration of African traders to Guangzhou, located in southeastern China, started in the 1990s and has ever since been increasing. During the last years, the neighbouring city of Foshan has become a second centre of African migration. While a growing number of migrants have been moving from Guangzhou to Foshan, an increased direct migration to Foshan can also be witnessed. The aim of this contribution is to demonstrate how the dynamics and complexity of processes in and between the two cities as well as regional, national and global conditions and developments are influencing this new migration flow. Urban development aims, their interpretation and implementation at various levels are considered, together with the reactions, flexibility and adaption strategies of the migrants. Furthermore, this example highlights how global developments (e.g. global economic crisis 2008/2009, China's entry to the WTO), national conditions (e.g. national immigration regulations, development strategies, hosting of major international events) and interurban competition and co-operation are interrelated and have diverse impacts on the migration flows. The analysis is based on a quantitative survey of 253 African migrants, four expert interviews and 14 qualitative interviews with African migrants.

Keywords

China • International migration • Urbanisation • Global change

T. Bork-Hüffer (✉)
Asia Research Institute, National University of Singapore, 469A, Tower Block, Bukit Timah Road, 259770, Singapore
e-mail: t.bork@uni-koeln.de

B. Rafflenbeul
Department of Geography, University of Cologne, Albertus-Magnus Platz, 50923, Cologne, Germany

F. Kraas
Institute of Geography, University of Cologne, Albertus-Magnus Platz, 50923, Cologne, Germany

Z. Li
School of Geography and Planning, Sun Yat-sen University, No. 135 Xiangangxi Road, Guangzhou, 510275, China

F. Kraas et al. (eds.), *Megacities: Our Global Urban Future*, International Year of Planet Earth, DOI 10.1007/978-90-481-3417-5_10,

10.1 Global Change, Urbanisation and International Migrants in the "World's Workshop"

Following the foundation of the People's Republic of China on 1 October 1949, foreign immigration and urbanisation were strictly regulated and limited. Only as a result of the reform policy pursued since, 1978 a process of extensive catch-up urbanisation begun, especially in China's coastal provinces. Likewise entry and exit regulations for foreigners were relaxed; nevertheless, migration of foreigners to China remained limited and subject to strict regulations. However, one of the consequences of China's tremendous growth in significance in the global economy and community was an increasing influx of foreigners to China, especially in the last decade and mostly to the booming cities of the coastal provinces. Estimates assume that a total of up to two million international migrants are currently living in China (see Pieke 2010: 20).

One densely inhabited region that has grown very rapidly since 1978 is the Pearl River Delta in the southeastern Chinese province of Guangdong, which was declared a special economic area in 1985 and is one of the first and largest urban development corridors in China. Over the last three decades, a multinodal, mega-urban agglomeration has developed here (with the megacities Guangzhou, Shenzhen and Dongguan and the emerging megacities Foshan and Zhuhai), where one city merges into the next (Kraas 2004). With the largest concentration of export-oriented industries worldwide, the Pearl River Delta has become known as the "world's workshop" (Sun et al. 2006: 28). Cross-border trade relations have a long tradition in the region of the Pearl River Delta due to its proximity to Hong Kong, Macau and Taiwan and the activities of overseas Chinese stemming from the region (Cartier 2001). As the provincial capital, Guangzhou is one of China's most important trading cities, and many immigrants have settled here (at least temporarily) since the nineteenth century (Cartier 2001). Most of the few foreign business people allowed into the country during the period between 1949 and 1978 stayed in Guangzhou, in connection with the trade fair (Brady 2003: 2). The fair continues to attract hundreds of thousands of foreign traders today.

The aim of this contribution is to demonstrate the dynamics and complexity of processes in and between urban areas as well as their interdependence with local, regional, national and global conditions and developments using the example of African migration to Guangzhou and Foshan. The focus of the analysis is on migrants' relocation from Guangzhou to Foshan and direct migration to Foshan. The example highlights the diversity and interrelatedness of influences of global developments and agreements (e.g. global economic crisis 2008/2009, China's entry to the WTO), national conditions (e.g. national immigration regulations, development strategies, hosting of major international events) and interurban competition and co-operation. Regional, urban and local development aims, their interpretation and implementation at various levels are also considered, together with the reactions, flexibility and adaption strategies of urban actors. The analysis is based on a quantitative survey of a total of 253 African migrants,[1] 179 of whom live in Guangzhou and 74 in Foshan. Four interviews were carried out with experts (president of one African migrant association, representatives of

[1] All of the data used were obtained as part of the research project "Internal and International Migrant Communities in the Pearl River Delta/China – Linking Informal Migration Dynamics, Global Change and Urban Health", which is part of the German Science Foundation's programme 1233 "Megacities – Megachallenge: Informal Dynamics of Global Change".

In this article, the term "migrant" is used to describe individuals residing primarily in China for longer periods of time or planning to do so – even if this cannot be realised, for example, because of rejected visa applications. The term also includes individuals who repeatedly spend time in China over a longer period. Thus, the use of the concept is adapted to the group studied here, African citizens resident in the Pearl River Delta, some of whom demonstrate a very high level of transnational mobility. However, more recent definitions of migrants also take into account the growing mobility of international migrants (see Castles 2000). The UNESCO (2010) defines a migrant as "any person who lives temporarily or permanently in a country where he or she was not born, and has acquired some significant social ties to this country".

local self-administration in Foshan, representatives of an international organisation in Guangzhou, and estate agents), augmented by 14 qualitative interviews with African migrants in both cities. Because of the sensitive nature of the issues discussed, only a small number of interview partners agreed to an audio recording of their interview. Thus, most of the interviews could only be recorded in writing and not taped or transcribed. For the most part, the quantitative survey and interviews were carried out between March and May 2010. Numerous field walks were carried out between 2006 and 2010, and field observations were recorded.

10.2 Changes in Entry and Exit Regulations and the Monitoring of Foreigners in China Since 1949

For nationalist and historical reasons – including the Opium Wars, the system of trading ports, the Taiping Rebellion, the Treaty of Washington in 1919, the Japanese invasion, the "Cold War" and the reaction of the West to the massacre on Tiananmen Square in June 1989 – foreigners occupy a precarious position in China (Brady 2003). In the years following the foundation of the People's Republic, an extremely restrictive entry and exit system was established, and the influx of foreigners thereby kept to a minimum (Liu 2009). Furthermore, most foreigners were requested – either directly or indirectly – to leave the country in the early 1950s (1949–1952). Only a tiny, selected proportion of foreigners trusted by high-ranking CPC representatives were invited into the country by the party or permitted to remain (Brady 2000). Since 1954 foreigners also had to apply for an exit visa or certificate before leaving the country, in addition to their entry visa (Liu 2009). From the 1950s to the 1970s, the number of foreigners living in China (students, diplomats, journalists, business people and foreign experts working for the Chinese government) was therefore very low (Brady 2003: 2).

A cautious reform of entry and exit regulations was begun in 1979. Perceptible liberalisation for foreigners – including the abolition of the exit visa – was only implemented since 1985. The regulations governing the entry of foreigners into the country are still strict today (Liu 2009). The segregation of the foreign community from the local Chinese population by means of the strict regulation of their places of residence,[2] job opportunities and political taboos was only gradually relaxed in the 1990s – initially when both population groups began to circumvent official regulations, before a partial official relaxation of the rules (Brady 2000; Farrer 2010).

Nevertheless, the number of foreign residents grew rapidly as China opened up. These residents include migrants coming to work in China as well as students, business people, journalists and diplomats (Brady 2000). The Chinese Statistical Yearbook (NBS 2009: 767) provides information about the numbers of foreign visitors[3] arriving in China: Their numbers increased fourfold between 1995 and 2008.[4] The largest numbers came, in order of frequency, from South Korea, Japan and Russia. Previous international research on international migrants in China concentrated primarily on the personnel management of specialised foreign staff and adaptive strategies employed by individual managers and their families (see Stening and Yu 2006: for an overview up to 2006). Any research that went beyond this narrow focus tended to concentrate on migrants from a few countries (Korea, Taiwan, Britain and African migrants not differentiated according to nationality) and a few issues (the formation of ethnic enclaves and enclave economies, globalisation processes and immigration, migrants' spaces, gender aspects, employment networks) (e.g. Li et al. 2008, 2009; Wang and Lau 2008; Willis and Yeoh 2002; Yeoh and Willis 2005; Zhang 2008).

[2] Restricted to hotels for foreigners (shewai binguan), student dormitories for foreigners (liuxuesheng sushe), apartments for expats (waiguo zhuanjia lou) or diplomatic quarters (waijiao gongyu) (Brady 2003).

[3] Foreign visitors recorded include tourists, students, diplomats and migrants. The statistical yearbooks also include Chinese living abroad and Chinese from Hong Kong, Macau and Taiwan.

[4] There were 5,886,700 foreign visitors in 1995, and the figure rose to 24,325,300 in 2008.

The Chinese term *waishi*, which refers to external affairs in the broadest sense, describes the whole spectrum of the Chinese strategy for controlling or influencing foreigners and foreign cultures and technologies in China and beyond (Brady 2000: 943). An understanding of *waishi* is essential in order to understand and evaluate the position of international migrants in China. Brady (2003: 2) explains: "China's *waishi*, or foreign affairs system includes state-to-state and people's diplomacy, laws and guidelines regulating where and how foreigners are allowed to live, do business, marry, give birth, and go to school, as well as covering foreign propaganda, tourism and the promotion of old and new foreign friends".

Since 1978 it has been the aim of the *waishi* strategy to attract foreign technology and investment to China in the interest of modernisation and a better position in global economy and policy. In this context the entry of desirable foreigners, for example, specialised Western workers, is considered necessary and is encouraged, while the entry of undesirable foreigners is simultaneously restricted (Brady 2000: 946; Farrer 2010). As well as the control of foreigners, *waishi* strategy works towards a deliberate indoctrination of the native population with regard to their dealings with foreigners. This includes repeated reminders to the native population to regard foreigners as "other", that is, as outsiders whose ways are fundamentally different from Chinese culture and the Chinese character: "insiders are to be trusted; outsiders are to be feared" (Brady 2000: 954).

The *waishi* strategy is clearly reflected in the selective allocation of visas as well as short-term changes in the requirements for visa applications, which are adapted to the political strategies of the day: While the types of visa[5] available have remained almost constantly the same, allocation criteria and requirements have been subject to change, for example, before the Olympic Games in Beijing in 2008. The Chinese government feared protests that could have overshadowed the games and reached the attention of the world press. Thus, for example, applications for multi-entry visas were turned down, extra documents were called for and, depending on nationality and/or status, visa extensions were only possible in the applicant's home country and could not be granted as hitherto in China, Hong Kong or Macau (Cheng and Chao 2008).

Furthermore, since May 2009 (Qin 2009) and increasing since the beginning of 2010, extensive visa checks have been carried out in Guangdong and with greater intensity in Guangzhou. The aim of these checks is to contain the "three illegalities" (*san fei*) (see Pieke 2010): illegal entry, illegal residence and illegal labour. In particular the growing numbers of illegal immigrants and those who outstay their visas are to be reduced. Evidently the Asian Games, which are to take place in Guangzhou in November 2010, are a further reason to implement these measures (Coloma 2010; Pomfret 2009). So far, these more recent trends and Chinese visa policy overall are poorly documented in international scientific publications; newspaper reports tend to be based on the personal experiences of individual migrants or cite sparse government information or the state news agency Xinhua. This makes evaluation considerably more difficult.

10.3 Urban Development in China and the Pearl River Delta Since 1949

The Mao era was characterised by the "three pillars of socialist control": the single party state, household registration and state work units. While the former ensured total control by the party at all administrative levels, access to all welfare rights was linked to the place in which households were registered (e.g. right to employment in the cities or land to farm in the rural areas, food rationing and other areas of welfare). This system also effectively stopped the migration of farmers

[5] Tourist visa (L visa), business visa (F visa), student visa (X visa), work visa (Z visa), residence permit (D visa) and special visas for crew members (C visa), for journalists (2 types: J1 and J2 visa) and for individuals in transit through China (G visa). The D visa has only been available since the reform of entry and exit regulations in 2001. From the introduction of the regulations up to 30 September 2005, only 649 foreigners were granted this type of visa (Liu 2009: 323), a tiny figure.

to the cities and urban growth. At the same time, the fact that urban residents were tied to work and housing units in the cities created a stable, controllable urban society (Wu 2002). Liu (2009: 314) sees a strong connection between the household registration system and entry and exit permits in China: "[...], persons residing in urban areas were subject to control. [...] These regulations essentially laid down the limits of freedom of movement". The almost total control over the movement of the native population achieved by these means also made it easier to control the movements of foreigners.

In the last 30 years, this relatively rigid situation was overridden by the introduction of the market economy as well as the allowance of (at least temporary) residential moves (Gu 2001). China's cities are experiencing booming economic growth, especially in the industrial sector, and massive in-migration from rural areas. The cities have been substantially restructured (modernisation of the central business districts, development of new city centres and axes as well as new shopping and services districts) and have expanded dramatically into their hinterlands. Particularly strong economic and spatial growth have taken place and are continuing in the cities of the special economic zone, in the form of large-scale development projects on the urban fringe, infrastructural development for industry – especially export-oriented industries – and concentrated capital investment in property, made possibly by urban land reforms (Cartier 2001, 2002; Shen et al. 2002; Wu et al. 2007).

The introduction of a market economy and accelerated economic growth combined with institutional decentralisation have greatly increased the influence of local governments, which in turn has fanned interregional and interurban competition for investors and subsidies (Gu 2001; Wu et al. 2007; Yao et al. 2008). This includes competition to attract desirable migrants, to enhance cities' cosmopolitan image (see Farrer 2010), as well as the exclusion or monitoring of less desirable migrants, reflected in numerous local strategies for the control and management of foreigners: "Competition has become so intense and intrinsic to local development that cities have to take a more entrepreneurial stance in order to remain at the top of a region and enhance their attractiveness to the footloose capital, residents and visitors" (Xu and Yeh 2005: 283).

Looking at the role of Guangzhou in China's urban system and that of the Pearl River Delta, the city has lost its function as the essential controlling centre since 1978. It has lost the primary position it held in the controlled socialist system, as well as its economic role as China's "southern gateway to the world", caused by the reintegration of Hong Kong and Macau. However, it has been able to retain its international significance as a trading centre thanks to its role as the location for the trade fair. At the same time, new (mega)cities have developed very rapidly in the Pearl River Delta. After Guangzhou, Shenzhen and Dongguan, Foshan has emerged as the fourth largest city in the Pearl River Delta (Guangdong Statistical Yearbook 2010: Table 4–5). The city has become known outside the immediate region because of its porcelain industry and is now the headquarters for many large private companies. Foshan's city administration is working to attract foreign companies and hence foreign direct investment. The district of Nanhai to the east of the city, directly adjacent to Guangzhou, has benefited especially.

The relationship between Guangzhou and Foshan has been strengthened by both municipal governments in recent years: Since 2005 a special exchange programme has been developed with the aim of intensifying co-operation in the areas of industrial development and water management as well as creating a shared transport network. By now China's first interurban underground line opened, the Guangfo Line, which links Guangzhou directly with Foshan – with the aim of creating a direct link to the Foshan High-Tech Industrial Development Zone, for example. Foshan's advantages include its comparatively low land and labour costs, its strong industrial basis and its strategic locational advantages through its proximity to Guangzhou, Hong Kong and Macau.

10.4 African Migration to Guangzhou and Foshan

Reliable figures on international migration to China are difficult to find, as they are not published (Brady 2000; Li et al. 2008). The Chinese Statistical Yearbooks do not offer much help in determining the number of African migrants (China, Guangdong or Guangzhou), as they only record the numbers of foreign visitors[4,5] without differentiating according to whether these are migrants, students, tourists or other groups of foreigners. According to the Chinese Statistical Yearbook (NBS 2009: 767), a total of 378,400 foreign visitors from Africa entered China in 2008. This is a very small proportion, barely 1.6%, of all foreign visitors to China.[6]

With regard to the number of African migrants in Guangzhou, the following figures can be found in the scientific literature: Up to 2008 researchers, civil servants, Chinese residents and migrants agreed on an estimated figure between 15,000 and 20,000 persons (Le Bail 2009: 6; Li et al. 2009: 709). This included illegal migrants without a passport or valid visa. Media reports on Africans in Guangzhou cite highly divergent figures. The People's Daily Online (2010), for example, reports that "official figures from the provincial public security department showed that in 2008, about 80% [130,400] of the 163,000 Africans with temporary residence in Guangdong live in the capital [Guangzhou]". A (somewhat) different picture is painted by the *Southern Weekly* (2009), a national daily newspaper with its headquarters in Guangzhou: According to this source, the annual growth rate of the African population in Guangzhou since 2003 is between 30 and 40%. In 2009 about 50,000 registered foreigners are believed to have lived in Guangzhou, more than 20,000 of whom came from Africa. These figures do not include the illegally resident Africans, who are believed to account for a further 200,000.

These marked divergences demonstrate how difficult it is to gain even a rough idea of the actual numbers of African migrants in China and Guangzhou. According to Cheng (2010), foreign residents were included in the National Census for the first time in 2010. This is supposed to give researchers and political decision-makers a better basis to plan a far-reaching reform of migration policy.

Although it has been difficult to determine the exact numbers of African migrants in China, more is known about how they came to China. The first African traders came from Hong Kong to Guangzhou in the 1990s. The influx of African traders rose rapidly in the early 2000s, largely influenced by the relaxation of trade barriers after China's entry into the WTO in 2001. Today three main groups of African migrants live in China: a small number of diplomats, a growing group of students and a large majority of business people. The latter group can be subdivided into tradesmen and salesmen.[7] The former usually reside long term in the city and generally have offices or shops in China where they sell goods for which there is a high demand (clothing, electronics, etc.). Salesmen generally spend shorter periods of time in China, as they move regularly between China and their home country. This group in turn can be subdivided into three categories, namely, (1) those who travel regularly between their home country and China and usually have a business visa, (2) those who usually only spend a few days in China for specific business and (3) those entering the country without previously established aims or business in order to "try their luck" – usually with a tourist or business visa. However, some of these remain in the country for longer than their visa permits. Tradesmen and salesmen tend to be concentrated

[6] About 24.3 million foreign visitors were reported for China as a whole in 2008 (NBS 2009: 767). The largest group is from Asia, accounting for 59.82%, followed by Europe with 25.17%, North America with 9.54%, Oceania with 2.83% and Latin America with 1.07%. The number of actual migrants is considerably lower (Pieke [2010: 20] estimates that about 2 million foreigners are living in China for a longer period of time).

[7] For reasons of brevity, the masculine term is used here; men and women are included.

in the two Chinese wholesaling cities of Guangzhou and Yiwu in Zhejiang province (Le Bail 2009: 1).

African business people are to be found in two main locations in Guangzhou: the area around Guanyuanxi Street, where many African migrants from English-speaking countries work, and the area around Xiaobei Street, favoured by Africans from French-speaking countries. The area around Xiaobei Street is in the Yuexiu District along the Inner Ring Road in the northern part of Guangzhou. Here several high-rise buildings (mansions) are primarily used by the African population. The best known is Tianxiu Mansion, a high-rise complex consisting of three blocks, which initially served as the only centre for the African population in Guangzhou, but has since been augmented by other locations (Li 2008: 388). Guangyuanxi Street is a multilane traffic axis (at the intersection of the three districts of Liwan, Yuexiu and Baiyun), whose favourable location close to Guangzhou's central railway station and within easy reach of the international airport plays a major role in attracting tradesmen and salesmen.

There are marked differences in the commercial structure of Xiaobei and Guangyuanxi Street: Along Guangyuanxi Street larger trading centres dominate, such as Canaan Wholesale Trading or Tangqi Export Clothing Center, while Xiaobei Street is home to smaller shops, showrooms and many offices. The built fabric of the two streets is also quite different: Guangyuanxi Street is dominated by warehouses, which contain many small showrooms and shops on several levels. Xiaobei Street is dominated by high-rise buildings whose lower floors contain showrooms, small shops and offices, while the upper floors are residential.

The urban district of Nanhai in Foshan has emerged as a new location for African migrants in recent years: The main foci are the neighbouring areas of Binjiangdasha and Hepanyuan, Bizhong, Dasha as well as Nanhai's old shopping street which has experienced a far-reaching trading-down process in the last few years. A small number of African "ethnic businesses" (hairdresser, restaurant, call centre) have since opened in the street. There are also Chinese-owned restaurants, cafés and clubs that are mainly frequented by Africans.

In the quantitative survey, Nigeria emerged as the most important source country with 52% of the respondents. Nine percent of those surveyed come from South Africa and 6% from Mali. However, if the residential areas are looked at separately, a different picture emerges: In Foshan Nigerians dominate with 85%, and a further bare 10% come from South Africa. Nigerians also form the largest group among those surveyed in Guangzhou but only account for 38% here. Nine percent come from Mali, 8% from South Africa and 7% from Ghana. Thus, the migrants in Foshan are more homogenous with regard to their country of origin than those in Guangzhou. Most of the African migrants surveyed were men aged 25–45 (85%). 90.5% of those surveyed work as tradesmen and salesmen; only 3.5% are students, 2% are looking for work and 1% employed in the services sector (hairdresser). The remaining 3% are employed in other occupations.

A largely negative, stereotyped image of African migrants is usually presented in the Chinese media; they are often associated with the "three illegalities" (see Sect. 10.2) and drug dealing (Le Bail 2009). Africans are more frequently confronted with these prejudices than other migrant groups – some of whom also include individuals without the necessary documents (see Epstein et al. 2010).[8] In 2009 African migrants drew the attention of the political leadership as well as national and international media: Two African textile retailers were seriously injured when attempting to escape a police raid in Guangyuanxi Street in Guangzhou in July 2009. One was badly cut on broken glass; the other was in a coma for several days (Coloma 2010). Immediately after the incident, about 100 angry African traders assembled outside the local police

[8] Epstein et al. (2010) report on illegal immigrants to China from Southeast Asia (in particular from Vietnam, Cambodia, Myanmar) who work in low-paid jobs on sugarcane plantations, in the clothing industry and on building sites.

station to protest against the Chinese police's rough treatment of Africans (Pomfret 2009). Both victims recovered, but rumours of their supposed death could be read even in the international press (D'Inka et al. 2009).

10.5 Findings: The Background to the Settlement of African Migrants in Foshan

African migrants only began to settle in Foshan in greater numbers quite recently. The data presented here suggest that many factors have had an influence on the choice of Foshan as a new location. 75.0% of the Africans surveyed in Foshan (see Table 10.1) stated that they first lived in Guangzhou, evidence for a recent and ongoing shift of the African migrant population towards Foshan. However, this figure also shows that the process involved is not only a locational shift: Some migrants living in Foshan migrated directly to Foshan, that is, a simultaneous process of new African migration to Foshan is also taking place. All of those surveyed who live in Foshan, and are in employment, gave Guangzhou as their place of employment. Thus, the shift affects only the migrants' place of residence and not their place of employment. At the same time, this means that Guangzhou's position as a location for African tradesmen and salesmen is not experiencing competition through some African migrants' choice of Foshan as a residential location.

The length of time spent in China by the migrants surveyed here varies considerably; the vast majority (90.7%) of those surveyed had come to China since 2005. While some of the migrants living in Guangzhou had been in China since the early 1990s, the first individuals surveyed had come to Foshan in 2007 (Table 10.1). Experts as well as representatives of the African migrant association believe that the number of African tradesmen and salesmen in Guangzhou has declined sharply since the onset of checks prior to the Olympic Games (since May 2008); the situation worsened again since early in 2010 (09/GZ/81/E,[9] 10/GZ/87/E, 10/GZ/89/M, 10/GZ/101/M, 10/GZ/103/M). One expert (10/FS/99/E), who carried out research on the African migrant association in 2009, estimated in March 2010 that the number of African migrants had also declined in Foshan since early in 2010.

In the following the background to the current migration of African tradesmen and salesmen to Foshan will be discussed.

10.5.1 The Role of Global Conditions and Trends

As already mentioned above, China's entry into the WTO in 2001 led to a marked increase in the immigration of African traders. At the same time, the global economic crisis of 2008/2009 placed severe economic restrictions on the traders. In this survey, 51.5% of the migrants stated that their economic situation had deteriorated somewhat or very much because of the crisis; 34.5% stated nothing had changed because of the global economic crisis, and 10.1% felt that their situation had improved. The considerably worsened economic situation of some migrants could explain their preference for Foshan as a cheaper residential location (see Sect. 10.5.3).

Africans' choice of China as a migrant destination is influenced not only by China's cheap production for global markets, the large Chinese market and the favourable conditions for trade, but also by the restrictive entry policies of other countries which push the decision in favour of China. For example, two African tradesmen in Foshan (10/FS/106/M, 10/FS/107/M) report:

[9] The qualitative interviews and interviews with experts carried out for this project are coded. The first two figures indicate the year in which the interview was held. This is followed by the location (GZ for Guangzhou, FS for Foshan) and the interview number from the sequential interview list. The letters following this number denote the status of the interview partner: E for expert, M for migrant. Thus, the code "10/GZ/89/M" refers to an interview carried out in 2010 with an African migrant in Guangzhou.

Table 10.1 Year of migration to China and places of employment of those surveyed in Guangzhou and Foshan

Place of residence			Year of migration to China						
			1993–1995	1996–1998	1999–2001	2002–2004	2005–2007	2008–2010	Total
Guangzhou	Place of employment	Guangzhou	0	2	5	13	46	86	152
		Foshan	0	0	1	0	1	3	5
		No employment (students, unemployed etc.)	1	0	0	0	3	13	17
	Total		1	2	6	13	50	102	174
Foshan	Place of employment	Guangzhou	0	0	1	0	14	51	66
		Foshan	0	0	0	0	0	2	2
		No employment (students, unemployed etc.)	0	0	0	0	1	2	3
	Total		0	0	1	0	15	55	71
Total			1	2	7	13	65	157	245

Source: Authors' survey (2010)

Interviewee 1: And [I buy up] tyres, used tyres, motor tyres, used tyres in Germany. I haven't been in Germany. [...] If people have the visa to go to Europe, to Germany, not everybody can go to Germany, so the people have the visa I give them money to go and buy for me. So this people have been buying these things for me for maybe one year, two years I cannot make any profits [...] So I have tried to get a visa. I applied for the visa, go to the embassy but: denied. [...] They denied me to come to Germany to buy my goods, buy my stuff, you understand? [...During] 1998–2000. For three years I was looking for, applying for visas. Not only I but I am telling you about myself. So many people don't get it. [...] Before, someone told me to go to China, somebody go to China and buy goods. [...] I think this China might be a very good place. And I come to China and I can see that business here [...]
Interviewer: So was it easy for you to get a visa here?
Interviewee 2: Easy.
Interviewee 1: Europeans, Americans, Germany, those aren't easy. It is no more easy any more.

10.5.2 The Role of National Conditions

In qualitative interviews with African migrants, numerous interview partners reported that either they themselves or acquaintances had chosen Foshan as a place of residence because they did not have a valid visa or passport and fewer checks are carried out on the streets and in apartments there (10/FS/106/M, 10/FS/107/M, 10/FS/111/M, 10/FS/114/M). At the same time, an expert from the local administration in Foshan stated that in his opinion a large proportion of the African migrants within his administrative district did not have a valid passport (10/FS/96/E).

As already mentioned, it is difficult or impossible to quantify the proportion of illegal migrants, so that it is not possible to confirm the claim that illegal migrants in particular have settled in Foshan. Nevertheless, questions about residence permit status were integrated into the quantitative survey in order to at least gain some additional information. As an indicator of actual ownership of a visa or passport, the interview partners were asked to show their visa at the end of the interview. 31.3% of those interviewed showed their visa, 2.7% stated they did not have a visa (any more) and the remaining 66.0% either could not or would not show their visa or passport (in spite of the fact that it is obligatory to carry a passport). Of those who showed their visa, 70% had a business visa, 15% a tourist visa, 5% a student visa and 4% a work visa. Of those who did not show their passport, 76% claimed to have a business visa, 11% a tourist visa, 7% a work visa and 4% a student visa. Determination of legal status is made more difficult by the fact that it is apparently possible to acquire a new African identity following deportation from China. According to several Nigerian migrants, it is easy to get a forged/new passport in an African state other than one's home country and to use this passport to apply for a new visa for China (10/GZ/113/M, 10/FS/114/M).

Those with a tourist visa are working illegally in China: It is easier and cheaper to apply for a tourist visa before travelling, even if this is only valid for "sightseeing, family visiting or other private purposes" (Xu 2010). Applying for an F visa is more complicated: Besides the usual forms, Nigerians, for example, must submit a *visa notification form* issued by a Chinese ministry or an authorised Chinese firm or interest group. They must also submit a written invitation from the firm or organisation where they plan to work and evidence of a return ticket to their home country, and a separate interview is carried out at the Chinese embassy in the applicant's home country (Xu 2010).

Interestingly, a considerably greater proportion of interviewees showed their visa in Guangzhou than in Foshan (Table 10.2). At the same time, the migrants resident in Foshan demonstrated a considerably lower level of transnational mobility: 40% of migrants in Foshan claimed that they never visited their home country, in comparison with 18.3% in Guangzhou. There may be economic reasons for this pattern, or it may be related to legal status, as a future return to China would be impossible without documents (of course individual reasons may also play a role in migrants' failure to visit their home countries, but they do not explain the marked difference between the two locations). A further indicator is the difference between those who

Table 10.2 Selected parameters for comparison of the samples in Guangzhou and Foshan

Areas	Parameters		Interviewees with place of residence in Guangzhou	Interviewees with place of residence in Foshan
Background to migration	Main countries of origin	1	Nigeria (38.0%)	Nigeria (85.1%)
		2	Mali (8.9%)	South Africa (9.5%)
		3	South Africa (7.8%)	Algeria (1.4%)
		4	Ghana (7.3%)	Ivory Coast (1.4%)
		5	Guinea (6.1%)	Guinea (1.4%)
	Proportion of migrants who moved from Guangzhou to Foshan		–	75.0%
	Average annual number of journeys between China and home country		3,587	2.143
	Proportion who have not returned to their home country since entering China		18.3%	40.0%
Residence permit status	Proportion who stated that they did not have a visa		2.8%	2.9%
	Proportion who showed their visa		38.6%	14.3%
	Proportion who experienced difficulties extending a visa		32.2%	51.4%
Economic situation	Average monthly per capita rent		2,269.94 RMB	939.93 RMB
	Average monthly rent per apartment		2,785.71 RMB	1,490.27 RMB
	Average monthly per capita income		31,359.14 RMB	7,324.42 RMB

Source: Authors' survey (2010)

have already experienced difficulties extending a visa (Guangzhou: 32.2%, Foshan: 51.4%). These three differences are the only figures that may actually indicate that illegal status could play a role in the preference of Foshan as a place of residence.

An interesting finding in this context is the fact that parallel to the limitation of visas granted and increased checks prior to the Olympic Games in Beijing in 2008, informal channels for acquiring visas have increased: Foreigners buy their visas for higher charges, usually involving an agent (according to statements made by various interview partners this person may be located in China (10/GZ/87/E), Hong Kong (10/GZ/89/M) or the migrant's country of origin (10/FS/107/M)). This agent may organise the visa without the required documents (e.g. written invitation, birth certificate) or a type of visa (e.g. visa for multiple entry) which would not have been officially approved, either because of insufficient documentation or current changes in visa regulations, such as occurred before the Olympic Games when the number of visas granted was severely restricted.

Interviews with representatives of international organisations and firms show that these channels are not only used by migrants whose nationality or status is "undesirable" but also, for example, by Western expats and even representatives of high-ranking international organisations (10/GZ/87/E). The fees for the informal acquisition of a visa vary according to the applicant's nationality: While migrants from Western states generally pay a fee a little lower than the costs of acquiring the visa themselves (e.g. travel to Hong Kong to get a visa there, 10/GZ/87/E), migrants from countries for whom it is much more difficult to get a visa pay much higher fees. The figures cited by African interview partners range from US$2,000 to get a visa in Hong Kong (10/GZ/89/M) to US$5,000 for a visa from the home country (10/FS/107/M); these figures also depend on the individual agent. An African trader in Guangzhou (10/GZ/112/M) reports:

> I have had different visas [...] [shows his different visas in his passport]. This one here is for three months and this one is for one year. Three months is 10,000 RMB. [...] Yes, because I changed from

L [tourist visa] [...] L is 4.500 RMB. It is false. The government can charge maybe less than 200 RMB. But the agent, not the government. [...] It depends on the agency. For certain people it is more, it is not the fault of the government. The government cannot take more than maybe 500. [...] But one year is 30,000 RMB. 3 month is 10,000 RMB.

Some migrants who cannot afford the high costs of informal visa acquisition or whose application was refused remain illegally in the country. Proximity to Hong Kong and Macau plays an important role for foreign traders and therefore for Guangzhou, as several procurers of visas are established there, and also some types of visa require that the holder leaves the country once every 30 or 60 days. Travel to Hong Kong or Macau is simpler and cheaper than travelling home or to other neighbouring countries, which would not be possible without a visa for those countries anyway.

10.5.3 The Role of Regional and Local Conditions

Regional or local reasons for the shift towards Foshan can be found in interurban differences in the implementation of sanctions against migrants guilty of one of the "three illegalities" and different rent levels. A comparison of migrants' incomes and rent levels in both locations demonstrates that economic reasons play a role in (new) settlement in Foshan: There are marked discrepancies between the income of those surveyed in Foshan (average income 7,324 RMB or ca. 840 Euro) and Guangzhou (average income 31,359 RMB or ca. 3,580 Euro). Those interviewed in Foshan also paid considerably lower rents (1,490 RMB or ca. 170 Euro in Foshan and 2,786 RMB or ca. 320 Euro in Guangzhou).

As already explained in the preceding section, some migrants stated that they had moved to Foshan because of the laxer police checks there. Field observations and qualitative interviews with all of the migrants (10/GZ/88/M, 10/GZ/89/M, 10/GZ/101/M, 10/GZ/103/M, 10/GZ/108/M) in Guangzhou who were willing to discuss this issue did indeed show that inspections of Africans' documents (passports, visas, registration of residence, where applicable work permits and business licences) in Guangzhou had increased markedly following the African traders' protests in July 2009, and again since the beginning of 2010. The migrants reported that police raids were carried out as frequently as once a day.

At the end of January 2010, a major raid on most of the foreign shops as well as a comprehensive inspection of passports, visas, residential registrations, business licences and other documents was carried out. African migrants and the main locations of African offices and showrooms were the focus of these checks (10/GZ/87/E). During fieldwork in March and April 2010, the authors observed several raids in the main business locations for African traders in Guangzhou. In the context of the collection of the data presented here, Xiaobei was surveyed several times between 2006 and 2010, and since 2008 a marked reduction was observed in the number of shops and offices run by Africans.

The representative of the local self-administration organisation referred to above reports that increased checks on migrants have been taking place in Foshan since April 2009, and the local government is obviously trying to prevent them from settling in Nanhai (10/FS/96/E). Spot passport checks identify migrants without (valid) papers; however, according to this interview partner, the local government lacks the funding for deportations so that these migrants generally remain in the country (10/FS/96/E). However, the frequency of checks is reportedly much lower than in Foshan – this is also confirmed by migrants (10/FS/95/M, 10/FS/106/M, 10/FS/107/M). The apparently generally informal measures used by the local government in Foshan include verbal prohibitions instructing estate agents in Nanhai not to procure apartments for African migrants (10/FS/92/E).

Interview partners (10/GZ/88/M, 10/FS/106/M, 10/FS/107/M, 10/GZ/112/M) report that members of their group who did not have a (valid) passport were treated quite differently: Some only had to pay fines, others were imprisoned for varying lengths of time and some were deported at their own cost. According to these accounts, the size of the fines, the period of imprisonment or deportation depend on varying local practices in the cities. Thus, some migrants stated that it is better for those without (valid) papers to stay in Foshan than in Guangzhou as the punishments there are much milder (10/FS/106/M, 10/FS/107/M).

10.5.4 The Role of Social Networks and Migrant Associations

The African migrants are well organised in (informal) migrant associations and demonstrate strong ethnic cohesion as well as social and economic support networks. They are organised in associations according to nationality, each headed by a chief representative or "president". These provide their members with important information and contacts (e.g. business contacts, contacts with visa agents or estate agents willing to rent to Africans), financial support in special circumstances and legal advice (10/FS/95/M, 10/FS/107/M, Li et al. 2009; Zhang 2008). The increasingly restrictive environment for (undesirable) foreigners in China since mid-2008 makes the migrant associations as well as institutions and organisations from their home countries and individual social networks even more important.

Contacts with established members of their own migrant associations and/or Chinese mediators are very important for migrants in general, and essential for illegal migrants, in order to gain access to housing and offices as well as to develop business contacts. Without these migrants will experience major difficulties finding a place to live or renting hotel rooms (10/FS/96/E, 10/FS/95/M, 10/GZ/103/M, 10/FS/109/M). Such contacts can help African migrants to gain access to apartments in Foshan in spite of the verbal ban on renting to African migrants which has been imposed on the local estate agents. Chinese-speaking contacts arrange apartments for Africans, for a fee. They deal with the apartment owners and thus enable illegal migrants to rent an apartment without the necessary documents (valid passport, valid visa) (10/FS/96/E).

In this context it is interesting that the majority of those interviewed in Foshan came from Nigeria. In qualitative interviews it was reported that most of the Nigerians are members of the Igbo people from southeastern Nigeria (10/FS/106/M, 10/FS/107/M). Many Igbo have emigrated since the Nigerian civil war, and the Igbo diaspora has a strong influence abroad. This includes numerous migrant organisations operating in Nigeria, the migrants' destination or in numerous countries (Gordon 2003). It is evident that the migrant association in China is active across international borders, as a significant proportion of the Nigerians in Foshan are now coming directly to Foshan: They were informed about the location and directly supported by other members of the migrant association (10/FS/95/M, 10/FS/106/M, 10/FS/107/M).

10.6 The Settlement of African Migrants in the Context of Processes of Global Change and National Development Aims

The findings presented here demonstrate that in recent years a proportion of African migrants have moved their place of residence from Guangzhou to Foshan. At the same time, new migrants have been coming directly to Foshan. Because the change involved is a shift of residence and not of workplace, the proximity of the Nanhai district in Foshan to the two main employment locations for African migrants in Guangzhou was a precondition for the move. Guangzhou retains its significance for migrants as a location for trade and employment, although the overall number of Africans has declined, according to all statements and indications. Migrants' residential relocation to Foshan while retaining Guangzhou as a place of employment has created new patterns of circular mobility and altered the action spaces of the African migrants.

International migrants play an important and positive role in China's economy. This includes African traders who commission goods to be produced in China and/or buy them to sell on in their home countries. This strengthens the Chinese labour and sales markets and reinforces the role of Guangzhou as a trading city. Thus, it appears that an over-restrictive policy on visas and foreigners may actually damage China's interests. Thanks to its unparalleled economic growth, the country will remain attractive to migrants from many regions of the world in the future. In relation to China's total population, the number of international migrants is so low that there is no reason to fear "inundation" by migrants or serious competition from foreign workers on the Chinese labour market.

Nevertheless, the activities of African tradesmen and salesmen who are viewed as "undesirable" in the context of *waishi* strategy are currently being increasingly restricted. Brady (2000: 962) observes: "government-orchestrated xenophobia has always been selective and controlled, as has government-administration of foreigners and foreign things". The decentralisation of administration means that aspects of *waishi* are pursued and implemented differently at local level. There is regional and national competition between cities to attain functional primacy and hence also to attract capital, goods and human capital. The municipal governments of megacities such as Shanghai, Peking, Guangzhou and Shenzhen have developed (and control) labour markets directed primarily at qualified workers (Orban et al. 2003). This development, which has led to interurban competition to attract international qualified workers – and which according to Farrer (2010) should be seen as a symbol of China's globalisation and cosmopolitanism and is intended to give the cities a cosmopolitan image – also leads to the increasingly urgent need to exclude all foreigners who supposedly fail to represent this image or to direct and monitor their activities as far as possible.

The African migrants' protest demonstration in the summer of 2009 appears to have had a significant influence on their more stringent treatment; the protest attracted worldwide attention in the media and casts a negative light on Chinese treatment of foreigners. The Chinese government and the city of Guangzhou, which as a trade fair location is especially concerned about its image, intend to act more radically to prevent such protests. The tendency to control and direct everything that is unfamiliar or new has parallels in the treatment of rural–urban migrants: For a long time, municipal governments were very concerned to control their movement and activity by means of strict monitoring, regulation and sanctions. Only in recent years have they begun to realise that the desired order cannot be enforced by control and sanctions (see Zhao 2003).

The less co-ordinated measures of the city of Foshan indicate that this municipality has hitherto had less experience with managing foreign residents. This example of local experimentation with measures (see Heilmann (2008) and Bork et al. (2011) on "experimental informality"), with which there is no prior experience, is typical of the Chinese transformation process. It is not very long since the country was first opened to foreigners. Thus, lack of experience and poor strategies for dealing with (illegal) migrants are evident in many areas. The fact that the decentralisation of administration allows cities to apply different sanctions to punish illegal activity, thereby encouraging illegal migrants to relocate to other cities, is just one example.

A further approach to explaining the more stringent monitoring of African tradesmen and salesmen could be China's Africa strategy: The Chinese government is increasingly supporting the involvement of Chinese actors in African economies and societies (see Alden and Hughes 2009). This includes the replacement of African traders by Chinese counterparts. The extent to which the restrictive Chinese visa policy is an attempt to impede the entry of African traders and their employment in China in this context remains unclear.

Because the incentives for migration remain strong, more frequent checks and greater strictness on the part of the police are not adequate to stem illegal immigration or illegal residency. In the light of the growing number of informal visa purchases, improved monitoring of visa allocation is a more necessary precondition for the long-term reduction of illegal immigration than arresting illegal migrants already in the country. Ultimately the Chinese government must decide whether its current strategy for the prosecution of the "three illegalities" is genuinely effective, given the fact that these problems are at least to some extent "home-made", as has been demonstrated here: This study demonstrates that there are illegal migrants without a valid visa or passport (even if their actual numbers cannot be readily determined) and that other illegal strategies (false passport, tourist visa) are chosen in order to gain access to China. Yet, a not insignificant proportion of the Africans living illegally in China today were most probably only pushed into illegality by the tightening of entry regulations (the drop in numbers of visas approved since 2008) and thus overstayed after not being granted a visa extension and did not actually enter the country illegally.

10.7 Conclusion

The settlement of migrants in Foshan has many underlying causes. The example studied here demonstrates the variety and complexity of the influence of global developments and agreements (e.g. global economic crisis 2008/2009, China's entry into the WTO), national conditions (e.g. national immigration regulations, development strategies, hosting major international events), interurban competition and co-operation. It also shows how development aims formulated at different levels (in this case the national versus the municipal level) are interpreted and implemented in different ways at a local level and are compromised or counteracted by the reactions, flexibility and adaptive strategies of actors in the cities.

Thus, indications were found that much tighter documentation checks in Guangzhou and laxer sanctions in Foshan in the case of one of the "three illegalities" have drawn increased numbers of illegal migrants to Foshan. In this way both national immigration regulations and their local implementation could be undermined. At the same time, differences in rent levels in both cities are responsible for the movement of migrants with lower incomes to Foshan. The global economic crisis of 2008/2009 caused a massive deterioration in the economic situation of a significant proportion of migrants and thereby encouraged migration to "cheaper" Foshan. Ethnic and social support networks within the migrant communities and influential individual actors played a significant role in migrants' access to housing and thus the possibility to settle in Foshan in spite of preventive measures on the part of the municipal government.

However, since the protests organised by African tradesmen and salesmen in the summer of 2009, the more stringent monitoring of Africans in Guangzhou – which is now also incipient in Foshan – must also be re-evaluated in the context of interurban economic competition, primarily directed towards the development of a cosmopolitan image. The Chinese government and the city of Guangzhou, which as a trade fair location is especially concerned about its image, intend to act more radically to prevent such protests. A further explanation could be the likelihood that China has an interest in replacing African traders with Chinese counterparts. Either way, the tightened entry regulations prior to the Olympic Games in August 2008 are likely to have spurred on the "illegalising" of migrants; thus, part of the illegal migrant problem is "home-made", so that the aims of the monitoring must be questioned. Furthermore, informal channels for acquiring a visa increased in response to the tighter controls.

References

Alden C, Hughes CR (2009) Harmony and discord in China's Africa strategy: some implications for foreign policy. China Q 199:563–584

Bork T, Gransow B, Kraas F, Yuan Y (2011) Marketization and informalization of health care services in mega-urban China. In: Krämer A, Khan MMH, Kraas F (eds) Health in megacities and urban areas. Springer, Heidelberg/Dordrecht/London/New York, pp 173–188

Brady AM (2000) "Treat insiders and outsiders differently": the use and control of foreigners in the PRC. China Q 164:943–964

Brady AM (2003) Making the foreign serve China: managing foreigners in the People's Republic, Asia/Pacific/Perspectives. Rowman & Littlefield, Lanham

Cartier C (2001) Globalizing South China. Blackwell, Oxford/Malden

Cartier C (2002) Transnational urbanism in the reform-era Chinese city: Landscapes from Shenzhen. Urban Studies 39:1513–1532

Cheng Y (2010) Rising tide of incomers prompts start on China's first immigration law. People's Daily Online. http://english.people.com.cn/90001/90776/90882/6995005.html. Accessed 7 Oct 2010
Cheng J, Chao L (2008) Fewer foreigners visit Beijing amid tighter rules on visas. Wall Street J. 22 May 2008. http://online.wsj.com/article/SB121138172668010789.html?mod=hps_asia_whats_news, Accessed 10 June 2010
Coloma T (2010) 100,000 immigrants/$100-billion in trade/Africa Town: everything to gain. The Globe and Mail (Canada). 15 May 2010. http://www.theglobeandmail.com/news/world/g8-g20/opinion/in-africa-town-everything-to-gain/article1569451/. 17 Oct 2010
D'Inka W, Kohler B, Nonnenmacher G, Schirrmacher F, Steltzner H (2009) Afrikaner protestieren in Südchina. Demonstranten beklagen Tod eines Nigerianers. Behörden sprechen von "Verletzten". Frankfurter Allgemeine Zeitung. 17 July 2009:7
Epstein G, Yang L, Thuy DM (2010) China's immigration problem. Forbes 186:26
Farrer J (2010) 'New Shanghailanders' or 'New Shanghainese': western expatriates' narratives of emplacement in Shanghai. J Ethn Migr Stud 36:1211–1228
Gordon AA (2003) Nigeria's diverse peoples: a reference sourcebook, Ethnic diversity within nations. ABC-CLIO, Santa Barbara
Gu C (2001) Regional polarization under the socialist-market system since 1978: a case study of Guangdong province in south China. Environ Plan 33:97–119
Guangdong Statistical Yearbook (2010) Permanent population at the year-end by city. www.gdstats.gov.cn/tjnj/table/4/e4_5.htm. Accessed 12 Sept 2010
Heilmann S (2008) Policy experimentation in China's economic rise. Stud Comp Int Dev 43:1–26
Kraas F (2004) Urbanisierungsprozesse in China. Petermanns Geogr Mitt 148:58–59
Le Bail H (2009) Foreign migration to China's city-markets: the case of African merchants. Asie Visions 19. http://www.ifri.org/files/centre_asie/AV19_LeBail_GB.pdf
Li Z (2008) Ethnic congregation in a globalizing city: the case of Guangzhou, China. Cities 25:383–395
Li Z, Xue D, Lyons M, Brown A (2008) The African enclave of Guangzhou: a case study of Xiaobeilu. Acta Geogr Sin 63:207–218
Li Z, Ma LJC, Xue D (2009) An African enclave in China: the making of a new transnational space. Eurasian Geogr Econ 50:699–719
Liu G (2009) Changing Chinese migration law: from restriction to relaxation. J Int Migr Integr 10:311–333
NBS (2009) China statistical yearbook. China Statistics Press, Beijing
Orban E, Chen X, Koehn PH (2003) Great power decentralization and the management of global/local economic policy and relations: lessons in fluidity from the People's Republic of China. Int Rev Adm Sci 69:235–258
Pieke FN (2010) China's immigrant population. China Rev Summer:21–23
Pomfret J (2009) Out of Africa and into China, Émigrés struggle. Reuters. 20 Aug 2010. http://www.reuters.com/article/idUSTRE57K00K20090821. 12 Sept 2010
Qin Q (2009) Guangzhou to increase checks on illegal immigration. China.org. 17 July 2009. http://www.chinadaily.com.cn/china/2009-07/17/content_8439196.htm. Accessed 18 Jan 2010
Shen JF, Wong KY, Feng ZQ (2002) State-sponsored and spontaneous urbanization in the Pearl River Delta of South China, 1980–1998. Urban Geogr 23:674–694
Stening BW, Yu Y (2006) Expatriates in China: a review of the literature. School of Management, Marketing, and International Business Working Paper Series 1. http://dspace-prod1.anu.edu.au/bitstream/1885/46046/1/MMIB%20Working%20Paper%20Series%20Volume%201_Numer%201.pdf. Accessed 12 Sept 2009
Sun Q, Qiu LD, Li J (2006) The Pearl River Delta: a world workshop. In: Zhang KH (ed) China as the world factory. Routledge, Abingdon, New York, pp 27–52
UNESCO (2010) Migrant/migration. http://www.unesco.org/new/en/social-and-human-sciences/themes/social-transformations/international-migration/glossary/migrant/. Accessed 17 Oct 2010
Wang J, Lau SSY (2008) Forming foreign enclaves in Shanghai: state action in globalization. J Hous Built Environ 23:103–118
Willis K, Yeoh B (2002) Gendering transnational communities: a comparison of Singaporean and British migrants in China. Geoforum 33:553–565
Wu F (2002) China's changing urban governance in the transition towards a more market-oriented economy. Urban Stud 39:1071–1093
Wu F, Xu J, Yeh AG-O (2007) Urban development in post-reform China: state, market, and space. Routledge, London
Xu J (2010) Types of Chinese visa. http://ng.china-embassy.org/eng/hzqz/qzzl/t142016.htm. Accessed 17 Aug 2010
Xu J, Yeh AG-O (2005) City repositioning and competitiveness building in regional development: new development strategies in Guangzhou, China. Int J Urban Reg Res 29:283–308
Yao S, Chen S, Guan C, Wang S (2008) New characteristics of urbanization and strategies of regional spatial development in China. In: Keiner M (ed) Sustainable urban development in China: wishful thinking or reality? MV-Verlag, Münster, pp 65–84
Yeoh B, Willis K (2005) Singaporean and British transmigrants in China and the cultural politics of 'contact zones'. J Ethn Migr Stud 31:269–285
Zhang L (2008) Ethnic congregation in a globalizing city: the case of Guangzhou, China. Cities 25:383–395
Zhao S (2003) Peasant migration: order building and policy rethinking. Social Sciences in China. Winter, pp 168–176

Phobopolis: Violence, Fear and Sociopolitical Fragmentation of the Space in Rio de Janeiro, Brazil

11

Marcelo Lopes de Souza

Abstract

Against the background of the evolution of Brazil's economy especially in the last 20 years, this chapter analyses recent trends related to the worsening situation in terms of socio-spatial segregation and social conflict in Rio de Janeiro since the beginning of the 1980s. *Sociopolitical fragmentation of the space* (constituted above all by the formation of enclaves dominated by drug-trafficking organisations, on the one hand, and self-segregation of the elites in gated communities, on the other hand) and '*phobopolisation*' (i.e. an urbanisation process which is decisively influenced by increasing fear of crime and violence) are identified as key characteristics of Rio de Janeiro (along with many other big cities) today. What can be done to cope with the new challenges for state and civil society in a city in which violence, lack of public safety and fear have become central features of daily life? It seems that progressive local-level measures and policies can probably only, in the best of all cases, contribute to reducing the problem, rather than eradicating it.

Keywords

Urban space • Violence • Sociopolitical fragmentation • Phobopolis • Rio de Janeiro

11.1 Introduction: City of Fear

Phobopolis (Port. *fobópole*) is a new word; it means *city of fear*. I introduced the concept in order to emphasise the degree of intensity in terms of violence and fear prevailing in some cities today (de Souza 2006, 2008). Is there something new about 'criminality and violence in cities', however? Something which could justify a neologism?

M.L. de Souza (✉)
Department of Geography, Federal University of Rio de Janeiro (UFRJ), Prédio do CCMN, Bloco H. Cidade Universitária (Ilha do Fundão), 21.949-900 Rio de Janeiro (RJ), Brazil
e-mail: mlopesdesouza@terra.com.br

F. Kraas et al. (eds.), *Megacities: Our Global Urban Future*, International Year of Planet Earth,
DOI 10.1007/978-90-481-3417-5_11,

Let us have a look at the following description offered by Yi-Fu Tuan in his book *Landscapes of Fear*:

> Medieval European and early colonial American towns were small places with populations of less than 10,000. They could use the curfew because relatively few suspicious-looking strangers roamed the streets and their movements could be supervised. In a large metropolis, such as Rome in the second century A.D. or eighteenth-century London, curfew without the support of a well-manned police force would have been totally ineffective. Instead, as night approached, the citizens themselves recognized the need to withdraw into the security of their homes, abandoning the dark alleys to thieves and foolhardy revelers. On moonless nights the warrens of Rome no doubt looked extremely sinister. Respectable people barricaded the entrances to their houses; the shops fell silent, and traders drew safety chains across the leaves of the doors. If rich party-goers stepped into the streets, they did so protected by slaves who carried torches to light them on the way. (Tuan 1979:160)

Special attention was devoted by him to old London:

> Crime in eighteenth-century London was rampant. After dark, townspeople were reluctant to go out into the ill-lit streets. (...) In 1751, Fielding, a magistrate of Bow Street, reported: "The innocent are put in terror, affronted and alarmed with threats and execrations, endangered with loaded pistols, beat with bludgeons and hacked with cutlasses, of which the loss of health, of limbs, and often of life, is the consequence; and all this without any respect to age, dignity, or sex." (...) Many citizens went about armed. Friends, when they were to return from a coffeehouse or tavern, made up parties for mutual protection. Many families refused to go to the theater on account of the dangers of the homeward journey. (...) Criminals operated boldly in the heart of London. Moreover, large sectors of the metropolis were wholly given over to them. Not only ordinary citizens but officers of the law hesitated to venture into these blighted areas, known as 'Alsatias' in the eighteenth century and "rookeries" a hundred years later. (Tuan 1979:161)

Considering such information, it seems that 'there is nothing new under the sun' regarding criminality, violence and fear in urban spaces. From ancient Rome to medieval European cities to eighteenth-century London to nineteenth-century New York, as Tuan and several other authors (see, for instance, Schwerhoff 2000 and Schuster 2000) show, history offers many examples of criminality and of public concerns about violence in the cities of the past. The point, however, is obviously not related to the *presence* of these phenomena but – as I have already suggested above – to their *intensity*, more precisely, to a particular intensity in a certain historical and geographical context and to the (socio)political consequences of it.

A 'phobopolis' is a city whose inhabitants experience a situation which could be described as quasi-'Hobbesian' in terms of fear of violence, as the socio-spatial consequences of this have become a daily experience for many people. In other words, a 'phobopolis' 'is a city in which violence has become a very widespread concern, and in which fear has become an almost omnipresent feeling' (de Souza 2009:29). The question of the particular intensity of this almost omnipresent feeling' is related to a certain historical and geographical context – that is, to a specific space-time – in which the (socio)political effects of violence and fear are dramatic to the point of influencing various different kinds of decision on the part of the state apparatus, capitalists and families. Among the types of decisions which are meant here, the following can be mentioned: the decision to move to a gated community or to go shopping only in presumably 'secure' shopping malls, the decision to avoid new investments in certain cities and metropolises, the decision to migrate (inside a city or to another city) in order to escape notorious so-called risk places, the governmental decision to send in the army itself to fight drug traffickers, etc.

Surely, every comparison must take into consideration the historical/situational relativity of *intersubjective* feelings. People always compare the gravity of their present-day situation to their own condition (and subjectivity) in the past and/or to the situation in other places, which is *supposed* to have been or be better. The picture of London which emerges from Tuan's pages may seem very familiar to the inhabitants of today's Rio de Janeiro, Mexico City, Johannesburg or Los Angeles, but a closer inspection could reveal some important differences. It is symptomatic enough that despite all horror of robbery, murder and rape, fear of crime and violence in the past did not seem to dominate public concern as intensely as it does nowadays in São Paulo or

Rio de Janeiro: people did not collectively think about leaving Rome or London because of fear; violence and crime did not play an important part in affecting patterns of migration; and an expression like *civil war* was of course related to politically motivated conflicts and not to ordinary criminality.

In contrast to this, the German essayist Hans Magnus Enzensberger coined the suggestive expression *molecular civil war* at the beginning of the 1990s and said that 'not only in Lima and Johannesburg, in Bombay and Rio, but also in Paris and Berlin, in Detroit and Birmingham, in Milan and Hamburg' such a kind of *civil war* was ongoing (Enzensberger 1993:18–19). It is clear that in many cities in the world, crime is no longer perceived in the sense of 'business as usual'. Especially from the perspective of upper and middle classes – and of many governments as well as of the media – something seems to be 'out of control'; actually many people have the impression that the state apparatus has simply failed in guaranteeing 'law and order' (due to 'structural corruption', lack of resources and sometimes lack of organisational flexibility – for instance, in comparison with modern networks of organised criminality). Violence has become an important part of the (geo)political concerns at the national level (from presidential speeches to interventions of the army in order to 'keep the streets safe' and occupy favelas, as occurred several times in Rio de Janeiro since 1992 – see de Souza [2008]).

In a 'phobopolis', *militarisation* of the urban question is a common phenomenon. Both from a progressive, left-wing perspective (such as in the case of Enzensberger's) and from a conservative or even right-wing viewpoint (from many politicians to most mainstream media to many military and criminal experts: *war on drugs*, *war on terror*, etc.), the use of the word *war* in a context which is more or less different from the traditional one has become increasingly common. From discourse to reality, it seems that established views regarding the border between *policing* and *warfare* are becoming rapidly obsolete in an age of 'new wars'. The traditional conservative idea according to which *policing* is something to be implemented at home (or in the cities of consolidated, formal colonies), with the aim of controlling civilians, whereas *warfare* is something to be carried out preferentially abroad, on battlefields in a proper sense, against foreign enemies of the nation-state, presupposes a conceptual border between the *internal* and the *external* which is now blurring or at least being relativised. On the one side, we are witnessing right-wing politicians (as well as scholars, journalists and generals) supporting military interventions in other countries (such as former colonies, but not necessarily) increasingly as a kind of 'international policing' (increasingly with help of mercenaries), as the case of Iraq tragically illustrates; on the other side, right-wing politicians (again: as well as scholars, journalists and generals) from Brazil to Germany are increasingly considering or have already implemented the view according to which internal 'disorder', be it due to ordinary criminality, protest marches, strikes or riots, is no longer a matter of policing purely but actually a challenge which demands military methods and personal (de Souza 2009). In this framework, it is understandable that the metaphor of *war* becomes more and more widespread, and it is also understandable that even the nonconservative discourse cannot avoid completely the seduction of this metaphor, as Enzensberger's essay on a *molecular civil war* or as Ulrich Beck's notion of a *gefühlter Krieg* – that is, a situation in which 'it feels like a war', although there is formally 'peace' (Beck 2007), as well as the use of the expression *global civil war* by Hardt and Negri (2005) and Agamben (2004). Nevertheless, this metaphor should never be treated without caution, as it can be instrumentalised to reinforce militarisation – as it has largely been by the three pillars of the contemporary capitalist *industry of fear*: the *mass media*, the *political system* and the *security market* (i.e. production and sale of different security 'tools' and 'apparatuses', from guns to armoured cars to surveillance cameras to whole gated communities), as it was examined in de Souza (2008).

In historical terms, the period between the end of the nineteenth century and the middle of the twentieth century (i.e. from the Victorian age to the immediate post-war years) was a *relatively* calm one, at least in the big cities – the situation was often quite different in the countryside and in the small cities, particularly in Latin America.

A synergy of factors was responsible for that comparatively *positive* image. Among these factors, we can mention the long-term trend which led to both *war* and *public safety* becoming very centralised in the hands of the state apparatus, for example, through the replacement of *militias* (composed of mercenaries) by professional armies between the sixteenth and the eighteenth century (see Münkler 2004), as well as the consolidation of police as an institution with the purpose of enforcing 'order'. Another factor contributing to the *relative* public safety of the period was the stabilising role of culture (traditional moral values) and some institutions such as religion and family, which functioned as instances of *sociopolitical domestication* in a very effective way in the past.

That *positive* image, for many of us still to a large extent a legacy of the vivid testimonies of our parents and grandparents, emerges as such, not only because it is very often compared to present-day 'chaos' but also because a kind of ideologically embedded nostalgia induces us to underestimate many different types of oppression. Characteristic for those 'well-ordered times', oppression was a factor leading to strikes, acts of terrorism (perpetrated especially by anarchists) and other forms of resistance and violent clashes between the oppressed and their oppressors in the context of class struggle. Nevertheless, as it was already shown, some rights conquered, and gains obtained by the working class, at least in certain countries, also contributed to the relatively low degree of violent criminality in the aforementioned period of relative calmness. However, as we consider the role of factors such as the gradual and partial centralisation of the instruments of violence (in the framework of the *legal monopoly on violence*) in the hands of the state, as well as the function of *traditional values*, we can see that that calmness had much more to do with repression and conformism – in a nutshell, with *heteronomy* – than with freedom and emancipation – that is, with *autonomy*.

Anyway, present-day 'chaos' is in reality the emergence of a 'new order' (or of new order*s*, both legal and illegal ones and at many scales) generated by contemporary capitalism: from '(hyper)precarisation' in terms of labour relations in the context of 'flexible' patterns of capital accumulation (large-scale diffusion of insecure, non-guaranteed working conditions, massive unemployment)[1] and the *erosion of the welfare state* in the '(Global) North' to the worsening of

[1] '*Hyper*precarisation' is related to the '*hyper*precariat', that is, to the workers who depend on – and often were expelled from – the informal sector in semi-peripheral countries (and who work and live under very vulnerable conditions). I have proposed the neologism *hiperprecariado* (Port.)/hyperprecariat (see de Souza 2008, 2009) in order to avoid the problematic term lumpenproletariat. Lumpenproletariat became a traditional term in Marxist jargon after it was used by Marx and Engels in the *Manifesto of the Communist Party* as well as in various later works. According to Marx and Engels, the *lumpenproletarians* (German: *lumpenproletarier*) are to be considered as an amorphous group of different kinds of people that, together, are always and above all potentially reactionary. Marx and Engels employed several discriminatory and even moralistic expressions to designate the *lumpenproletarians* – very much in contrast to a virtuous 'working class' (*proletariat, Arbeiterklasse*) which was supposed to be the real protagonist of social change towards socialism (under guidance of the Communist Party and the enlightened party leaders, of course). In contemporary Europe and North America, the debate regarding *precarity* (French: *précarization*; German: *Prekarisierung*) is about the effects of the erosion of the welfare state in the context of globalisation, neo-liberalism and 'flexible accumulation'. Authors such as French sociologist Bresson (2007:73) believe that this debate is pertinent only to the so-called industrialised world, being irrelevant to deal with the problems of the countries usually known as 'developing countries'. This oversimplifies the capitalist world-system. In fact, there is a whole set of countries – from Brazil to Mexico to South Africa – which are at the same time *significantly industrialised* (although this industrialisation is usually very concentrated in some regions) and *socially highly unequal and unjust* (the degree of inequity and social injustice is often even higher than in many typical peripheral countries). They are *semi-peripheral* countries, whose economies can be by no means described as typically 'nonindustrialised' (despite the importance of the primary sector). Of course, in contrast to Western Europe and the USA, at the semi-periphery of the capitalist world-system, *precarity* or even *extreme precarity* was *always* a chronic problem, due to the intensity of exploitation and oppression and the absence of a welfare state in a Western European sense. As a consequence, the concept of *hyperprecariat* does not have very much to do with the European discussion about *precarity, in spite of some relevant links at the global level.*

since ever 'precarious' working conditions, to rising unemployment rates and the collapse of the 'developmentist' (*desenvolvimentista, desarrollista*) state in many countries in the '(Global) South'; from the 'criminalisation of the world economy' (*criminalisation de l'économie mondiale*: so Riccardo Petrella in *Le Monde Diplomatique* [1995]) on the basis of the increasing importance of many types of illegal trade and of money-laundering to the 'corrosion of character' (Sennett 1999) in the framework of 'flexible capitalism'. Both aspects – the intensity of contemporary problems and changes and the comparison with an historical period which was relatively 'calm' in terms of public safety for ordinary urban, middle-class citizens – are crucial in order to understand the rise of widespread fear as an intersubjective phenomenon. That is, they are fundamental for the understanding of the concept of 'phobopolis'.

Already in the 1950s, Hannah Arendt (Arendt 1998) offered a major contribution to the analysis of what Richard Sennett would term the *fall of public man* in his homonymous book (Sennett 1977), firstly published in 1974 – at a time when problems related to crime and violence in big cities were already very relevant, especially in the United States, although Sennett did not mention these questions in his analysis. Today it is no longer possible to ignore this topic. The *fall of public man* has been increasingly related also to *fear and lack of public safety in the public spaces* in many countries and cities around the world, both in the '(Global) North' and in the '(Global) South', and especially (*but not exclusively*) in big cities and metropolises.

In a typical 'phobopolis', fear and the lack of confidence in the police are so widespread that many people simply renounce visiting certain places at certain times – and the number of these places and times is increasing. Urban dwellers navigate the streets and enter banks and shops with an almost constant feeling of fear (the exception usually being shopping malls). There is a growing acceptance of measures such as monitoring by surveillance cameras (at the cost of privacy, maybe of some degree of freedom), and the urban population is increasingly welcoming conservative types of discourse such as *zero tolerance*. There is a deepening desire to live self-segregated lifestyles (in *edge cities, fortified enclaves, gated communities, citadels*, etc.). As the leader of one of Rio de Janeiro's neighbourhood associations said to me already at the beginning of the 1990s, 'Rio's [middle-class] citizens live in 'prisons', and only in 'prisons' they feel themselves free' ('prisons' is the metaphor he used for 'gated communities'). This kind of statement gives us a taste of the social-psychological atmosphere in a 'phobopolis'. However, the problem is not exclusively related to upper- and middle-class citizens, since even the poor is experiencing fear as an increasing problem – surely in a form which is very different from that experienced by privileged residents – as I hope to show in the next section.

Rio de Janeiro is very much like a type of synthesis of both the positive and the increasingly negative elements of urban Brazil, as the funk-rap *Rio 40 graus* (Rio 104 degrees) indicates:

> Rio 104 degrees
> Marvelous city
> Purgatory of beauty and chaos
> Hot-blooded capital of Brazil
> Hot-blooded capital
> Of the best and worst of Brazil
> (From the song *Rio 40 graus*, written by Fernanda Abreu and others)

Moreover, Rio de Janeiro is one of the best examples of a 'phobopolis' in the contemporary world. In fact, Rio is not only an impressive example of some problems which are becoming increasingly important in so-called megacities in (semi-) peripheral countries; it also depicts a reality that displays aspects which are becoming increasingly relevant *worldwide*, including in the (big) cities in Europe and especially in the USA, so that *fragmentation* and phobopolisation cannot be confined to a specific academic pigeonhole called *problems of Third World cities*. Unfortunately, fear, violence, criminality, and strategies of control, prevention and repression implemented by the state or by private firms correspond very much to a *global* phenomenon, in spite of the obvious differences between countries as well as between cities in terms of intensity and specific forms of the problems. From a '(Global) North'-based perspective, terrorism and the experiences of certain countries such as the USA and the UK have been often privileged (although ordinary criminality still is a crucial component of the discursive landscape and a key factor for the 'security market', especially in the USA), as Stephen Graham did it in his recent book *Cities under Siege* (Graham 2010); in contrast to this, from a '(Global) South'-based viewpoint, ordinary criminality and its connections with the above mentioned three pillars of the *industry of fear* have understandably deserved much more attention (although terrorism is sometimes not completely absent as a real or potential threat to the existing socio-spatial order), as I did it myself in my book 'Phobopolis' (de Souza 2008) as well as in previous works since the beginning of the 1990s.

11.2 Rio de Janeiro as a 'Fragmented Metropolis'

I would like to begin with two remarks, in order to avoid any misunderstandings: first of all, it must be stressed that poverty and unemployment do not determine urban violence *directly*, since institutional factors (such as the degree of efficiency of the police) as well as cultural and socio-psychological factors (religious values, cultural change, and so on) play a very important role (de Souza 2000:83–85, 2006:446–447, 2008:176–187); that is, the socioeconomic factor alone cannot explain why Latin American metropolises like Rio de Janeiro and São Paulo are much *less* violent than, say, Calcutta, where poverty is much *more* evident. Moreover, there are different types of violent crime; some of them are undoubtedly related to inequalities and poverty (even if only in an indirect way), like street robbery perpetrated by poor teenagers, while some others are not. *Nevertheless*, economic problems such as poverty and inequalities, 'mediated' by cultural and institutional factors (for instance, dissolution or weakening of some 'traditional' values, increasing hedonism and consumerism and 'structural' corruption of the police), do represent a background against to which we can understand the increasing magnitude of many aspects of violent criminality in Rio de Janeiro as well as in other (semi) peripheral cities.

Since the beginning of the 1990s, Brazil has experienced a *productive restructuring* which is characterised by both the selective introduction of new technologies and by patterns of *flexible accumulation* (as well as by privatisation of state enterprises) – all this in the framework of an increasing opening of Brazil's economy to the global market. On the one hand, this restructuring has led to an increasing productivity of the industrial sector; however, on the other hand, it has decisively contributed to an increase in terms of unemployment and the rate of informality (including employees not officially registered and self-employed people not covered by social security) in the labour market.

A process of exclusion of workers from the formal sector and of *precarisation* of working relations was registered during the 1990s. In the six biggest metropolises (São Paulo, Rio de Janeiro, Belo Horizonte, Porto Alegre, Salvador and Recife), the percentage of employees without a work permit in relation to the total number of employees increased from 18.4% in December 1989 to 26.9% in June 1999, and the percentage of self-employed people increased in the same period from 17.7 to 23.5% – while the percentage of employees with a work permit decreased from 59.5 to 44.7% (Mattoso 1999:15). Between 1999 and 2004, the trend of steady growth of the informal sector (as well as of unemployment in the industrial sector) was reversed, but no consistent pattern can be identified (Rocha and Albuquerque 2006). This 'reversion' seems to have been above all a kind of 'stabilisation' at a very mediocre level in terms of structure of employment.

And then the USA came to the financial mess in 2007–2008, which began as a liquidity crisis in the banking system and resulted in the collapse of large financial institutions in that country, contaminating the whole world. At this moment (2010), it seems that the Brazilian economy begins to recover from the present-day global economic crisis, but even the modest evolution in terms of higher rates of employment and formal work was largely eroded as a result of the financial crisis in 2009. Probably we need to be very cautious. At this juncture, maybe it is appropriate to remember Peter Ustinov's joke: 'The point of being an optimist is to be foolish enough to believe the best is yet to come'.

In contemporary Brazil, Rio de Janeiro as well as other metropolises are still places par excellence for the concentration of wealth; however, as stressed by Rocha (1995, 2003), urban poverty in Brazil is highly concentrated in the metropolises, especially in São Paulo and Rio de Janeiro. 30.24% of the Brazilian population (i.e. 51.3 million inhabitants) lived in the nine largest and oldest metropolitan areas in 2000 (IBGE *apud* Rocha 2003:130); 34.19% of all poor people in Brazil (regarding absolute poverty), that is, 11.2 million people, lived in these few metropolitan regions in 1999, whereas 20.41% of the poor people lived in the countryside in the same period (see Rocha 2003:83). 16.66% of all poor people who lived in the nine metropolitan regions were located in the metropolis of Rio de Janeiro in 1999 (Rocha 2003:127).

The tremendous magnitude of informal labour has several negative implications. First of all, the fact that an enormous number of employees have been excluded from legal guarantees, such as social insurance, unemployment benefits and paid holidays (additionally, one must not forget the trend towards a reduction in the contribution of the state to social welfare). From a *sociopolitical* standpoint, the most remarkable aspect of this evolution is the increasing importance of criminal strategies of survival, especially drug trafficking. It is very difficult even to estimate the number of those among the poor involved in drug trafficking; however, it is not difficult to perceive that this criminal part of informal-sector employment is not irrelevant, above all in Rio de Janeiro and São Paulo (see de Souza 2000, 2001).

Although Enzensberger provoked his readers by mentioning not only Lima, Johannesburg, Bombay and Rio but also Paris, Berlin, Detroit and Birmingham as examples of cities related to the phenomenon of *molecular civil war*, the most impressive examples can be found in some big cities and metropolises of (semi)peripheral countries. Surely Rio de Janeiro is one of them. As a first illustration of this, let us remember the phenomenon of *mass robbery* popularly known as *arrastão* (= dragnet), which has occurred several times in Rio since 1992. The most famous of all *arrastões* in fact began as a conflict between gangs, but it was reduced by the press to a threat against the middle class:

> Yesterday the Zona Sul became a battle zone, with *arrastões* carried out by gangs of adolescents from the shanty towns of Baixada Fluminense, armed with sticks. The Military Police, with 110 men armed with revolvers, machine guns and rifles, had difficulty in putting down the violence of the various groups involved in the attack. Even a parallel police force, constituted by the Guardian Angels – a voluntary group aiming to defend the population – entered the fray.
>
> Panicked beachgoers and inhabitants of the area had to take refuge in bars, bakeries and street stands. The attack began about midday, on Arpoador Beach [between Copacabana and Ipanema], where several bus lines from the periphery make their final stop. As the gangs got off, they began to form *arrastões*, spilling over Copacabana, Ipanema and Leblon. Angry inhabitants demanded the death sentence and army patrols on the streets.
>
> (*Jornal do Brasil*, October 19, 1992)

Of course, not all types of violence in the context of Rio de Janeiro's everyday life are drug-related. *Arrastões*, for instance, are not directly related to the drug trade. However, drug trafficking has very relevant and general implications in terms of violence and feelings of fear. Nowadays, shanty town formation (which is – along with poor, peripheral semi-legal settlements – the most pervasive spatial symbol of urban poverty in Rio de Janeiro as well as in other Brazilian metropolises) is impressive not only because of their growth rates (1,092,783 people – corresponding to 18.6% of the total population of the municipality of Rio de Janeiro – lived in favelas in the year 2000; moreover, the demographic growth rate of the population of the favelas was 2.4% in comparison with only 0.38% of the *formal city* for the period 1991–2000 [Prefeitura da Cidade do Rio de Janeiro 2002]). Their complexity and sociopolitical significance – largely on account of drug trafficking – are also crucial aspects.

The most impressive aspect of Rio de Janeiro's '*phobopolisation*' since the 1980s has been the increasing *territorialisation* of shanty towns (in the sense of a process of exercising territorial control over them) by drug-trafficking gangs and organisations. (Under 'drug-trafficking organisation',

I do not mean a single 'gang', which usually controls only one or a few favelas, but the very loose organisations known as *comandos*.) Although the excessive highlighting of favelas by the media leaves retail drug dealers who are not based in favelas in the shadows (not to mention the role of the wholesalers who operate at a regional, national and even international level and who – it goes without saying – do not live in favelas), it cannot be denied that favelas have a great logistical importance for drug dealers who operate within the circuit of retail drug trafficking in Rio as well as in other Brazilian cities. Three typical characteristics of favelas can explain this relevance: (1) their *location* (many favelas are located very close to middle-class districts and neighbourhoods, that is, very close to the main drug consumers); (2) their *internal spatial structure* (small streets and a 'chaotic' spatial pattern which makes the task of invading and taking them under control difficult for the police); and (3) their *socioeconomic characteristics* as a 'reservoir of cheap labour' (abundant poor young people to be recruited for different activities) (see Zaluar 1994; de Souza 1996, 2000, 2005, 2008).

The end of the 1970s and the beginning of the 1980s can be seen as a historical benchmark regarding this territorialisation process. Since then Rio de Janeiro has been experiencing a transformation in the context of which favelas which until the end of the 1970s were more or less 'open' – that is, individuals who lived in different favelas could visit each other without significant problems or restrictions – became increasingly 'closed', and the mobility between favelas controlled by rival drug-trafficking organisations became increasingly difficult. Rivalry between criminal gangs as well as between *comandos* influences the spatial mobility of favela inhabitants, in as far as they take the risk of facing hostility and menace when they visit people in other favelas (see, for details on this issue, de Souza 1996, 2000, 2008).

Favela-based retail drug trafficking combines a strong hierarchy at the scale of the favela with a decentralised, network-based form of organisation at a larger scale (the *comandos*). In terms of organisation and hierarchy, there are different functions in the context of a single drug-trafficking crew (that controls one or more shanty towns and belongs to a *comando*), from the *dono do morro* ('owner of the hill') over the *gerentes* ('managers' = those who control the selling places) and soldados ('soldiers' = security staff) to *vapores* ('vapours' = street sellers) and *aviões* ('aeroplanes' = go-between sellers). While *donos* ('owners') who belong to the same *comando* usually respect each other's territories, bandits belonging to rival *comandos* often try to take possession of 'enemy territories' – leading to *guerras* ('wars') between rival groups which occur very often and which can take several days or even weeks. In the context of these 'wars', several specific drug-trafficking crews belonging to the same *comando* can help during the invasion of an 'enemy territory', very much in the sense of 'mutual help' (de Souza 1996, 2000, 2005, 2008).

It is necessary to stress the ambivalence of the role drug traffickers usually play in the favelas in which they live.

On the one hand, one cannot deny that drug dealers are 'useful' for favela inhabitants in several respects. For instance, they do not allow that ordinary crimes such as robbery or rape are committed within their territories (and they do so not primarily because of their 'sense of justice' but fundamentally for the sake of their 'business', since unnecessary violence would lead to social instability as well as to excessive public exposure of the community where they operate). Even more important is the fact that drug dealing represents an important source of income for many *favelados* nowadays – either directly or indirectly (de Souza 2000, 2005, 2008).

On the other hand, the relevance of drug dealers as 'safety guarantors' for the other favela residents is relative. Although rapists and thieves who commit crimes inside the favela or even in its surroundings are usually punished by drug dealers, drug traffickers themselves often behave in a brutal manner towards ordinary *favelados* (for instance, taking women by force, although drug dealers usually condemn rape if it is committed by 'ordinary criminals', even implementing 'exemplary punishment' of transgressors in the form of murder or castration). One of the many impressive examples was reported by the press in 2003, informing that a teenager who lived in a certain favela was publicly humiliated by local drug dealers (she was forced to walk naked in the streets of the favela), then raped, tortured and finally murdered, only because local drug traffickers discovered that the girl's boyfriend lived in another favela which belonged to the network of a rival *comando*. It seems that both the protection of the drug business and symbolic aspects such as demonstrations of virility (see Zaluar 1994, 2002a) have contributed not only to an increasing use of violence among criminal crews as well as to an increasing 'tyranny' of drug traffickers towards favela inhabitants (de Souza 2000, 2005, 2008).

Favela-based drug bosses also go quite often beyond their own 'criminal business' to control or influence several aspects of favela life, from residents' associations to favela-upgrading projects, a fact that can be interpreted as another aspect of their 'tyranny'. As far as the attempt to control residents' associations is concerned, it is important to note that in many shanty towns, drug dealers already started to influence the *associações* in the 1980s. Leaders of residents' associations who refused to obey orders (which are usually related to requests to permit the use of the association's infrastructure) are menaced, often evicted and in some cases murdered (de Souza 2000, 2005, 2008; Zaluar 2002b; see also Leeds 1996). In fact, not only have community leaders been menaced and evicted: the relatives and friends of the 'enemies' are usually evicted too, when in the course of a 'war' a criminal organisation takes control over a favela which was until then dominated by another one. By virtue of this problem – and often simply as a consequence of feelings of insecurity – around 20% of dwellers have already left their respective shanty towns in Rio de Janeiro, according to an estimation made by the local Federation of Dwellers' Associations (Zaluar 2002a:149). Furthermore, local bosses often try to manipulate residents' associations more directly, for instance, during the elections for the *associação* – advancing or supporting candidates linked or subordinated to them.

As far as the interference through favela-upgrading projects is concerned, Rio de Janeiro's community-upgrading programme *Favela-Bairro* has served as a very interesting 'laboratory' since 1994 (when it begun to be implemented by Rio de Janeiro's administration): drug-trafficking bosses have repeatedly prohibited the implementation of specific aspects of a certain favela-upgrading project in the framework of the programme; in other cases, they modified aspects of the project after its implementation. This kind of interference, imposed by drug traffickers because they are not interested in certain types of physical improvements that could threaten their security, is usually 'tolerated' by the state. Such improvements could, for example, include a better integration of the small streets of the favela and those of the surrounding districts. Problems like these (along with plain arbitrary behaviour on the part of armed dealers wishing to demonstrate power or extort money from *Favela-Bairro* teams) have often been reported by the press, and my research team conducted interviews in the framework of research projects between 1994 and 2007 and especially between 2004 and 2006, in the framework of which similar experiences have been reported by community leaders, architects working for the municipality and so on (see de Souza 2005, 2008).

Nevertheless, drug dealers who operate in favelas are only the poorest part of a long chain of agents and interests. As a matter of fact, *they are themselves oppressed people who (quite often) oppress other oppressed people* (de Souza 2005:7, 2008:61, 2009:33). In fact, the power of drug traffickers has been challenged in recent years in many favelas in Rio de Janeiro – but by illegal means. In many favelas, dealers were already replaced by paramilitary groups (death squads and militias integrated by policemen or ex-policemen), who expel (or kill) the criminals and 'sell protection' to the community – in a nutshell, another kind of tyranny.

Let us now consider the *self-segregated spaces*, which are also relevant in order to present an integral picture of Rio de Janeiro as a 'phobopolis'.

Although the state apparatus has the legal monopoly on violence, this monopoly is steadily challenged by *microlocal warlords*, who have been responsible for the formation of many *territorial enclaves* in the city. Moreover, the state is no longer able to offer safety in the 'ordinary' neighbourhoods; public spaces are particularly vulnerable. From the standpoint of several members of the urban elites and middle class in Rio de Janeiro, *condomínios exclusivos* (i.e. gated communities) seem to be at least a partial solution for urban violence. As far as the metropolis of Rio de Janeiro is concerned, most gated communities are concentrated in the district of Barra da Tijuca. These antipodes of the favelas are luxury apartment or house complexes which are in essence similar to the gated communities of other Brazilian cities (as well as those of other countries).

The escapism of Rio de Janeiro's urban elites is not restricted to *condomínios exclusivos*: huge, sophisticated shopping malls are also a part of their typical way of life. Surely not even luxury shopping malls are so closed to strangers as gated communities; low-income people can often be found at least in some of them (for instance, young people looking for diversion), although they obviously do not have money to buy hardly anything. It is hardly possible to keep 'undesirable visitors' out openly, since shopping malls are not private property in the same sense as gated communities are. However, the attractiveness of those shopping malls to middle- and upper-class people depends on their 'exclusiveness' as well as on their image of being absolutely safe (in contrast to the city outside them). Since the prejudices of the Brazilian urban middle- and upper-classes associate poor people, especially *favelados* and blacks, with potential robbers and 'unpleasant' people in general, one of the expectations on the part of the urban elites regarding shopping malls is that these 'undesirable people' are kept outside. Consequently, people who are regarded as 'undesirable visitors' are steadily monitored, sometimes threatened and humiliated by private security agents.

Citizenship is usually seen as a set of formal political and civil rights (such as the right to vote, the right to express one's opinion freely and the right to freedom of movement inside the boundaries of a specific country). Nevertheless, the mere existence of formal rights is not enough, and even the introduction of some 'social rights' in addition to these older, civil and political rights has failed in preventing concrete inequalities, for example, in terms of income, from undermining the exercise of certain liberties, such as the freedom of movement (which is not independent of the financial capacity of a citizen to travel or even to have access to certain places where payment is required for admission). As I have pointed out in previous publications (de Souza 2000, 2005, 2008), a further problem arises if some formal political, civil and social rights cannot be fully exercised and enjoyed due not (only) to material inequalities but because of the *fear of violence*. The room for manoeuvre required for the full exercise of certain basic liberties that are tied to the notion of citizenship shrinks under circumstances marked by increasing prejudice and fear. For instance, in many favelas of Rio de Janeiro, the right to freedom of movement and the right to freedom of association have been severely limited by drug traffickers, be it directly (through the orders they give) or indirectly (due to the fear of being considered an enemy or a possible spy by the dealers of a favela controlled by a rival group). However, in the public spaces outside the segregated and self-segregated areas, the fear of becoming the victim of assaults and robbery also restricts the full use of these rights.

Sociopolitical fragmentation of the urban space can be understood as the process in the wake of which excluded enclaves dominated by drug-trafficking organisations (or paramilitary militias) proliferate, whereas self-segregation (gated communities), on the other side, becomes increasingly complex. Public life as well as public safety is declining in a dramatic way, due to the fact that public spaces and 'ordinary' districts and neighbourhoods at large are being transformed into highly unsafe areas – particularly from the perspective of the middle-class inhabitants. *Sociopolitical fragmentation of the urban space* is not just a new term for the old phenomenon of residential segregation but a complex phenomenon which comprises both the transformation of an increasing number of poor, segregated areas into illegal *territorial enclaves* under the control of drug traffickers and self-segregation, which is to a large extent a response of privileged residents to growing feelings of insecurity.

Rio de Janeiro is becoming an increasingly 'unsafe' and militarised city – like many other cities in Brazil and throughout the world. However, in the case of Rio, the problem has been particularly intense. Although the metropolis of Rio de Janeiro still is the second most important central place in Brazil after São Paulo, its economic and social decline is remarkable – a decline which is becoming more and more connected to factors such as urban violence and criminality. The sociopolitical fragmentation of the urban space is the context in which the urbanisation process becomes decisively influenced by increasing fear of crime and violence. In other words, this fragmentation is the very spatial expression of phobopolisation.

11.3 'Sociopolitical Fragmentation of the Urban Space' and 'Phobopolisation': Can We Cope with Them?

Is it possible to overcome the sociopolitical fragmentation of the urban space and phobopolisation by means of state-led urban planning measures and local-level public policies? To be realistic, the obstacles are simply tremendous.

As far as the attempt to introduce mechanisms of popular participation in urban planning and management is concerned, it is sufficient to consider the difficulty (not to say impossibility) of 'empowering' favela residents who live in spaces controlled by local drug-trafficking bosses. (It must be stressed that as far as Rio de Janeiro is concerned, the introduction of participatory urban development mechanisms has also faced political obstacles since the beginning of the 1990s, because the city has been governed in a very technocratic and authoritarian manner since then.) However, it is not just a question of 'empowering' but sometimes of plain implementation of public policies and planning projects in general, as was mentioned in the previous section, illustrated by the example of the *Favela-Bairro* programme.

Unfortunately, it would not be realistic to expect that the police can prevent drug dealers' interference in 'their' territories – for instance, regarding the implementation of public policies and upgrading projects such as *Favela-Bairro*. The police in Rio de Janeiro as well as in other Brazilian states are both structurally corrupt – many policemen extort money from drug dealers as well as from ordinary citizens in different ways – and very ineffective (Soares 2006). However, the army itself is understandably not able to cope with the challenge represented by drug traffickers (or at least not yet – *fortunately*, we should add) who not seldom posses heavy weapons (from grenades to automatic rifles to machine guns) and who operate inside favelas as their *territorial enclaves*, simply because it is not prepared for this type of mission. In fact, troops have already intervened several times in Rio's favelas since 1992, as I have already mentioned, but the army has been challenged itself by drug traffickers: for instance, soldiers have been often co-opted by criminals to facilitate the theft of weapons and ammunition from army barracks. The militarisation of the urban question (de Souza 1996, 2000, 2008) is not only a reactionary strategy; it should be regarded as a to some extent nonrealistic strategy even from a conservative standpoint.

Not only the formation of *territorial enclaves* controlled by criminals operating within the circuit of retail drug trafficking challenges the state apparatus and the formal legal system. Many gated communities are simply *illegal*, since the 'closure' of public streets and denying or restricting access to public spaces is not allowed by federal law. As far as the gated communities are concerned, the local state in Rio and elsewhere in Brazil has been unable to monitor the implementation of the urban law; in fact, local administrations have usually closed their eyes to this kind of illegality perpetrated by the elites and middle classes. (Tragically, the practice of 'closing' streets has been to some extent imitated since the 1990s even in lower-middle-class districts.)

It is not difficult to see that, from a pragmatic viewpoint, alternative responses to urban problems in a metropolis like Rio de Janeiro must integrate progressive public safety strategies – something which is not easy to develop and even less to implement – with (and above all) progressive, social-justice-oriented 'urban development' policies. In such a city, the integration between urban policies (which aim at the reduction of inequalities, at the improvement of the conditions of life and so on) and public safety strategies (which should be designed in order to enhance accountability of and citizens' control over the police as well as in order to support the *demilitarisation* of the urban question) is a matter of *interdependence*, since progressive urban policies must often be supported by some public safety measures, and on the other hand, progressive urban policies can decisively contribute to public safety (see more about this in de Souza 2005, 2008). However, the efficacy and in some cases even the existence of local-level policies is highly dependent on measures and processes at the national and international scale. More precisely, it is very difficult if not impossible to overcome the problem of socio-political fragmentation of the urban space within the existing political-institutional and economic framework which is typical of a semi-peripheral country. Even in European cities, where drug trafficking is by far not as sociopolitically problematic as in metropolises such as Rio de Janeiro or São Paulo (and where financial and institutional resources to cope with the problem are much more available than in Brazilian cities), dealing successfully with drug trafficking and its socio-spatial effects is still very difficult.

Furthermore, measures implemented by the state apparatus should never be considered from a truly critical point of view as a *non plus ultra* in terms of socio-spatial development strategies, not to mention the public safety measures – even those implemented by progressive governments in specific, favourable conjunctures. *Structurally* seen, the *state apparatus* is ultimately always a *heteronomous* instance of power, no matter how sincerely committed to (re)distributive measures and 'popular participation' particular *governments* can be. Civil society itself, especially in the form of social movements and their organisations, must develop and propose and as far as possible try to implement a more radical 'public safety' agenda (especially in those countries and cities in which even progressive governments cannot avoid police's corruption and brutality completely, so that activists must learn to defend themselves without becoming unnecessarily violent themselves), as I have stressed in several works (de Souza 2008, 2009).

Surely, state-led local-level policies oriented towards socio-spatial development can play an important role in empowering poor people by means of measures and programmes aiming at income generation, income redistribution and social inclusion sometimes – and this can contribute to a weakening of drug traffickers' power. However, state-led local-level public policies are by far not enough to *overcome* the problem, no matter how creative and consistent they are. *And the same is even more true in relation to typically conservative measures*. Just as an exercise of imagination, suppose it could be possible to incarcerate all drug dealers very easily, having as a consequence of this the elimination of drug trafficking in the short run. What would happen then? Since drug trafficking nowadays represents an important source of income for many *favelados*, either directly or indirectly, it is reasonable to expect that they would become desperate. Probably crime would 'migrate' even more strongly and very rapidly from the favelas to the *formal city*, and the number of cases of robbery, kidnapping and other types of violent crime which constitute the nightmares of upper and middle classes would increase dramatically.

At least as far as the majority of the favelas' inhabitants involved with the drug trade is concerned, it would be an exaggeration to say that violence can serve as 'an instrument for social mobility, or even an instrument for reversing traditional social hierarchies', as was assumed in the introductory chapter of a book on the legacy of civil war, violence and terror in Latin America (Kruijt and Koonings 1999:11). However, involvement with violent criminality can certainly be a strategy of satisfying material and some immaterial basic needs. Considering this, even if poverty is not the only factor causing criminality, it is necessary to provide an attractive *economic and social alternative* for the poor; more jails and more policemen (as well as more gated communities) only represent a *pseudosolution* – perhaps a *suicidal* solution in the long run. However, offering better employment and social opportunities to the poor is a thing which depends decisively on processes and institutions related to the *national* and even the *international/global* level, and expecting significant changes in this regard for the coming years would be simply unrealistic. Moreover, public policies and all sorts of strategies which are conform to the capitalist status quo in its essence can – in the best of all cases – *mitigate* the problem, provided they are sufficiently intelligent and progressive. But this is surely not enough to exorcise phobopolisation once and for all.

Another obstacle lies in the fact that the legal and the illegal parts of the economy and society are becoming more and more interconnected with each other within contemporary capitalism, from the local to the global level. Contemporary capitalism is a social model (and not just a mode of production) which is becoming increasingly 'criminogenous' for a number of reasons,[2] and in this framework, the powers behind the illegal economy should not be underestimated. From 'structural corruption' in Rio de Janeiro's police to money-laundering worldwide, many important economic sectors and social groups (or parts of institutions) are becoming increasingly dependent of illegal activities. Considering the complexity of processes discussed, one can easily come to the conclusion that the implementation of a consistent solution (even of a partial one) is a challenge which goes beyond the possibilities of the 'official' side of the state apparatus in a country like Brazil, a tendency that will probably increase in the future.

11.4 Final Remarks

In the aforementioned introductory chapter, Kruijt and Koonings pointed out that

> [i]t is by no means certain that stable, civic forms of rule and social integration will prevail or endure in Latin America. This caution is inspired by the legacies of repressive dictatorships and civil wars, as well as by the persistence of poverty, inequality, and social and political exclusion. The latter form the backdrop to new and perturbing forms of violence that seem to be on the rise in post-authoritarian Latin American societies. (Kruijt and Koonings 1999:3)

Metropolises such as São Paulo, Rio de Janeiro, Mexico City and so on – and big cities in general – are playing a very important role as 'laboratories' of violence in *post-authoritarian* Latin America. Sure, in spite of some Latin American peculiarities, phobopolisation and militarisation of the urban question is very much a *global* phenomenon, as it was already stressed in this chapter. Nevertheless, the Latin American and Brazilian big cities in general, and Rio de Janeiro in particular, are very good examples of this problem. Unfortunately, in the face of all the obstacles, it seems that hypothetical progressive, state-led, local-level measures and policies can probably only, *at best*, contribute to reducing the problems of violence and sociopolitical fragmentation of the urban space, rather than eradicating them. Especially under the conditions of globalisation, phobopolisation is not a mere *local* challenge but also a *national* and *international* one – in a very strong sense, because the 'criminogenous' dynamics of the existing social order are not a mere local phenomenon either. What is

[2] As I put it elsewhere, capitalism 'shows itself as 'criminogenous', especially these days, as it (a) wakes in many people a desperate desire to consume, while it simultaneously give to only few people the opportunity to satisfy (even) their (basic) needs in a legal way; (b) generates and disseminates competitive, individualistic and hedonistic values, so placing 'me' clearly above 'us,' 'to have' above 'to be,' and property above life; (c) generates a 'cultural industry' for which violent criminality is a major types of 'food' – and which reinforces values such as individualism and the primacy of force through the powerful influence of several ways (newspapers, movies, television, videogames) and through superficial and a-critical narratives on violence and crime; (d) facilitates, by means of deregulation of the financial system, money laundering and corruption worldwide; (e) disseminates the belief according to which everything can be transformed into a commodity and everybody has a price – and the feeling that the difference between being sent or not being sent to overcrowded and inhuman prison depends (more or less according to the country) on whether or not one can spend the necessary sum of money' (de Souza 2009:47–48, n. 7).

more, 'phobopolisation' is a challenge that the capitalist state is very probably not able to overcome, even in the framework of 'progressive' political conjunctures; in reality, structurally seen, the state apparatus seems to be part of the problem rather than part of the solution.

References

Agamben G (2004 [2003]) Estado de exceção. Boitempo, São Paulo

Arendt H (1998 [1958]) The human condition, 2nd edn. The University of Chicago Press, Chicago

Beck U (2007) *Weltrisikogesellschaft: Auf der Suche nach der verlorenen Sicherheit*. Suhrkamp, Frankfurt am Main

Bresson M (2007) Sociologie de la précarité. Armand Colin, Paris

de Souza ML (1996) As drogas e a 'questão urbana' no Brasil. A dinâmica sócio-espacial nas cidades brasileiras sob a influência do tráfico de tóxicos. In: Castro I et al (eds) Brasil: questões atuais da reorganização do território. Bertrand Brasil, Rio de Janeiro, pp 419–468

de Souza ML (2000) O desafio metropolitano: Um estudo sobre a problemática sócio-espacial nas metrópoles brasileiras. Bertrand Brasil, Rio de Janeiro

de Souza ML (2001) Metropolitan deconcentration, socio-political fragmentation and extended suburbanisation: Brazilian urbanisation in the 1980s and 1990s. Geoforum 32:437–447

de Souza ML (2005) Urban planning in an age of fear: the case of Rio de Janeiro. Int Dev Plan Rev 27(1):1–18

de Souza ML (2006) Clima de guerra civil? Violência e medo nas grandes cidades brasileiras. In: Albuquerque ES (ed) Que país é esse? Pensando o Brasil contemporâneo. Globo, São Paulo, pp 101–140

de Souza ML (2008) Fobópole. O medo generalizado e a militarização da questão urbana. Bertrand Brasil, Rio de Janeiro

de Souza ML (2009) Social movements in the face of criminal power: the socio-political fragmentation of space and 'micro-level warlords' as challenges for emancipative urban struggles. City 13(1):26–52

Enzensberger HM (1993) Aussichten auf den Bürgerkrieg. Suhrkamp, Frankfurt am Main

Graham S (2010) Cities under siege: the new military urbanism. Verso, London

Hardt M, Negri A (2005 [2004]) Multidão: Guerra e democracia na era do Império. Record, Rio de Janeiro

Kruijt D, Koonings K (1999) Introduction: violence and fear in Latin America. In: Koonings K, Kruijt D (eds) Societies of fear: the legacy of civil war, violence and terror in Latin America. Zed Books, London/New York, pp 1–30

Leeds E (1996) Cocaine and parallel polities in the Brazilian urban periphery: constraints to local-level democratization. Lat Am Res Rev 31(3):47–84

Mattoso J (1999) O Brasil desempregado. Fundação Perseu Abramo, São Paulo

Münkler H (2004) Die neuen Kriege. Rowohlt, Reinbeck (bei Hamburg)

Petrella R (1995) Le retour des conquérants. Le Monde Diplomatique 494, May

Prefeitura da Cidade do Rio de Janeiro (2002) Os dados mais recentes sobre a população de favelas na cidade do Rio de Janeiro. [Originally published in Rio Estudos, n.° 46, fevereiro/2002] http://www.armazemdedados.rio.rj.gov.br/index.htm. Accessed 15 Sept 2003

Rocha S (1995) Governabilidade e pobreza: o desafio dos números. In: Valladares L, Coelho MP (eds) Governabilidade e pobreza no Brasil. Civilização Brasileira, Rio de Janeiro, pp 221–265

Rocha S (2003) Pobreza no Brasil: afinal, de que se trata? FGV Editora, Rio de Janeiro

Rocha S, Albuquerque RC de (2006) A questão do emprego no Brasil (1996–2004). http://www.iets.org.br/biblioteca/A_questao_do_emprego_no_brasil(1996–2004).pdf. Accessed 20 Dec 2006

Schuster P (2000) Hinter den Mauern das Paradies? Sicherheit und Unsicherheit in den Städten des späten Mittelalters. In: Dinges M, Sack F (eds) Unsichere Großstädte? Vom Mittelalter bis zur Postmoderne. UVK (Universitätsverlag Konstanz), Konstanz, pp 67–84

Schwerhoff G (2000) Insel des Friedens oder Brennpunkt der Gewalt? Die Reichsstadt Köln ca. 1470-1620. In: Dinges M, Sack F (eds) Unsichere Großstädte? Vom Mittelalter bis zur Postmoderne. UVK (Universitätsverlag Konstanz), Konstanz, pp 139–156

Sennett R (1977) The fall of public man. Cambridge University Press, Cambridge and other places

Sennett R (1999 [1998]) A corrosão do caráter. Conseqüências pessoais do trabalho no novo capitalismo. Record, Rio de Janeiro

Soares LE (2006) Segurança tem saída. Sextante, Rio de Janeiro

Tuan Y-F (1979) Landscapes of fear. University of Minnesota Press, Minneapolis

Zaluar A (1994) Condomínio do diabo. Revan and Editora UFRJ, Rio de Janeiro

Zaluar A (2002a) Violence related to illegal drugs, easy money and justice in Brazil: 1980–1995. In: Geffray C et al (eds) Globalisation, drugs and the state, vol 1. UNESCO/UNODCCP, Neuilly-Sur-Seine, pp 142–163

Zaluar A (2002b) Drug trafficking and the informal market in Rio de Janeiro. In: Geffray C et al (eds) Globalisation, drugs and the state, vol 2. UNESCO/UNODCCP, Neuilly-Sur-Seine, pp 74–91

Dar es Salaam, Megacity of Tomorrow: Informal Urban Expansion and the Provision of Technical Infrastructure

12

Alexandra Hill, Tanja Hühner, Volker Kreibich, and Christian Lindner

Abstract

In the emerging megacity of Dar es Salaam, Tanzania, rapid urban growth is progressing informally without statutory planning. Suppliers of technical infrastructure are operating without reliable information about the distribution of future demand, while settlements are starting to grow without basic services.

Household surveys and focus group discussions conducted in new settlements in the urban periphery confirmed the dominance of the shelter first principle in the locational decisions of poor settlers but also the high importance assigned to infrastructure provision during the consolidation process.

Expert interviews with representatives of planning authorities and utility providers have revealed a serious coordination deficit and the need for integrated urban development strategies and tools.

A simulation model capable of demonstrating patterns of future urban growth and of ensuing demand for public utilities proved to provide a powerful tool for the coordination of development policies and the efficient provision of trunk infrastructure (Hill and Lindner 2010).

Keywords

Dar es Salaam • Urban growth • Informal planning • Technical infrastructure

Dar es Salaam, Tanzania's largest city, is expected to become one of the future megacities on the African continent. Since the 1970s, the city has been growing rapidly without adequate planning regulations. The majority of unplanned settlements in the urban periphery are unserviced by

A. Hill (✉) • T. Hühner • V. Kreibich • C. Lindner
Institute of Spatial Planning, Technical University Dortmund, Germany
e-mail: alexandra.hill@tu-dortmund.de; tanja.huehner@tu-dortmund.de; volker.kreibich@tu-dortmund.de; christian.lindner@tu-dortmund.de

F. Kraas et al. (eds.), *Megacities: Our Global Urban Future*, International Year of Planet Earth, DOI 10.1007/978-90-481-3417-5_12, © Springer Science+Business Media B.V. 2014

technical infrastructure. This chapter[1] presents findings from empirical research on the mutual linkages between informal urban sprawl and the provision of technical infrastructure in Dar es Salaam.

12.1 Urbanisation Under the Conditions of Poverty

Recent population projections by the United Nations assume that the number of inhabitants of Dar es Salaam will increase from about 3 million today to about 5 million by the first half of the 2020s (UN 2008). This means that by the year 2025, Dar es Salaam is estimated to achieve megacity status.

Although the annual population growth rates have declined from more than 10% in the 1970s to half of that rate today, Dar es Salaam is still one of the most rapidly growing cities on the African continent with an average annual increase of 4.7% (from 2000 to 2005). Only the cities of Lagos, Nigeria and Kumasi, Ghana, have exceeded this pace of growth with rates of 5.1 and 4.9%, respectively, in the same time period (UN 2006). Today, Dar es Salaam has to accommodate about 150,000 new dwellers each year with natural growth displacing migration gains as the driving force of population increase.

Urbanisation under the conditions of poverty in the least developed countries differs significantly from rapid urbanisation in the nowadays developed countries at the time of their peak urban growth during the period of early industrialisation starting in the middle of the nineteenth century. At that time, rapid population increase and spatial expansion of urban areas were a consequence of economic growth which attracted migration into cities. Epidemics caused by the subsequent deterioration of hygienic conditions combined with increasing problems of the transport system paved the way for the introduction of new planning concepts, institutions and tools. In Germany, for example, the responsibility for urban and infrastructure planning was handed over from the police to the municipality following the enactment of the so called *Fluchtliniengesetz* (Urban Alignment Act) in 1875. Its main goal was the delineation of road reserves and block layouts, which had to comply with fire protection rules. As a consequence, the municipalities were equipped with additional budgets and technical staff.

Today, cities in poor developing countries are experiencing rapid urban expansion despite the fact that their economic growth by far does not enable them to bear the follow-up costs of urbanisation. As a result of the circumstances of urbanisation in poor countries, national and municipal authorities lack financial and human resources as well as appropriate planning visions, concepts and instruments. Due to these constraints, public authorities are barely able to guide urban growth in accordance with the requirements of existing planning legislation and to assure at least a basic supply of social infrastructure and functioning utilities, in particular roads, water and drainage systems. These deficits can be observed throughout sub-Saharan Africa, where on average, two thirds of the urban population live in informal settlements (UN-HABITAT 2006: 21) and less than one third have access to fresh water (UN-HABITAT 2006: 77).

In the rapidly growing cities of sub-Saharan Africa and other regions of the Global South, informality therefore appears no longer as a residual category but has to be accepted as a 'generalized mode of metropolitan urbanization' (Roy 2005: 147). As such, it is not only the disempowered public sector and its excessive bureaucracy which obstruct the entrepreneurial

[1] The research project 'Trunk infrastructure and urban growth – managing rapid urbanisation under poverty in Dar es Salaam, Tanzania' was conducted from 2005 to 2008 at the School of Spatial Planning, TU Dortmund University, Germany, in close cooperation with Prof. Dr. Wilbard Kombe, Director of the Institute of Human Settlement Studies, Ardhi University, Dar es Salaam. It was funded by the Federal Ministry of Education and Research within the programme 'Research for a sustainable development of megacities of tomorrow'. Besides the authors of this chapter, the project team comprised Timo Basteck, Dr. Wolfgang Scholz and Prof. Dr. Hans-Peter Tietz.

potential of informal settlers (De Soto 2000). In addition, urban informality 'can be seen as structured through 'extra-legal' systems of regulation' (Roy and AlSayyad 2004: 26) and, thus, as a newly emerging mode of urban governance.[2]

It is the intention of the research performed in Dar es Salaam and other large cities in Tanzania (Kombe and Kreibich 2000, 2006) to understand the settlement patterns which are evolving in the context of urbanisation under conditions of poverty. The poor land seekers are confronted with an unregulated private land market which is extremely fragmented and not very transparent, resulting in high transaction costs which are, paradoxically, a prerequisite for affordable plot prices. While tenure security is not an issue of imminent concern in the prevailing informal mode of governance (Kombe and Kreibich 2000, 2006), the supply with trunk infrastructure, especially piped water, road access and electricity, is considered a critical factor contributing to uncertainty in the allocation process. The research project therefore focused on the mutual linkages between locational decisions of informal settlers and investment strategies of utility suppliers.

12.2 Informal Urban Development in Dar es Salaam

About two thirds of residential plots in Dar es Salaam have informal status meaning that they were built upon without planning designation, surveying, land titling, provision of services and building permit (Kombe and Kreibich 2006). Based on the observation that informal settlements develop higher population densities than settlements in formally planned areas, Kironde estimates that more than 80% of the population in Dar es Salaam are accommodated in informal settlements (Kironde 2006: 463).

The dominance of informal settlements and building activities is a direct result of the political, technical and administrative deficits of local public authorities. Out of 261,668 applications to acquire a building plot[3] that were handed in to the responsible authority between 1978/1979 and 1991/1992, only 17,751 resulted in the assignment of plots, which corresponds to merely 6.8% of all applications (Kombe 1994: 30). For 2001, the demand for plots was estimated to be about150,000, whereas only an average of about 8,000 plots were assigned per year throughout the preceding decade (Silayo 2004: 354).

The urban structure of Dar es Salaam resembles a four-finger-type model pointing from the coast of the Indian Ocean towards the interior of the country (see Fig. 12.1). While urban development is restricted to the southern part of the city by a natural barrier in the form of a major creek, the northern and western parts of the city have experienced the growth of informal settlements over the course of the last three decades, particularly along the coastline and the arterial roads towards the hinterland.

Informal urban sprawl in Dar es Salaam can be described as a diffusion process progressing in wave-like concentric rings from the city centre towards the periphery in combination with broadening ribbons following the arterial roads (see Figs. 12.1 and 12.2). The negative influence of the congested arterial roads on the overall structure of the city is amplified by the fact that the frequency of *dalla dalla* minibuses, the local public transport system, is restricted by the bad condition of the feeder roads which are often not passable in the rainy season.

The increasing land prices alongside the major development axes push new settlers both to the outer periphery of the city – already within a distance of 30 km from the centre – and to the interstitial areas. There, poor dwellers can afford to

[2] This research topic is at the centre of the priority programme 'Megacities – Megachallenge. Informal dynamics of global change' of the German Science Foundation (Kraas and Hansjürgens 2008) which contains a focus on informal urban land management (Hackenbroch et al. 2009).

[3] Officially, only lease titles are assigned, but title holders as well as informal occupiers treat their plots like freehold property.

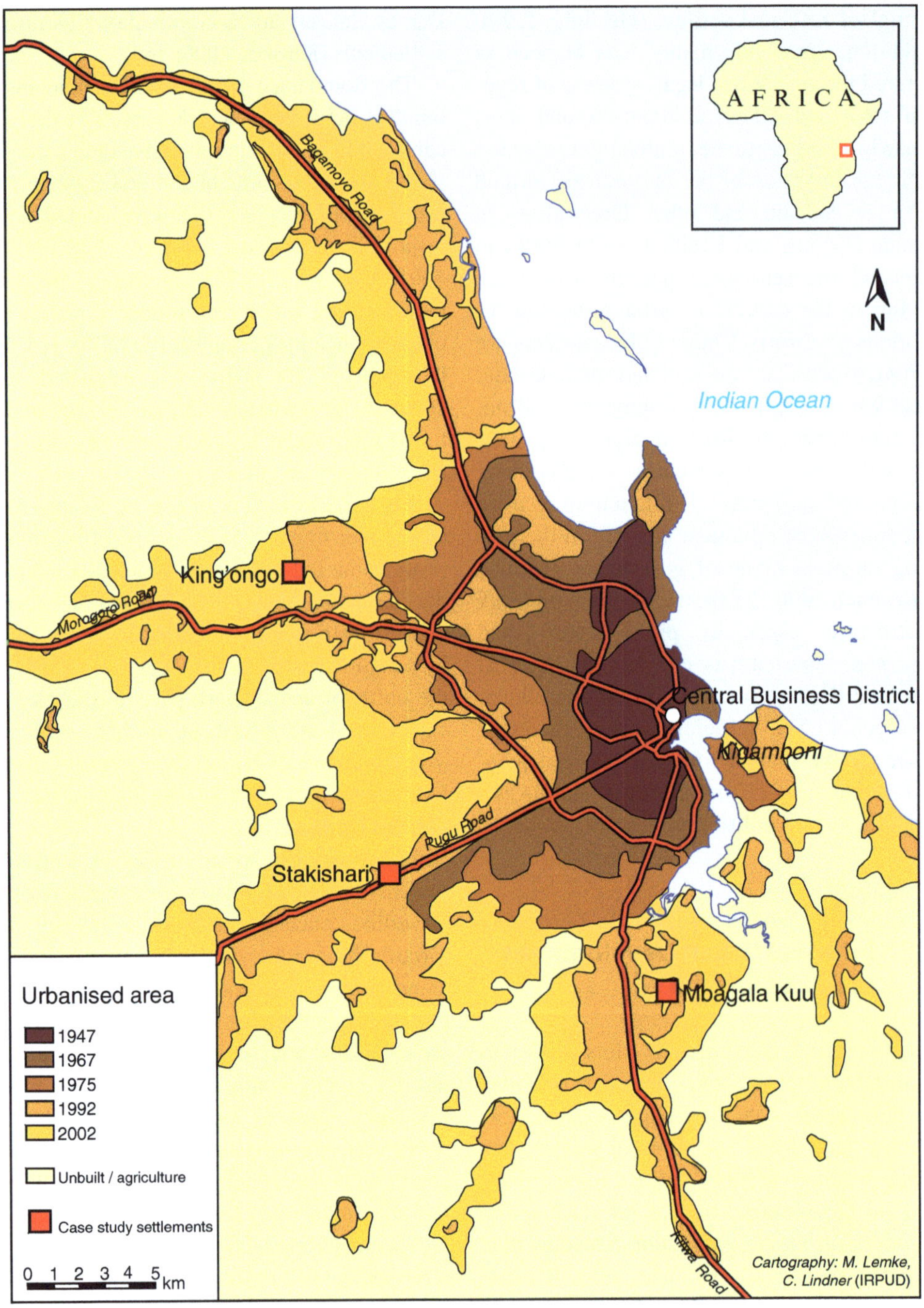

Fig. 12.1 Dar es Salaam, urban growth 1947-2002 (Source: Based on land-use and transport network data provided by ITC, Enschede; updated for 2002 by IRPUD; DCC, Dar es Salaam 2000)

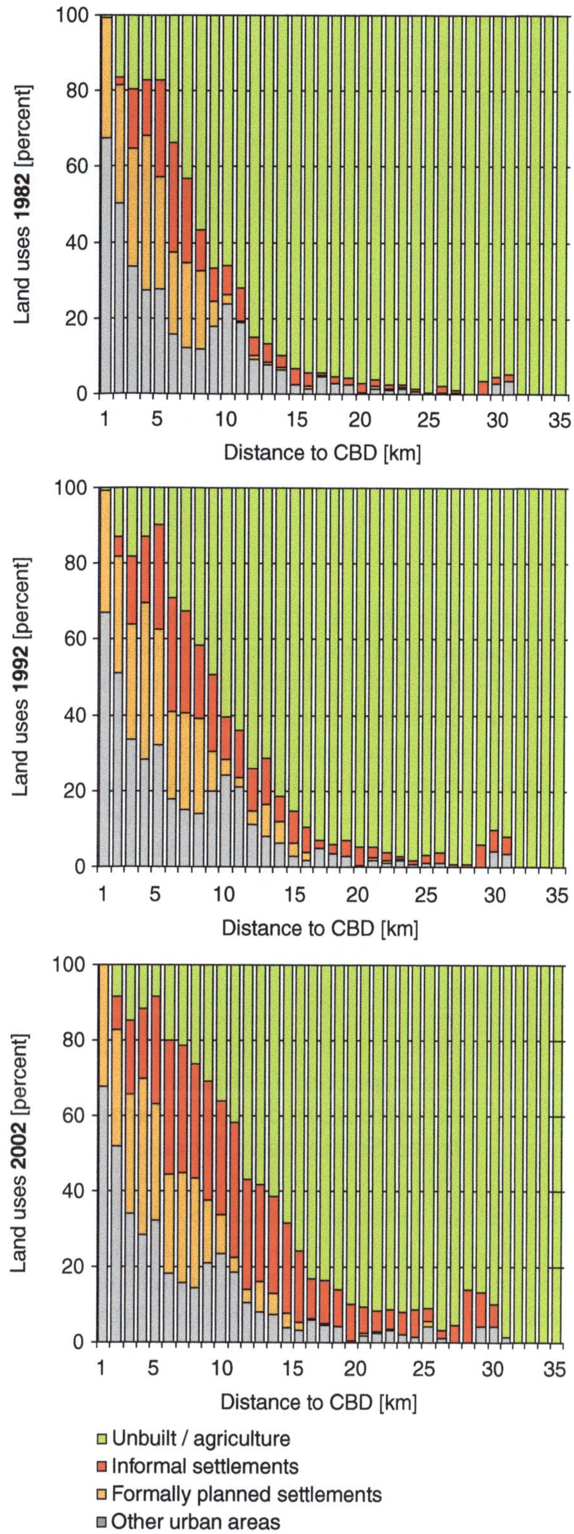

Fig. 12.2 Land uses and distance to central business district (CBD) 1982, 1992 and 2002 (Design: Alexandra Hill, Christian Lindner (IRPUD); Source: Based on own calculations and land-use and transport network data provided by ITC, Enschede; updated for 2002 by IRPUD)

Photograph 12.1 Focus group discussion in King'ongo (Source: own photograph)

buy affordable plots large enough to conduct subsistence farming. For some years, the ever-moving peri-urban frontier of the poor has interfered with suburbanising high-income earners who can afford a private car and seek to escape urban agglomeration disadvantages by resettling on large plots in a quasi-rural environment. The spatial structure resulting from this double-layered urban sprawl can be characterised as fragmented at low density and, thus, highly inefficient in terms of infrastructure supply costs and energy economics.

12.3 The Impact of Technical Infrastructure on the Informal Urban Settlement Process

Unregulated urbanisation under conditions of poverty leads to both urban land management and utility supply being almost entirely up to market forces. The mutual linkages between unregulated urbanisation and market forces are, however, complex. On the demand side, poor settlers following a 'shelter first' strategy will seek unplanned and unserviced plots in order to avoid the costs associated with obtaining planning consents. On the supply side, consolidating settlements will become attractive for the extension of private sector utility networks, especially electricity and public transport, and for a decentralised supply of drinking water. Both sides face serious risks when the informally growing settlements are located in areas which can later be serviced only at high or even excessive costs. However, statutory planning which could direct peripheral urban growth into serviceable locations and reduce risks, and thus costs for both parties, is absent.

In this environment of high uncertainty, little is known about the locational decisions of informal dwellers and the investment strategies of utility suppliers. It is the aim of the research presented here to open the two 'black boxes'. In the first part, the importance of utility supply for the locational decision of informal dwellers looking for a plot in the urban periphery of Dar es Salaam has been examined. In three typical informal settlements located in the peri-urban zone of Dar es Salaam, a survey of 200 households was undertaken by way of semi-structured interviews. The survey was complemented with focus group discussions in the settlements (see Photograph 12.1).

The fieldwork aimed at understanding the settlers' rationale behind acquiring land in an unplanned settlement with limited access to utilities, especially to important trunk infrastructure such as potable water, electricity and roads.

The case study areas are informal and rapidly growing settlements located in the peri-urban

Photograph 12.2 Water transmission main along Morogoro Road (Source: own photograph)

zone of Dar es Salaam. Services like roads, drinking water, schools, and health-care facilities are at least partially available, making it possible to analyse their influence on the locational decisions of residents.

In order to investigate the increasing perception that there is a higher need for trunk infrastructure during the consolidation of settlements (and with the overall goal of improving housing affordability for settlers), three settlements were selected which either represent the primary or the consolidating stage of development (see Fig. 12.1).

The first case study settlement, King'ongo, is situated in Kimara Ward at the western periphery approx. 14 km outside of the city centre. It is located to the north of Morogoro Road, one of the main arterial transport routes to the inner city centre. One of the city's three water transmission mains runs almost parallel to Morogoro Road (see Photograph 12.2). Although the settlement is not located directly at the main road, it has been rapidly growing since 2003 when a new bridge at N'gombe River was constructed, connecting the settlement with the main road.

The second case study, the settlement of Mbagala Kuu, is located in the hinterland of Kilwa Road, approx. 15 km south of Dar es Salaam city centre. Even though it is not connected to this main transport route by a tarmac road resulting in poor public transport, this settlement has been growing continuously since the end of the 1990s.

The third case study settlement, Stakishari, is part of Kipawa Ward. It is located along Pugu Road about 10 km from the city centre. The settlement has been experiencing rapid growth since the mid-1970s. In the decade from 1992 to 2002, the number of households increased from 292 to 742 (Kombe and Lupala 2004: 54). One aspect that makes the settlement attractive for dwellers is good public transport. Especially along the main road, the building density is increasing.

The results of the household survey revealed some general factors that will influence urban growth in the near future. Natural population increase has overtaken rural-urban migration as the main driver of population growth. Only 10% of the interviewed residents originate from outside Dar es Salaam. In fact, peri-urbanisation is mainly driven by internal migration. Especially young families move to the urban fringe searching for an affordable plot. Many interviewed settlers even came from neighbouring settlements.

The availability of an affordable plot turned out to be the dominant reason why settlers moved to a settlement at the urban fringe. One third mentioned it as their first and most important motive. Especially in the recently growing settlements of King'ongo and Mbagala Kuu, where cheap plots are still available and building density

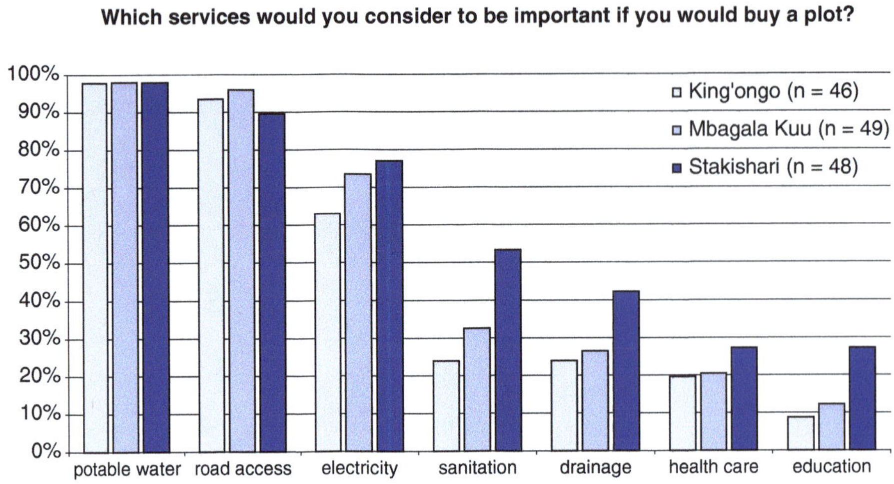

Fig. 12.3 Importance attached to infrastructure supply (Source: own survey 2006/2007)

is still low, the majority of settlers mentioned this motive. Most residents explained that they could afford a sufficiently large plot only in informal settlements at the urban fringe.

Only one out of ten respondents regarded the level of technical infrastructure available in a settlement as a reason to acquire a plot. For this minority, access to a main road and public transport is necessary to commute to the inner city. Water supply was just mentioned by two interviewed residents as being an influencing factor. Other motives of importance were social relations such as existing connections to friends and relatives already living in the settlement. Furthermore, 10% mentioned good environmental conditions and low building density as favourable factors.

The question directly addressing technical infrastructure availability as a motivation for moving to a settlement triggered a variety of responses. One third of the settlers interviewed affirmed that the existing infrastructure was one factor for them to move to a specific area. This particularly applies to the settlement of Stakishari, where the supply of utilities has already been improved. Consequently, in King'ongo and Mbagala Kuu, where water supply is still inadequate, infrastructure was not a relevant motive.

The differences between settlements in distinct stages of development are therefore significant. An older settlement like Stakishari in a consolidating stage of development with a water supply system in place is generally more attractive to urban dwellers. Those who cannot afford a plot there move to areas without services but with cheaper plots. They hope that a growing population in their settlement will be a motivation for the supply companies to provide utilities: 'Services will come sooner or later'. This expectation was confirmed by the ranking of services according to their importance when buying a plot in a specific settlement. In all settlements, road access and reliable access to potable water were prioritised (see Fig. 12.3). Access to electricity was important to three out of four respondents.

The provision of utility services, like water supply and electricity, is ranked as being more important by settlers than disposal services, such as sanitation facilities or drainage system (see Fig. 12.3). There are, however, significant differences between the settlements. The importance of disposal services increases with the development of the settlement. The case of Stakishari once again illustrates that the stage of development of a settlement is an important factor; significantly more respondents compared to the other two case study settlements considered the availability of disposal systems as very important. The higher building density in Stakishari generates problems with sanitation and seasonal flooding. People's awareness of the importance of disposal systems in mitigating these problems is therefore higher than in the younger settlements.

Photograph 12.3 Environmental pollution along an open sewer with water transmission lines (Source: own photograph)

Overall, the results of the survey indicate that the factors influencing settlement decisions of informal dwellers will vary depending on the development stage of a settlement. It appears that settlers do not recognise the relevance of infrastructure until acquiring an affordable plot and moving into the settlement. Once water shortages occur or expenditures and commuting times rise, residents realise the importance of well-functioning infrastructure. Later on in the consolidation phase of a settlement, when environmental problems arise, disposal systems gain more attention (see Photograph 12.3).

The changing awareness and its impact on the locational decisions of informal urban dwellers could be further differentiated in the focus group discussions. Particularly in densely populated settlements like Stakishari, conversations with elderly settlers confirmed the priorities indicated by the household survey. Initially, the decision to move to the area primarily depended on the price of the plot. After settling down, the improvement of supply services became important. Nowadays, the settlement is so densely populated that settlers recognise the need to improve the disposal systems.

This order of priorities reflects the hierarchy of poor settlers' basic needs which are not met adequately when public planning and services are lacking. In their locational decisions, poor settlers follow the principle of 'shelter first', hoping that public utilities and social infrastructure will be improved later, either by utility companies or municipal authorities or through their own initiative.

12.4 Inefficient Trunk Infrastructure Supply Linked to Coordination Deficits

It is the task of urban planning to reduce insecurity, risks and transaction costs associated with settlement decisions of peri-urban dwellers and with investment strategies of utility providers by regulating land-use change. The supply of trunk infrastructure should follow network logics in addition to economies of scale and needs to be based on reliable expectations concerning the future distribution and growth of settlements at customer thresholds. In the absence of efficient land-use planning, it is left to the sectoral utility suppliers to establish at least a basic consensus about the future direction of the extension of their road, water and electricity networks.

The second part of the research conducted in Dar es Salaam served to establish the level of information exchange and cooperation between municipal planning authorities and utility providers and between the various stakeholders. Detailed personal interviews were conducted with urban planners, experts from the central and local government and representatives of electricity and water supply companies in Dar es Salaam.

The findings revealed that every institution develops its own plans and operates based on individual assumptions concerning the future growth and distribution of the urban and regional population. Only informal mechanisms to share information among the different stakeholders involved in the urban development process are in place. There is practically no cross-sectoral cooperation.

The main reason for this coordination deficit is the lack of a legally binding and coherent spatial development concept (like a land-use or master plan) designating favourable and constraint areas for peri-urban residential development. Due to the lack of integrated (and integrating) planning, the informal growth of new settlements depends on the availability of new building land and is, thus, guided exclusively by interests and mechanisms of the private land market. The outcome is an inefficient dispersion of residential areas in terms of transportation costs, economic supply of utilities, protection of agricultural land and environmental sustainability.

The lack of strategic information forces utility suppliers to passively react to urban growth. This means that they either have to invest in the construction of unprofitable network extensions or refuse the provision of services until the demand is high enough to ensure a profitable operation. Investments in water mains or electricity transmission lines are postponed until new settlements have grown large enough to guarantee adequate thresholds. Depending on their location with regard to a network and on the suitability of the local topography, it may occur that settlements cannot be connected at all and have to be served in a decentralised mode at much higher unit costs.

With increasing distance to the city centre (see Fig. 12.2), the investment and service costs for the provision of trunk utilities rise, and the time until the investments become profitable increases. Unplanned urban sprawl can therefore amplify the expenses for vital utilities – especially drinking water, public transportation and electricity – to a degree that the cost benefits which poor settlers gained by acquiring cheap unserviced plots following the 'shelter first' strategy are quickly eliminated.

When informal settlements enter into the consolidation stage, land reserves for the construction of reticulated utilities (piped water, sewage, drainage, roads, even electricity) become scarce because the population density is high and the land required for construction belongs to many different owners. Land pooling and restructuring are time consuming in Tanzania like in most developing countries because the required legal instruments, financial and human resources and practical experience are lacking. Furthermore, settlers are reluctant to make land available for the provision of infrastructure since they do not trust in public bodies. Under the inefficient planning system, the ex post supply of utilities causes delays in the construction of new roads, and costs increase for the installation of desperately needed potable water and electricity supply services.

With inadequate services, especially a lack of cheap and safe potable water, the incidence of diseases increases causing, for example, higher infant mortality. Furthermore, girls and young women are forced to spend precious time on fetching water over long distances. With augmenting building density in unserviced settlements, public health is likely to deteriorate and environmental hazards to increase resulting in additional opportunity costs.

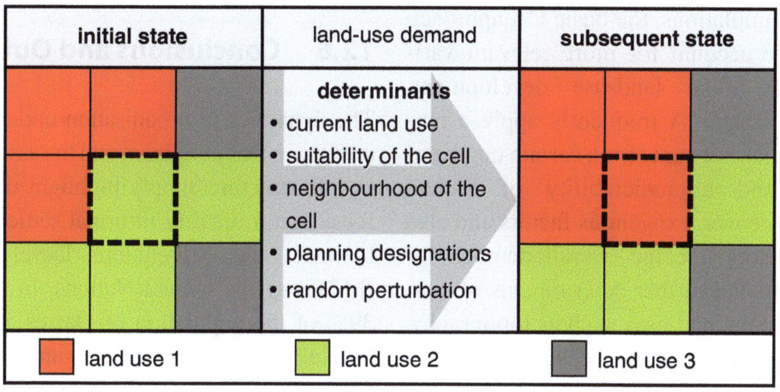

Fig. 12.4 Illustration of the basic modelling principle (Design: C. Lindner (IRPUD))

The majority of informal settlers who cannot afford a private borehole, a generator and off-road vehicles have to pay high additional costs as a consequence of their decision to acquire an affordable plot at the urban fringe. These adverse effects of peripheral urbanisation under the conditions of poverty could be avoided with a better coordination of urban growth and trunk infrastructure provision.

12.5 Coordinating Urban Development by Simulating Settlement Dynamics

In the absence of efficient planning regulation, land owners, settlers and utility suppliers are operating in an environment of severe information deficits, high transaction costs, insecurity and unforeseeable risks. Poor settlers, obviously the weakest party, with their 'shelter first' strategy are banking on later improvements as a consequence of settlement growth without reliable information on the logics and strategies which the suppliers of trunk infrastructure are forced to follow. The latter largely operate in isolation which prevents them from optimising the use of their scarce resources.

The second part of the research conducted in Dar es Salaam revealed that new and alternative tools are required to overcome the deadlock caused by disempowered public authorities and ineffective planning regulation and the resulting information deficit on behalf of public utilities suppliers, particularly trunk infrastructure providers. It was assumed that plausible information on the future course of urban growth, especially its spatial distribution, could provide a common platform for the exchange of information and subsequent cooperation of stakeholders. A model to simulate urban development capable of responding to strategic variables seemed to provide the most adequate tool.

The simulation model used to forecast land-use development and its spatial distribution was designed and implemented based on the assumption that informal land-use changes are predominantly driven by local-scale processes on the neighbourhood level. The model uses the basic principles and techniques of cellular automata (CA) which model change dynamics as decentralised bottom-up processes (see Benenson and Torrens 2005; Batty 2007). Due to their inherent ability to account for neighbourhood relations, CA models are particularly suited for modelling spatial diffusion processes of urban growth.

For the purpose of modelling land-use development, the urban space is partitioned into a regular grid of uniform cells. Each of these cells is assigned a specific land use, that is, the main land use found in this cell. Future cell states are then calculated based on the present land use of the cell and a function considering a predefined cell neighbourhood and a set of transition rules (see Fig. 12.4 for an illustration of the modelling principle).

For urban simulations, the basic CA approach is expanded to account for more relevant variables driving urban land-use development. Additionally, urban CA frequently apply a random factor which accounts for a certain degree of irrationality and unpredictability of human behaviour. Moreover, exogenous factors and also exogenous factors like the overall demand for residential land and further determinants of suitability for specific land uses such as topography, accessibility and planning designations are included (see White et al. 1997; Barredo et al. 2003). Furthermore, constraint areas may be defined which should be excluded from any urban development. For the purpose of estimating the overall demand for residential land, the Dar es Salaam model uses demand figures which are calculated on the basis of UN population projections for Dar es Salaam.

The presentation of a prototype of the Dar es Salaam land-use model (Hill and Lindner 2007) and its discussion with local experts, including representatives of most utility suppliers and municipal as well as regional authorities, supported the underlying assumption that the information generated is of strategic value. The option of simulating the impact of planning measures and utility extension on peri-urban growth has raised the interest of local planners and utility companies. The virtual example of constructing a new ring road in the north-western periphery of the city provided evidence on the location of future hot spots of infrastructure demand. A lively discussion on envisaged measures and their likely impacts produced valuable insights into the practices and policies of the public and private sector. What seems even more important in the light of the detected information deficits is the observation that the mutual discourses on the linkages between model assumptions and simulation output have established a new basis for an improved exchange of information and a better cross-sectoral cooperation. The final version of the simulation model gave empirical insights into the mechanisms of informal urban dynamics and demonstrated the potential impacts of strategig transport infrastructure measures (Hill and Lindner 2010).

12.6 Conclusions and Outlook

The dynamics of urbanisation under conditions of poverty in Dar es Salaam and the associated deficits in infrastructure supply highlight the urgent need for action to qualify informal settlements in order to meet the Millennium Development Goals (MDG) of the United Nations. In the year 2000, 39% of the population of Dar es Salaam did not have access to safe drinking water (UN-HABITAT 2003: 28). To meet the MDG, this rate has to be halved by 2015, requiring an upgrading of existing settlements and, additionally, supplying the majority of about 150,000 new dwellers per year with technical infrastructure.

In order to optimise the use of the scarce resources available for planning purposes, to reduce costs for investment and maintenance and to achieve affordable prices for the supply with water, electricity and public transport, settlement growth has to be directed to suitable locations. Urban structures with appropriate densities are necessary to create adequate threshold populations for the economically viable supply of trunk infrastructure. This requires public and private stakeholders to break new ground for cooperation.

A model simulating future peri-urban growth served as an attractive platform to bring together representatives from all relevant institutions and to establish a task force for continuous information exchange and cross-sectoral cooperation. Time will tell if the task force will be able to continue its work beyond the duration of external funding.

References

Barredo JI, Kasanko M, McCormick N, Lavalle C (2003) Modelling dynamic spatial processes: simulation of urban future scenarios through cellular automata. Landsc Urban Plan 64:145–160

Batty M (2007) Cities and complexity – understanding cities with cellular automata, agent-based models, and fractals. MIT Press, Cambridge/London

Benenson I, Torrens PM (2005) Geosimulation – automata-based modeling of urban phenomena. Wiley, Chichester

DCC, Dar es Salaam City Council (2000) Strategic urban development plan for the city of Dar es Salaam – city growth vision, vol I. DCC, Dar es Salaam

De Soto H (2000) The mystery of capital: why capitalism triumphs in the West and fails everywhere else. Basic Books, New York

Hackenbroch K, Baumgart S. and Kreibich V (2009): The Spatiality of Livelihoods - Urban Public Space as an Asset for the Livelihoods of the Urban Poor in Dhaka, Bangladesh. In: Die Erde 140, 47–68

Hill A, Lindner C (2007) GIS-based simulation of the impact of trunk infrastructure provision on rapid urban growth in developing countries – a CA application for the city of Dar es Salaam, Tanzania. In: Proceedings 10th international conference on computers in urban planning and urban management (CUPUM), Foz de Iguacu, Brazil, 11–13 July 2007

Hill A, Lindner C (2010) Modelling informal urban growth under rapid urbanisation - A CA-based land-use simulation model for the city of Dar es Salaam, Tanzania. Doctoral thesis, TU Dortmund University, http://hdl.handle.net/2003/27283

Kironde JML (2006) The regulatory framework, unplanned development and urban poverty: findings from Dar es Salaam, Tanzania. Land Use Policy 23:460–472

Kombe WJ (1994) The demise of public urban land management and the emergence of informal land markets in Tanzania – a case of Dar-es-Salaam city. HABITAT Int 18(1):23–43

Kombe WJ, Kreibich V (2000) Informal land management in Tanzania, vol 29, SPRING research series. SPRING-Center, Dortmund

Kombe WJ, Kreibich V (2006) Governance of informal urbanisation in Tanzania. Mkuki na Nyota, Dar es Salaam

Kombe WJ, Lupala J (2004) WSS practices and living conditions in the peri-urban interface of Metropolitan Dar es Salaam: the cases of Tungi and Stakishari. Unpublished report produced for the research project 'Service provision governance in the peri-urban interface of metropolitan areas', Development Planning Unit, University College London

Kraas F, Hansjürgens B (2008) Megacities – urban dynamics of global change. In: Höll B, Mauser W (eds) Global change research in Germany 2008. German National Committee on Global Change Research, Munich, pp 21–22

Roy A (2005) Urban informality: towards an epistemology of planning. J Am Plan Assoc 71(2):147–158

Roy A, AlSayyad N (eds) (2004) Urban informality: transnational perspectives from the Middle East, Latin America, and South Asia. Lexington Books, Lanham

Silayo EH (2004) The role of the cadastral system in poverty alleviation in Tanzania. Paper presented at the CASLE conference held at Danbury Park Conference Centre, Chelmsford, UK, 21–25 Apr 2004

UN, United Nations, Department of Economic and Social Affairs, Population Division (2006) Wallpaper urban agglomerations 2005. United Nations, New York

UN, United Nations, Population Division of the Department of Economic and Social (2008) World urbanization prospects: The 2007 revision population database. http://esa.un.org/unup. Accessed 21 June 2008

UN-HABITAT, United Nations Human Settlement Programme (2003) Water and sanitation in the world's cities. UN-HABITAT, London

UN-HABITAT, United Nations Human Settlement Programme (2006) The state of the world's cities 2006/07: the millennium development goals and urban sustainability. UN-HABITAT, Nairobi

White R, Engelen G, Uljee I (1997) The use of constrained cellular automata for high-resolution modelling of urban land-use dynamics. Environ Plan B Plan Des 24:323–343

13 Bus Rapid Transit: The Answer to Transport Problems in Megacities? The Example of TransMilenio (Bogotá, Colombia)

Jan Marco Müller

Abstract

In recent years, the Colombian capital Bogotá has become an internationally renowned showcase of how an innovative concept of public mass transport, embedded into a long-term urban development strategy, is able to completely transform a large metropolis. Basically, Bogotá's TransMilenio system relies on the Bus Rapid Transit concept developed for Curitiba (Brazil) in the 1970s but adapted to the needs of a megacity. The overwhelming success of TransMilenio has led to a "renaissance" of Bus Rapid Transit, revolutionising the way many metropolises, especially in developing countries and transformation countries, organise their public transportation systems. At present, the system is being implemented in dozens of cities around the world, primarily in Latin America, but increasingly also on other continents. This chapter presents the TransMilenio system, its institutional framework, implementation and accompanying measures and its strong points and drawbacks, and discusses the system's applicability to other megacities.

Keywords

Urban transportation • Bus Rapid Transit • Bogotá • TransMilenio • Sustainable development

13.1 Introduction

In recent years, the concept of Bus Rapid Transit has revolutionised public mass transport in many metropolises of the developing countries and transformation countries. Its basic principle is simple: separate bus lanes are introduced along major urban corridors, making it possible for articulated buses to circulate in extremely short intervals. Access to the system is provided through covered bus stations, usually placed in a distance of some 500 m of each other. Once in the system, the user can travel as far as he or she likes and back on a single ticket, until leaving the system again at one of the bus stations. Bus Rapid Transit provides

J.M. Müller (✉)
Bureau of European Policy Advisers, European Commission, BERL 08/364, Rue de la Loi 200, 1049, Brussels, Belgium
e-mail: jan.mueller@ec.europa.eu

F. Kraas et al. (eds.), *Megacities: Our Global Urban Future*, International Year of Planet Earth,
DOI 10.1007/978-90-481-3417-5_13, © Springer Science+Business Media B.V. 2014

a cheap and easy way to implement means of improving the mobility in large urban areas.

Bus Rapid Transit as such is nothing new: the Brazilian town of Curitiba, for instance – then still a middle-sized city – already introduced the system in the 1970s under Mayor Lerner. Grown to a metropolis in the meantime, the system continues to serve Curitiba's transport needs. However, this forerunner was initially not replicated in many cities outside Brazil; one of the few examples is the trolley bus system introduced in 1995 in Quito (Ecuador). Only since the introduction of Bogotá's TransMilenio system in the year 2000 has Bus Rapid Transit experienced a real international breakthrough. Bogotá showed for the first time that Bus Rapid Transit was also applicable as an integrated solution to the transport problems of a megacity, serving as the backbone for a complete reorientation of urban planning in the context of sustainable development.

Enrique Peñalosa, an independent technocratic but popular politician who governed Bogotá from 1998 to 2000, is considered the "father" of TransMilenio. He declared finding solutions to the transport problems threatening the future of the city a flagship initiative of his administration (see Peñalosa 1997). After decades of failed attempts to establish a metro in Bogotá (see Ardila Gómez 2004), Peñalosa managed the miracle of planning, building and putting into service the TransMilenio system within a period of 3 years, inaugurating it on 18 December 2000, 2 weeks before leaving office.

13.2 The Transport Situation in Bogotá before TransMilenio

Located in an altitude of 2,600 m, the Colombian capital Bogotá with its nearly 8 million inhabitants – 10 million in the whole metropolitan area – can be regarded as the highest megacity of the world. More than two thirds of the population belong to the low- and lower-middle-income classes (Müller 2002: 290). Like many cities in developing countries, before the introduction of TransMilenio, Bogotá was characterised by a completely chaotic, inefficient and environmentally unsustainable transport system. In 1998, 22,000 buses transported 72% of the urban population on 639 different bus lines. At the same time, 670,000 private cars moved only 19% of the population (all data: TransMilenio S.A. 2000: 9). The public transport system was unable to cope with the rapid population growth of the metropolis of up to 3% per year, with the urban area now covering more than 500 km^2.

Moreover, the transport system was dominated by the informal sector, with hundreds of "one-man companies" fighting a daily battle for each single passenger, a phenomenon that became known as the "Penny War". This resulted in constant traffic jams, a rapid deterioration of the infrastructure, enormous environmental pollution and a high number of accidents (see Fig. 13.1). The average time spent on urban transport before TransMilenio has been estimated at 2 h and 20 min per person per day, the average speed being just 10 km/h (TransMilenio S.A. 2000: 9). The transport sector was largely responsible for air pollution in Bogotá, amounting in 1991 to 290,000 tons of CO, 11,000 tons NO_x, 8,000 tons SO_x and 2,200 tons of particulate matter (see Japan International Cooperation Agency 1992). In this context, it should be taken into account that air pollutants in an altitude of 2,600 m are more likely to cause respiratory diseases than under normal circumstances due to the lower oxygen concentration in the ambient air.

The number of traffic deaths in Bogotá rose from 1,089 in the year 1991 to 1,387 in 1995, most of the victims being pedestrians and cyclists (Departamento Administrativo de Planeación Distrital 1997: 69–71). The buses were not suited for people with physical disabilities and operated with low occupancy rates, even during peak hours, the latter contributing to high crime rates, including incidents of sexual violation in running buses. Many bus drivers were obliged to work 16–18 h per day in order to gain enough money to support their families. Due to the low profit margin, hardly any investments were made into new buses: in the year 2000, approximately 35% of all buses were more than 20 years old (http://www.kolumbien-aktuell.ch/monatsberichte/2.03.html).

Fig. 13.1 The "Penny War" characterised many Colombian cities before the introduction of Bus Rapid Transit systems (Photo: Jan Marco Müller)

13.3 General Presentation of the TransMilenio System

The Bus Rapid Transit system TransMilenio has been implemented on selected large boulevards in the Colombian capital (so-called trunk lines). The space for the bus lanes was taken away from the road space, i.e. capacity formerly available to individual motorised traffic. In the first phase, which was initiated on 18 December 2000, two trunk lines with a length of 41 km were built. Since then, the network has been doubled to 84 km with a total of 114 bus stations. A fully canopied bus station was installed every 500 m along the trunk lines. Like the entire system, they are compliant with the needs of disabled persons and protect users from the rain and wind.

In contrast to Curitiba, TransMilenio bus stations are built in the middle of the street and are accessible via pedestrian crossings or bridges (see Fig. 13.2). This solution has been chosen in order to save costs, both in terms of staffing and infrastructure (instead of one bus station per direction, there is only one serving both directions). At the bus stations, users can buy a contactless smart card, which can also be precharged for several rides, that allows using the system without distance and time limits until leaving the system through a bus station. The inner area of each bus station can be accessed through turnstiles, while the buses can be accessed, depending on the direction to the left or right, through transparent doors that open automatically after the arrival of a bus.

Fig. 13.2 TransMilenio stations are mostly bidirectional, located in-between the bus lanes (Photo: Jan Marco Müller)

At present, approximately 1,300 red articulated buses are operating on the trunk lines, either as so-called "corriente" buses that stop at every station or as express buses that only stop at the most busy ones. As most parts of the carriageway provide two bus lanes for each direction, the TransMilenio buses can pass each other. The system is complemented by more than 500 green feeder buses, which, on a total length of more than 600 km, connect selected stations of the trunk lines on the outskirts with the surrounding residential areas (see www.TransMilenio.gov.co). At the end of each trunk corridor, large bus stations have been built, connecting TransMilenio with the regional and national bus services.

On average, express bus lines operate on a schedule based on a 2-min interval on working days, while the "corriente" buses circulate every 3 min. The combination of these two lines makes intervals as short as 13 s possible at the busiest stations. The green feeder lines operate based on a 5-min interval. The average speed of the buses is currently 27 km/h – a much higher speed than the old system would ever have been able to offer. TransMilenio operates every day between 5:00 a.m. and 11:00 p.m. (on Sundays and holidays 6:00 a.m.–10:00 p.m.). The system is continually monitored by a control centre, which receives a GPS signal with a 2-m precision from each bus every 6 min. In the case of an incident, supplementary buses are immediately added into service, and/or the bus drivers are advised to go faster or slower.

With the initiation of the second phase in April 2006, TransMilenio now circulates along nine corridors, connecting different parts of Bogotá (see Fig. 13.3). With the "Norte-Quito-Sur" trunk corridor, for the first time, a tangential line has been put into service, complementing the other routes that depart radially from the existing Caracas trunk corridor which serves as the central backbone of TransMilenio. When completed, TransMilenio is supposed to encompass 388 km of trunk roads connecting the whole urban area (TransMilenio S.A. 2000: 56). The ultimate goal is for 80% of the urban population to use TransMilenio on their way to work. The third phase (Carrera 10 as well as Avenida Eldorado) is under construction. In this context there is also a plan to enlarge the system by extending TransMilenio beyond the Capital District towards the neighbouring municipality of Soacha.

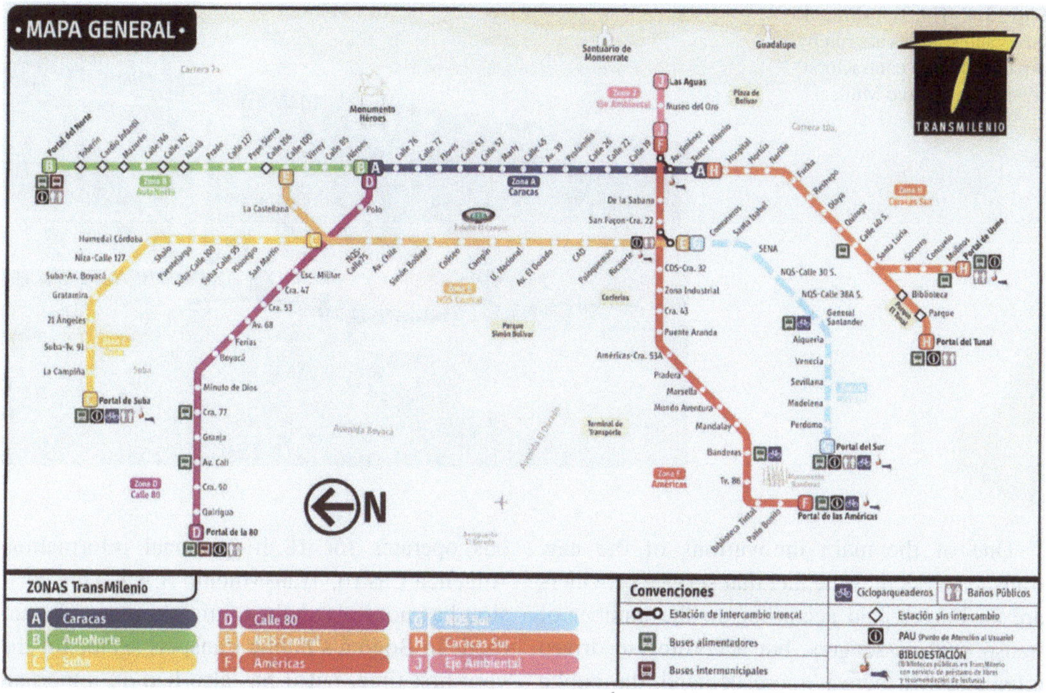

Fig. 13.3 TransMilenio network 2012 (Source: TransMilenio S.A.)

13.4 Administrative and Financial Framework

TransMilenio follows a model of public-private partnership: the public domain provides the infrastructure, while buses and fare collection are operated by several private contractors. Seven contractors operate the trunk buses and several others the feeder buses, and two of them are in charge of collecting fares (Fig. 13.4). This approach has been chosen in order to avoid being dependent on one single contractor, at the same time allowing a certain degree of competition within the system (personal information, Angélica Castro, TransMilenio S.A.).

The costs of all development phases of TransMilenio are expected to add up to 2.3 billion US$, not counting the rolling material. The first phase, which was initiated in 2000, had a total cost of 240 million US$ (i.e. 5.9 million US$ per km). 46% of these costs have been financed by private car users through a local surcharge on fuel in Bogotá. Further sources included the national government (20%), the World Bank (6%) and numerous other sources (28%) (all data: Federal Transit Administration 2006: VII). The second phase of TransMilenio, beginning in 2006, had significantly higher costs (545 million US$ = 13.3 million US$ per km) mostly due to higher investments into accompanying infrastructural measures, such as the creation of public parks and the beautification of public space.

Interestingly, 66% of the costs of the second phase were financed by the national government (compared to 20% of the costs in the first phase), indicating a change of attitude in the government following the visible success of the first phase (Federal Transit Administration 2006: VII). The government obviously noticed that TransMilenio provided a quick solution to the problems preventing a proper functioning of the national capital. This was in the national interest, since Bogotá generates about one quarter of the Colombian GDP (Müller 2001: 43).

Fig. 13.4 The TransMilenio buses and stations are run by separate private contractors (Photo: Jan Marco Müller)

One of the main innovations of the new transport system is the fact that service providers are no longer paid according to the number of transported passengers, but according to driven kilometres. If there is more demand, the public company administrating TransMilenio gives the contractors the instruction to drive more kilometres. The total duration of the contracts corresponds to 850,000 driven km, which is more or less equivalent to ten years of service. Two thirds of all revenues generated by TransMilenio go to the contractors of the trunk lines, 20% to the contractors of the feeder buses and 11% to the contractor of the fare collection. Only about 5% of the revenues remain with the public TransMilenio S.A., which is in charge of the overall operations (all data: Graftieaux 2006).

In most cases, the TransMilenio contractors are consortia of local bus companies that already offered transport services before the TransMilenio system began operating. They often have national and international partners. According to TransMilenio S.A., 94% of the traditional transport companies are involved in the consortia now running the Bus Rapid Transit (www.TransMilenio.gov.co). They were preferred in the public call for tenders in order to ensure more "social peace" by transferring traditional bus drivers into the new system.

The maximum working time for each driver has been limited to six hours per day, corresponding to three drivers per bus per day, since each bus operates for 18 h (personal information, Angélica Castro, TransMilenio S.A.). This decision has not only defined minimum social standards in Bogotá's public transport sector for the very first time, but it has also had an important effect on employment, replacing the jobs which threatened to break away after the old system was abolished. In its first year of existence, TransMilenio alone has created 3,000 new jobs in the formal sector, mainly benefitting bus drivers formerly employed in the informal sector (Perea 2002). In addition, an estimated 13,000 persons have been involved in the construction work for TransMilenio (www.TransMilenio.gov.co).

TransMilenio S.A., which is owned by the Capital District and which operates the control centre and is in charge of the overall system development, determines the fare. After buying a ticket, passengers can ride all TransMilenio lines as often as they wish until they exit the system at one of the stations. The fare includes the use of the feeder buses, which therefore can be used for free when purchasing a trunk line ticket, and is the same for all users. An introduction of tariff zones, stepwise fares or the like has been avoided in order to make the use of the system as simple and cheap as possible, also for illiterates and people with low level of education.

In 2011, the fare for one single ride was 1,700 Colombian pesos (= 0.48 euro/0.61 US$), which equals roughly 0.3% of the monthly minimum income in Colombia. TransMilenio revenues totalled 172 million US$ in 2005, of which 7 million stayed with the public TransMilenio S.A., while the rest was distributed among the contractors (Federal Transit Administration 2006: 21). With a net profit of approximately 30% (statement of a TransMilenio manager quoted in Federal Transit Administration 2006: 32), TransMilenio is highly profitable and can be run without any state subsidies (not counting the provision of the infrastructure).

TransMilenio is also able to make a profit because ridership has exceeded all initial projections. At the beginning, it was expected that TransMilenio would be used by 350,000 passengers daily after 1 year of service (personal information Angélica Castro, TransMilenio S.A.). In 2009, after extending the network to twice its original length, TransMilenio carried an average of up to 1.4 million passengers every day, already reaching the projected passenger load during three peak hours alone. Between December 2000 and February 2012, TransMilenio was used 3.7 billion times (Data: www.TransMilenio.gov.co), showing that Bus Rapid Transit can definitely compete with rail-based solutions in large urban agglomerations. When fully completed, TransMilenio is envisaged to carry 5.5 million passengers every day (www.TransMilenio.gov.co).

13.5 Accompanying Measures and Long-Term Strategy

The astonishing success of TransMilenio would not have been possible without a series of accompanying measures. In the first place, the restrictions imposed on motorised individual traffic during peak hours have to be mentioned: daily traffic load was reduced by forbidding the circulation of cars with certain digits on their licence plates on certain days (in Bogotá known as "Pico y Placa"). This tool, also used in many other metropolises like Mexico City and Santiago de Chile, removes about 40% of the individual traffic during peak hours. It has been extended by a "Pico y Placa Ambiental", forbidding the circulation of cars older than ten years during peak times. A survey conducted by TransMilenio S.A. shows that 9% of the TransMilenio users previously used private cars to commute to work (Federal Transit Administration 2006: IX). In addition, each year Bogotá celebrates an obligatory car-free day, which was approved by a referendum in October 2000.

In the same referendum, the population of Bogotá agreed to a probably worldwide unique "public deal" proposed by the city administration. Citizens agreed to completely prohibit individual traffic during peak hours as of 2016, while the city administration committed itself to providing full TransMilenio coverage by 2016 with every inhabitant having a station within 500-m walking distance. Surprisingly, this revolutionary proposal received a relative majority of votes; however, the decision did not become legally binding because it did not reach the absolute majority. Nevertheless, the result put strong pressure on the public administration to fulfil its promises while at the same time strengthening the citizens' commitment through the further development of TransMilenio.

The October 2000 referendum was only one element in the long-term strategy for the sustainable development of the Colombian capital. TransMilenio functions as the "backbone" in this strategy, which is characterised by a strong participatory approach and the embedding of all plans and measures into the local Agenda 21 process. TransMilenio was the cornerstone of the urban development and land use plan for the Capital District, finalised in 1999, which defined the development goals for the next 10 years (see Departamento Administrativo de Planeación Distrital 1999).

This synchronisation of TransMilenio with urban development measures can be seen in a number of examples, for example, the construction of social housing along the TransMilenio corridors ("Metrovivienda") or the creation of the "Tercer Milenio" Park in the context of TransMilenio construction work in what used to be El Cartucho, the most dangerous quarter in central Bogotá. In fact, it is amazing to see the speed and commitment with which the TransMilenio system is being implemented

Fig. 13.5 Demolition of buildings for the construction of the new TransMilenio corridor to Suba (Photo: Jan Marco Müller)

throughout the city, some TransMilenio corridors literally being carved through the urban fabric, expropriating and demolishing entire blocks of buildings in order to make room for the new transport system (see Fig. 13.5).

Further measures accompanying TransMilenio have been a 20% increase in the local surcharge on fuels, which contributes directly to the construction of TransMilenio, the introduction and the increase of parking fees and the removal of parking space and traffic lanes along major streets, replaced by pedestrian lanes and bicycle tracks. 300 km of new bike lanes have been built since, boosting the popularity of bicycles as means of transport and leading to an increase of cycling in the modal split during the Peñalosa administration alone from 0.5% in 1997 to 4% in 2001 (personal information Germán Camargo, Capital District Environment Agency DAMA). Dozens of new pedestrian crossings and bridges built to provide access to the TransMilenio stations at the same time helped to reduce the isolating character of major boulevards for pedestrians and cyclists. Thanks to all these measures, the percentage of motorised individual traffic in the modal split of Bogotá has been reduced from 20% in 1999 to 10% in 2006, meaning a 50% decrease (www.TransMilenio.gov.co).

The importance of TransMilenio's promotional campaign should not be underestimated as a key to success. From the very beginning, TransMilenio started with a clear branding and an intelligent approach to reaching people. In the first three weeks of its existence (18 December 2000 until 6th January 2001), TransMilenio could be used by everybody for free, which proved to be an excellent idea to make everybody familiar with the system. TransMilenio continues to foster cultural activities and the interaction with citizens, providing a broad spectrum of services online (www.TransMilenio.gov.co), including a special website for children (see Fig. 13.6).

13.6 Drawbacks of TransMilenio

One of the major criticisms of TransMilenio is its limited capacity. In fact, buses seem to be overloaded at all times, and long queues can be seen in front of the stations, provoking protests from users. Except for very few double-articulated buses with a capacity of 270 persons like in Curitiba (so-called ligeirinhos), TransMilenio only uses mono-articulated buses with a capacity of 160 passengers. This design has primarily been chosen to save costs (smaller buses, smaller stations). The capacity argument against Bus Rapid Transit is put forward especially by those favouring rail-based solutions. Nevertheless, one must also take into account the extremely short intervals Bus Rapid Transit can provide. In Bogotá, during peak hours, buses arrive every 13 s at the most busy stations, which is probably a world record!

Fig. 13.6 TransMilenio offers a special website for children (Source: http://www.TransMilenio.gov.co/WebSiteInfantil/index.html)

The US Federal Transit Administration (2006: 11) comes to the conclusion that TransMilenio at the end of the day offers as much or even more capacity than a rail-based solution: a subway train certainly can transport up to ten times more passengers than a TransMilenio trunk line bus, but due to technical reasons, the intervals between trains need to be at least ten times longer than in the case of TransMilenio – and, besides, investment costs are ten times higher than for Bus Rapid Transit. In the end, Bogotá is creating a mass transport system that covers the entire agglomeration with the investment costs of one single metro line.

Although the development of TransMilenio is characterised by a strong participatory approach, its implementation does not escape protests by affected lobby groups. For example, the system has been blocked repeatedly by "threatened" bus drivers of traditional transport companies, mainly those that were not successful in the call for tenders for operating TransMilenio. Such a protests took place on the 2nd of May 2006, when hundreds of bus drivers protested against the shutdown of traditional bus routes along the TransMilenio corridors and the allegedly poor reimbursement for the obligatory dismantling of old polluting buses. According to a regulation of the Capital District, for each new trunk bus between 2.7 and 8.9 old buses must be taken out of service, depending on their environmental performance (so-called chatarrización).

More than 2,000 old buses have already been taken out of circulation since TransMilenio started (www.TransMilenio.gov.co). Nowadays, even the taxi companies have understood that their revenues are not endangered by TransMilenio, but, on the contrary, increased due to reduced traffic jams (personal information Germán Camargo, Capital District Environment Agency DAMA). Nevertheless, there is certainly some truth in the statement by Hidalgo and Hermann (2004: 14): "Winning over the traditional transport sector is perhaps the biggest challenge to the successful implementation of BRT systems in developing cities".

Without doubt, TransMilenio and measures accompanying the new system have considerably improved the environmental conditions in Bogotá. Between 1998 and 2002, CO emissions have

Fig. 13.7 The trunk buses have special technical requirements (Photo: Jan Marco Müller)

been cut by 28% and NO_2 emissions by 16%. Also, ozone levels dropped by 6%. However, at the same time, SO_2 emissions increased by 15% and particulate matter (PM10) by 12% (Graftieaux 2006). This may also be related to the fact that TransMilenio does not use the latest technologies. Although the system itself contributes to an improved air quality in Bogotá, the diesel-driven TransMilenio buses of different manufacturers (Mercedes, Scania, Volvo, Marco Polo etc.) usually only comply with the Euro-2 emissions norm.

Moreover, the quality of the utilised diesel fuel is poor (sulphur content 1,000 ppm), especially considering that the buses run 350 km daily in an altitude of 2,600 m and most of the time fully loaded (Federal Transit Administration 2006: 11). So far, no buses running on biodiesel, natural gas or hydrogen are being used, thus leading to an underperformance of the system in terms of environmental impact, compared to solutions based on electrical power (e.g. trolley bus). It should also be noted that the dense traffic on the TransMilenio trunk lines is leading to a rapid deterioration of the road surface, causing higher investments into road maintenance.

Another drawback is the fact that the purchase of the buses is relatively expensive as these have special technical requirements due to the unusual position of the doors which are located 90 cm above the ground, on the left-hand side of the bus, instead of on the right-hand side, as is usually the case (see Fig. 13.7). However, this disadvantage has been accepted because the "same level" boarding not only ensures full suitability for disabled persons but also increases the capacity of the system, with a boarding time per passenger of 0.1–0.3 s. At the same time, the risk of people accessing the system without paying the fare, for example, ambulant salesmen, has been reduced. As indicated earlier, the higher purchasing costs for the buses were in part compensated by lower investment costs for the stations, since bidirectional stations in the middle of the street – instead of one station for each direction – were adequate, meaning that fewer stations had to be built.

Finally, the impact of TransMilenio on urban sprawl and land use as well as on the socioeconomic dynamics in Bogotá still needs to be examined. A first study conducted by Barrios (2002) indicates that real estate values have sharply increased along the trunk lines, declining by 0.8–1.1 US$ per m^2 with each block (100 m) of distance from the TransMilenio corridors. Targa (2003) proved that out of 28 city quarters that experienced increasing real estate values between 2001 and 2002, 15 were located within 1 km of a TransMilenio corridor. Clear indicators are also new shopping malls in close vicinity to the

TransMilenio stations, such as the "Centro Comercial Portal de la 80". Further geographical case studies are needed to complete the picture, especially on the impact that TransMilenio has on urban growth.

Likewise, there needs to be further investigation on whether Bus Rapid Transit actually contributes to stronger social cohesion. A recent master's thesis comes to the conclusion that TransMilenio did not contribute significantly to the social inclusion of lower-income inhabitants of Bogotá (Hörmann 2008: 95), at least not in the short run, mainly because of shortcomings regarding its affordability, availability, accessibility and acceptability. Still, it is seen as a useful instrument for a long-term urban planning, especially for mitigating social inequalities, while accepting that the implementation of Bus Rapid Transit is a learning process for all actors involved (Hörmann 2008:97).

13.7 Applicability of the TransMilenio Concept to other Cities

Bogotá has shown that Bus Rapid Transit can be a tool for solving transport problems in megacities. This refers especially to developing countries and transformation countries, where urgent mobility problems need to be resolved with limited financial and institutional resources. The example of TransMilenio shows that Bus Rapid Transit can provide citywide solutions at relatively low costs, much cheaper than any rail-based solution. Moreover, TransMilenio has proved that Bus Rapid Transit can be implemented within a very short time frame, addressing urgent needs. In Bogotá implementation only took 3 years, from the planning stage to full operation – a time frame short enough to pay off in the next elections!

In general, the implementation of rail-based solutions takes longer than one electoral term. Furthermore, Bus Rapid Transit can be implemented quite easily using existing transport corridors (see Fig. 13.8) and therefore does not necessarily result in "construction nightmares" annoying affected business owners and commuters. Another advantage is the fact that Bus Rapid Transit can be operated with relatively limited technical knowledge and without depending on imports from industrialised countries. Bus Rapid Transit systems are very flexible, depending on resource availability. They can be developed in a modular way, making it possible to implement several systems simultaneously in different parts of a megacity.

In the case of Bogotá, the implementation of the TransMilenio system abolished the "Penny War" in public urban transport and introduced an efficient, reliable and well-organised system that maximises the benefit of limited resources. It has also transferred informal jobs to the formal sector and significantly improved the state of the environment and the quality of life, as well as dramatically reduced traffic accidents: along the trunk corridors, accident rates have dropped 90% since TransMilenio was initiated (www.TransMilenio.gov.co). However, there is another aspect that goes beyond bare statistics: TransMilenio has fundamentally turned around the way citizens see their city.

It has become a symbol of identification and of pride, particularly in a country usually known as a source of bad news. This becomes clear on the official website www.TransMilenio.gov.co, which has the title "TransMilenio – Orgullo Capital" ("capital pride"). TransMilenio teaches respect and civil rights, especially towards people who need help. It is not just a transport system, but an integral part of a long-term philosophy aiming at the sustainable development of a megacity that is defending itself against becoming uninhabitable.

TransMilenio's overwhelming success has turned Bogotá into a real "Mecca" of transport planners from all over the world and its father Enrique Peñalosa into a worldwide requested expert in the field of urban sustainable development. Bus Rapid Transit is being actively promoted by organisations of technical cooperation such as the German GIZ (e.g. see http://www3.giz.de/E+Z/content/archiv-ger/01-2004/inw_art4.html). While in the past many politicians in developing countries followed the "myth of the

Fig. 13.8 Bus rapid transit systems are relatively easy to implement (Photo: Jan Marco Müller)

metro" which was aggressively pushed by the railway industry, TransMilenio showed that bus-based solutions can be inexpensive and modern at the same time.

Thanks to TransMilenio, Curitiba's concept of Bus Rapid Transit is celebrating a remarkable revival. Many large cities in developing countries are now introducing Bus Rapid Transit systems, interestingly enough even cities that already have a rail-based metro system, such as Santiago de Chile (Transantiago), Mexico City (Metrobús), Caracas (Metrobús) and Medellín (Metroplús). The Venezuelan cities of Mérida (Trolmérida) and Barquisimeto (Transbarca) are developing Bus Rapid Transit systems based on the TransMilenio model, so is the Peruvian capital Lima (ProTransporte). Public and political opinion in Colombia has almost entirely switched to being in favour of Bus Rapid Transit.

A second national Bus Rapid Transit system went into service in Pereira, a city with a population of 500,000, in 2006 (Megabús), followed by Colombia's third largest city Cali in 2009 (Mío), while further systems following the TransMilenio model have been developed in Cartagena (Transcaribe), Barranquilla (Transmetro) and Bucaramanga (Metrolínea). Bogotá's experience with TransMilenio is starting to pay off: one of the consortia running TransMilenio won the call for tenders for operating the Chilean counterpart Transantiago, which started operating fully in 2007 (see Gschwender 2006). Increasingly, cities on other continents are starting to invest into Bus Rapid Transit, especially African metropolises such as Johannesburg, Dar es Salaam and Dakar but also Asian megacities such as Guangzhou, Ahmedabad and Jakarta (see http://www.itdp.org/ for more information).

It is clear that every city has its own character, so individual solutions need to be found. Certainly, towns in the Americas offer certain advantages for the implementation of Bus Rapid Transit due to their basic morphology (chessboard layout, relatively little architectural heritage), while cities in other cultural contexts, for example, the Islamic world or India, may face other obstacles for the implementation of Bus Rapid Transit systems. Likewise, Bus Rapid Transit systems are easier to implement in bigger cities with their wide boulevards while creating major problems in smaller towns, for example, in Pereira where the space needed for the Megabús system has created major problems with traffic congestion (giving it the popular nickname MegaEstorbo, i.e. "major nuisance"). It should be remembered, however, that the situation in Curitiba was not any different in the beginning.

The institutional power of the mayor and his administration also plays a decisive role. It must be clearly stated that the success of TransMilenio

Fig. 13.9 TransMilenio has become part of the urban identity of Bogotá (Photo: Jan Marco Müller)

has only been possible thanks to a fortunate – and, in Colombia, extremely rare – continuity in municipal policies, despite the short mandates of mayors. Bogotá elected several independent and highly capable mayors in a row: Jaime Castro (1992–1994) restored the finances of the Capital District, Antanas Mockus (1995–1997) developed the city budget and fostered "civil education" and Enrique Peñalosa (1998–2000) spent funds wisely, implementing his vision of sustainable development. He was followed by Antanas Mockus in his second term (2001–2003) who continued this strategy of encouraging public savings and supporting sustainable development.

It should not be forgotten that during this period, changing national governments stayed committed to their promises. Although the goal of completing TransMilenio by 2016 has proved to be too optimistic (the TransMilenio website now mentions the year 2028), it is clear that without the continuity mentioned above, TransMilenio would never have become the success story it is today. Besides, the use of the unique "window of opportunity" that opened up under Mayor Peñalosa, during the presidential term of former Bogotá Mayor Andrés Pastrana, contributed to making TransMilenio a success story.

The importance of continuity in local politics can be demonstrated by the fact that the two recent administrations of Lucho Garzón (2004–2007) and Samuel Moreno (2008–2011) gave less importance to TransMilenio, thus considerably slowing down its implementation. While the implementation of the system is moving ahead in principle, the ambitious development plans of previous administrations have been cut back. Recent mayors revived the metro debate, although such a costly system, which certainly remains an option in the long term, would not help to solve any of Bogotá's most pressing transport problems in the short run. Despite these drawbacks, the professional implementation using a fully participatory and transparent approach has anchored TransMilenio so deeply in the Bogotans' hearts that the city will never go back to pre-TransMilenio times (see Fig. 13.9).

Like Curitiba in the 1970s, Bogotá can serve as an example of best practice. It is surprising to see that even the Federal Transit Administration, an operating administration within the US Department of Transportation, published a study entitled "Applicability of Bogotá's TransMilenio BRT System to the United States" in 2006, clearly demonstrating that the Bogotá model is seriously being considered, even in countries of the Northern Hemisphere. In fact, several North American cities have recently invested into Bus Rapid Transit, for example, Boston, Los Angeles and Las Vegas. The US study qualifies TransMilenio as "one of the world's premier

Bus Rapid Transit (BRT) systems" and comes to the remarkable conclusion (Federal Transit Administration 2006: XIII):

> Using average weekday ridership per mile as a measure of system performance, it was found that TransMilenio Phases I and II outperform the U.S. heavy rail systems by at least 3 times, U.S. Light Rail Transit by at least 12 times, and U.S. Bus Rapid Transit by at least 4 times. [...] It can be said that the capital cost effectiveness of TransMilenio is likely to be superior to most transit investments of any mode in the U.S.

Acknowledgement The author wishes to acknowledge the support provided by TransMilenio S.A. during the investigation.

References

Ardila Gómez A (2004) Transit planning in Curitiba and Bogotá. Roles in interaction, risk, and change. PhD thesis, Department of Urban Studies and Planning. Massachusetts Institute of Technology, Cambridge, MA

Barrios VP (2002) Influye la accesibilidad en la renta inmobiliaria? – Estudio del caso del sistema de transporte masivo TransMilenio Bogotá. Master's thesis. Facultad de Economía, Universidad de los Andes, Bogotá

Departamento Administrativo de Planeación Distrital (ed) (1997) Estadísticas Santa Fe de Bogotá. DAPD, Bogotá

Departamento Administrativo de Planeación Distrital (ed) (1999) Plan de Ordenamiento Territorial de Bogotá. Documento Técnico de Soporte. DAPD, Bogotá

Federal Transit Administration (ed) (2006) Applicability of Bogotá's TransMilenio BRT System to the United States. Final Report – May 2006. FTA, Washington, DC

Graftieaux P (2006) TransMilenio and Transantiago: Similarities and differences. http://web.worldbank.org/WBSITE/EXTERNAL/TOPICS/EXTTRANSPORT/0,,contentMDK:21069690~pagePK:210058~piPK:210062~theSitePK:337116,00.html

Gschwender A (2006) Transantiago: Das neue integrierte Bussystem für Santiago de Chile. Stadtverkehr 3: 30–39

Hidalgo D, Hermann G (2004) The Bogotá model for sustainable transportation: inspiring developing cities throughout the world. TRIALOG – Zeitschrift für das Planen und Bauen in der Dritten Welt 82(3): 11–15

Hörmann F (2008) Mobilität und soziale Exklusion in Megacities – Eine explorative Untersuchung am Beispiel des TransMilenios in Bogotá, Kolumbien. "Diplom" Thesis, Ludwig Maximilian University Munich (unpublished)

Japan International Cooperation Agency (ed) (1992) The study on air pollution control plan in Santafé de Bogotá city area – Final Report. JICA, Bogotá

Müller JM (2001) Struktur und Probleme des Verkehrssystems in Kolumbien – Ein integriertes Verkehrskonzept als Voraussetzung für eine dezentralisierte Regionalentwicklung unter neoliberalen Bedingungen. PhD thesis, Department of Geography, University of Marburg. Marburger Geographische Schriften 137: 264 pp. Marburg

Müller JM (2002) Das TransMilenio-System in Bogotá/Kolumbien: Ein innovativer Beitrag zur Lösung von Mobilitätsproblemen in Megacities? In: Kagermeier A, Mager TJ, Zängler TW (eds) Mobilitätskonzepte in Ballungsräumen. Studien zur Mobilitäts- und Verkehrsforschung 2: 287-304. Verl. MetaGIS-Infosysteme, Mannheim

Peñalosa E (1997) Programa de Gobierno para la elección del Alcalde Mayor de Bogotá, D.C. 1998–2000 por el Movimiento "Por la Bogotá que queremos". Bogotá

Perea R (2002) TransMilenio tendrá 15 buses más. El Tiempo, 25 July 2002

Targa F (2003) Examining accessibility and proximity-related effects of Bogotá's bus rapid system using spatial hedonic price models. Department of City and Regional Planning, University of North Carolina, Chapel Hill

TransMilenio SA (ed) (2000) TransMilenio – Un Sistema de Vida. Alcaldía Mayor de Bogotá, Bogotá

Further Sources

http://www3.giz.de/E+Z/content/archiv-ger/01-2004/inw_art4.html

http://www.itdp.org/

Interview with Angélica Castro, Head of Planning Department, TransMilenio S.A. (29 December 2001)

Interview with Germán Camargo, Head of Planning Department, Capital District Environment Agency DAMA (16 January 2002)

TransMilenio website: www.TransMilenio.gov.co

Part III

Governance and Management

Governance in Megacities: Experiences, Challenges and Implications for International Cooperation

14

Shipra Narang Suri and Günther Taube

Abstract

Megacities have moved from being associated predominantly with developed countries to a mainly developing-country phenomenon, which also reflects the overall trend and pattern of urbanisation in the last few decades. In this chapter, we explore major issues, experiences and challenges connected to megacity governance, with particular emphasis on megacities in emerging market economies and developing countries. While megacities in these countries are important drivers of national economies and international globalisation, they are also homes to rising inequalities and conflicts between different interest groups. Their governance assumes crucial importance in order to manage these increasing disparities and ensure that megacities become more equitable as well as efficient and sustainable as well as productive. This chapter also explores the evolution of the concept of urban governance, highlighting the governance challenges faced by megacities. It concludes by providing recommendations that could help making megacities in emerging markets and developing countries more inclusive and sustainable, emphasising the role of international cooperation in this regard.

Keywords

Megacities • Urban development • Governance • International cooperation • Developing countries

14.1 Introduction: The Urban Millennium and the Growth of Megacities

In the second half of the last century, the number of people living in urban areas increased from 750 million to 2.86 billion. In 2007, for the first time, more than half of the world's population was living in cities. By 2030, this proportion is estimated to rise to 61%, or nearly 5 billion

S.N. Suri (✉)
Department of Politics, University of York, UK
e-mail: shipra.narang-suri@york.ac.uk

G. Taube
InWEnt – Capacity Building International, Germany

F. Kraas et al. (eds.), *Megacities: Our Global Urban Future*, International Year of Planet Earth,
DOI 10.1007/978-90-481-3417-5_14,

people (United Nations 2004). Humanity's future seems to be urban, and the trend towards urbanisation irreversible. The number of megacities has also grown by leaps and bounds over the last three decades. Indeed, the very concept of a megacity was nebulous until the 1970s. It was then that the United Nations first defined a megacity as an 'agglomeration with a threshold of 10 million people'. In 1975, only five cities of the world passed this threshold. In the year 2000, this number rose to 19, with 8.8% of the world's population residing in these cities. Interestingly, only 4 of these 19 cities are in the developed north – New York, Los Angeles, Tokyo and Osaka. The other 15 belong to the developing world (UN HABITAT 2003a).

This trend in the growth of megacities is a reflection of the growth of the urban population in general. Urban populations are expected to rise much more rapidly in the less developed regions of the world, averaging 2.3% per year during the period 2000–2030. The most worrying aspect of this growth is the fact that it leads to the 'urbanisation of poverty'. Today, about one-third of the world's urban population – nearly one billion people – lives in slums. Asia has about 60% of the world's slum dwellers, Africa some 20% and Latin America around 14% (UN HABITAT 2003b). What these and other figures show is that megacity development is of particular relevance in Asia, and within Asia, and within in China and India in particular, where most megacities can be found.

14.2 Megacity Ambivalence

Megacities are of ambivalent character. On the one hand, they are powerful engines of economic growth and dynamic centres of globalisation. They are characterised by innovation, social change, action, modernity and – last but not least – improved living conditions. Cities are dynamic spaces, providing important economic, social and cultural opportunities for urban populations as well as the hinterland, often driving national economies (Narang and Reuterward 2006). In general, megacities have much higher productivity and economic growth rates than the national average, and cities like Lagos contribute nearly two-thirds of the national GDP (Prud'homme 1996; Taylor 2006). In all of the large Asian countries, the urban share of GDP ranges between 50 and 90% of the national GDP (ADB 2006). São Paulo in Brazil and Bangkok in Thailand, respectively, produce more than 40% of the national GDP (UN HABITAT 2006).

Reflecting the increasing importance of countries like India, China or Brazil in the world economy, a number of megacities from these countries are now also included in a category called 'global cities' or 'world cities' – denoting urban agglomerations that are global powerhouses. Beijing, São Paulo, Shanghai and Seoul, as well as Moscow, can be put into this category nowadays, next to New York, Tokyo, London and Paris. It may not be long before Delhi and Mexico City join this club. On the other hand, megacities pose severe challenges to their inhabitants, the environment and governments. Megacities leave 'megalarge', and often irreversible, ecological footprints in their surroundings and distant regions with huge natural resources consumption and disposing of their waste and other pollutants.

The environmental consequences of megacity development, including global warming, as well as the economic costs of environmental damage, are tremendous. Megacities also give rise to social tensions because of increasing income disparities, growing poverty and social exclusion. Megacities do not only grow; they also tend to become increasingly fragmented – gated communities for the small rich minority on the one extreme and slum settlements for the large poor majority on the other. Finally, local and national governments in many developing countries – including those that are doing well economically like Brazil, China, India and Mexico – have limited institutional capacity to cope with the ever-increasing demand for housing, infrastructure and services, including water, health, education and security. Moreover, this limited institutional capacity is often spread inefficiently over a large number of public bodies, often with overlapping jurisdictions and mandates that tend to lack coordination. Megacities do not pose challenges to those who seek to govern them only in terms of sheer size and growth but also in terms of increasing complexity. Taking this point further, Buehler (2003) argues that each megacity is unique due to a range of global, national and local influences which also increase a factor he calls 'uncertainty'. For example, complexity relates to an ever-increasing multitude of diverse actors – ranging from multinational companies or 'global investors' to poor households in slums – as well as to globalisation and rising mobility and competition.

In fact, what can be observed in many megacities (in the developing world) is that the room for manoeuvre for (national) governments appears to be shrinking due to limited institutional capacity and weak city finances, while the influence of private and informal actors seems to be on the increase. This can take the form of public-private partnerships (PPP) in providing infrastructure, or it can come about through self-help neighbourhood groups who join hands in providing services which the city government is unable to deliver. Furthermore, the power of national governments to rule megacities is shrinking because of the general trend towards decentralisation in many countries, involving the devolution of spending and taxation powers to local governments. At the same time, intergovernmental fiscal arrangements in many countries seem to be ill-equipped for the rapid changes and developments that megacities are facing.

14.3 The Shift from Management to Governance of Megacities

In this section, we explore the change in the conceptual thinking from urban management to urban, or metropolitan, governance. In 1996, Prud'homme argued that what counts is not the size of the city, but the quality of its management. He went on to emphasise that the most critical element of good urban management was coordination – internal (between the various city stakeholders), vertical (between the central government and local governments) and horizontal (between the various local governments that make up a megacity) (Prud'homme 1996). The Urban Management Programme (UMP) of UNDP and UN HABITAT, which ran from 1986 to 2006, played a major role in building up a constituency for improved urban management, particularly for addressing issues of poverty alleviation and environmental sustainability. At this time, the concept of urban management was seen as an innovation to replace the old-fashioned and ineffective urban planning approaches in cities and to make cities more productive and efficient (Biau 2005; Mehta 2005).

More recently, the term 'urban governance' has replaced 'urban management' in the international development lexicon. To some extent, this reflects that over the last decade, there has been a growing interest in the concept of (good) governance. Governance is defined in many ways, but all definitions focus on the relationships between the state, civil society and the private sector. The key aspects to be emphasised are that (a) governance is not government and (b) governance emphasises 'process' (UN HABITAT 2002; OECD 2000; ODI 2006). There is now broad international agreement on the notion that good governance is a crucial prerequisite for poverty reduction. For example, UNDP's 2000 Poverty Report calls good national governance the 'missing link' between antipoverty efforts and poverty reduction (see UN HABITAT 2002). ODI (2006) asserts on the basis of empirical governance assessments in 16 countries that better governance leads to greater socio-economic development, higher investments and growth rates. More broadly, Amartya Sen has argued that good governance, with its emphasis on participation, equity, transparency and accountability, is not just a means to economic development but also an end in itself (Sen 1999).

UN HABITAT's Global Campaign on Urban Governance, which ran from 1999 to 2007, put forth the following definition of 'urban' governance: "*Urban governance* is the sum of the many ways individuals and institutions, public and private, plan and manage the common affairs of the city. It is a continuing process through which conflicting or diverse interests may be accommodated and cooperative action can be taken. It includes formal institutions as well as informal arrangements and the social capital of citizens" (UN HABITAT 2002). The campaign also proposed seven principles of 'good' urban governance, described below, which are not very different from the OECD Principles of Metropolitan Governance (OECD 2000):

- Sustainability – balancing the social, economic and environmental needs of present and future generations
- Subsidiarity – assigning responsibilities and resources to the closest appropriate level
- Equity of access to decision-making processes and the basic necessities of urban life
- Efficiency in delivering public services and promoting local economic development
- Transparency and accountability of decision-makers and all stakeholders
- Civic engagement and citizenship – recognising that people are the principal wealth of cities and both the object and the means of sustainable human development
- Security of individuals and their living environment

14.4 Challenges for Megacity Governance

Against the background of these general principles of good governance, the above-mentioned megacity trends in emerging market economies and developing countries pose a number of challenges. In this context, key issues are associated with agglomeration or metropolitanisation, the challenges of ensuring participation and inclusion, problems related to urban planning in megacities and the challenge of developing governance indicators for measuring performance. In this section we will explore in detail only the first two of the above-mentioned factors. Issues of urban planning and urban governance indicators, though increasingly on the radar of practitioners and researchers alike, have a large range of sub-issues associated with them and are outside the scope of this chapter.

A major challenge for megacities is what has been called agglomeration, metropolitanisation (UN HABITAT 2004) or 'polycentric cities' (Leautier 2006). The World Bank research based on empirical data from more than 400 cities using indicators related to globalisation and governance has shown that globalisation and good city governance are related and that there appear to be dynamic pressures from globalisation and accountability that result in a better performance of cities. The same research also points out a growing importance of 'agglomeration economies', which would challenge the conventional view that larger cities exhibit a worse performance than smaller cities (Leautier 2006). At the moment, the largest agglomeration of this type exists in China, where the Pearl River Delta with its roughly 48 million inhabitants can also be characterised as an emerging 'world factory' (Kraas and Nitschke 2008).

Other megacities also consist of different municipalities whose activities, ranging from urban planning to infrastructure delivery and service provision, are not well coordinated. Even traffic rules and their enforcement, for example, differ significantly across the various municipalities that form Metro Manila in the Philippines. Clearly, aspects such as the above-mentioned call for less rather than more decentralisation of governance structures at the megacity, metropolis or agglomeration level. Establishing or strengthening intermediate government levels to deal with such problems is a political challenge as it would result in not only weaker local governments but also strong competition for regional or provincial and even national governments. The governance challenge clearly is to be able to adhere to the principle of subsidiarity described above, which emphasises the assignment of responsibilities as well as resources at the lowest appropriate level, which may not necessarily be a well-established government level. The aim should be to foster effective, contextually relevant, decentralised governance models in these agglomerations.

A second key challenge is to ensure participation, especially in the context of increasing informality and diversity of actors. The instrumentalist conception of participation argues that participation is important in order to increase efficiency, effectiveness and responsiveness of development decisions to citizens' needs, as well as improve and enhance implementation. The empowerment perspective views participation as a fundamental right of the citizen and an essential ingredient of democracy. Buehler (2003) argues that the two are intertwined and mutually reinforcing. However, it is extremely difficult to ensure participation of all stakeholders when the numbers involved are as mind-boggling as they are in megacities. The most important and the most frequently emphasised aspect of governance is the inclusion of new actors – city governments working in partnership with the private sector as well as with NGOs in slums (Herrle et al. 2006). Encouraging local initiatives in service provision offers an important opportunity for empowerment and democratisation, thus improving the quality of overall governance. Therefore, strengthening participation and ensuring inclusion of all population groups can be a key driver of better economic and political governance, in megacities as well as elsewhere.

14.5 Implications for International Cooperation

What can international cooperation contribute to strengthen the 'ingredients of successful megacities' (Taylor 2006) in emerging market economies and developing countries? In general, it would appear that simply by considering the numbers and the increasing problems related to urban poverty and exclusion, more attention to megacities in emerging market economies and developing countries is required. This can take a variety of forms. One is to step up aid for city development, for example, through well-established bilateral and multilateral channels but perhaps also through new initiatives that seek to exploit synergies between these two aid channels. Alongside those urban networks and initiatives that already exist (e.g. Cities Alliance), the City Development Initiative for Asia (CDIA), launched by the Asian Development Bank (ADB) and the German Development Cooperation, could play a role. The initiative seeks to facilitate international dialogue, training and advice with a view to enhance the problem-solving capacity of decision-makers in Asian cities while supporting them in preparing municipal infrastructure investment and mobilising financing. It is an innovative approach as it seeks to combine the relative strengths of one bilateral donor (capacity building through advisory services, training and dialogue) with that of a multilateral donor (financing).

In general, city-to-city cooperation across borders as well as peer learning would be a third avenue that should be used more extensively. There are many unexplored benefits that could be tapped through better networking. One such example would be to use existing national networks (say the Council of Mayors in China, CAM and the equivalent German Association of Cities – Deutscher Städtetag) which could potentially link city officials, including through the use of modern media such as chat rooms on the Internet. InWEnt, for example, seeks to increasingly use the several hundreds of alumni – Chinese mayors and other city officials who have been trained in town planning, municipal finance and other issues over the past 25 years – as facilitators and networkers through CAM to reach out to other city officials in China. Such networks as well as other channels allow learning from good or best practices, of which there exist an astonishing number and variety. A good collection of examples from Asia and elsewhere, plus many references to other good sources of good or best practices, can be found in a recent publication of the Asian Development Bank (ADB 2006).

Finally, megacity governance should also be given more prominence in the work of international agencies such as the World Bank and the UN. In case of the latter, UN HABITAT's World Urban Campaign, along with its network of global and regional partners, could be an appropriate vehicle to promote better governance in megacities of the south, mainly through policy reform and capacity building. Megacities are here to stay, and international cooperation can – and, indeed, must – play a major role in ensuring that they are not just successful for a few individuals, for a time, but sustainable in the long run and inclusive for all groups of society.

References

ADB (2006) Urbanization and sustainability in Asia. ADB, Manila

Biau D (2005) Two decades of urban management. Habitat Debate 11:8

Buehler R (2003) Urban development in megacities in developing countries: potentials of citizen participation in planning and managing urban development (unpublished)

Herrle P, Jachnow A, Ley A (2006) Die Metropolen des Südens: Labor für Innovationen? Mit neuen Allianzen zu besserem Stadtmanagement. Stiftung Entwicklung und Frieden, Policy Paper 25. Bonn

InWEnt (2007) Governing emerging megacities – challenges and perspectives. International Dialogue, Bonn

Kraas F, Nitschke U (2008) Megaurbanisierung in Asien: Entwicklungsprozesse und Konsequenzen stadträumlicher Reorganisation. In: Themenheft "Raum- und Stadtentwicklung in Asien". Informationen zur Raumentwickl 8–2008:447–456

Leautier F (ed) (2006) Cities in a globalizing world. Governance, performance and sustainability. World Bank, Washington, DC

Mehta D (2005) Our common past: the contribution of the urban management programme. Habitat Debate 11:6–7

Narang S, Reuterward L (2006) Improved governance and sustainable urban development: strategic planning holds the key. Eur J Spatial Dev

ODI (2006) Governance, development and aid effectiveness: a quick guide to complex relationships (ODI briefing paper). Overseas Development Institute, London

OECD (2000) The reform of metropolitan governance (Policy brief). OECD, Paris

Prud'homme R (1996) Managing megacities. Le courrier du CNRD 82:174–176

Sen A (1999) Development as freedom. Knopf, New York

Taube G, Nitschke U, Peters G (2006) Nord-Süd-Partnerschaften für urbane Entwicklungszusammenarbeit. Polit Ökol 101:68–69

Taylor P (2006) The role of megacities in development – opportunities and challenges, InWEnt international dialogue on "governing megacities: challenges and perspectives". Frankfurt

UN HABITAT (2002) The global campaign on urban governance: concept paper. UN HABITAT, Nairobi

UN HABITAT (2003a) Slums of the world: the face of urban poverty in the new millennium? UN HABITAT, Nairobi

UN HABITAT (2003b) The challenge of slums: global report on human settlements 2003. Earthscan, London

UN HABITAT (2004) State of the world's cities 2004/5. Earthscan, London

UN HABITAT (2006) State of the world's cities 2006/7. Earthscan, London

United Nations (2004) "Executive summary." World urbanization prospects: The 2003 Revision. Population Division, Department of Economic and Social Affairs, United Nations, New York

Managing the Metropolis: Planning in New York City

15

Werner Gamerith

Abstract

Planning and managing an urban cosmos like New York City face some tricky challenges. This chapter starts by analyzing some New York peculiarities in urban planning, for example, the fact that metropolitan management in the New York case extends across three different states. The situation is further complicated by the use of sometimes contradicting methods as well as different attempts to outline the area that could be termed metropolitan New York. Topography and physical shape were often responsible for tranquillity and isolation of different settlement niches, at least in the city's history. This may explain the deeply rooted localisms that are scattered all over metropolitan New York, making urban planning more difficult. In regard to planning, the chapter resumes there is sort of an organizational pandemonium in the New York area, with a sometimes stifling coexistence of both public and private agents. More often than not, New York planning has been dominated by individual persons, not always to the city's benefits. Finally, the chapter looks at some major fields of urban planning in recent decades, mainly architecture and transport. It concludes by shortly discussing the problems that are to be expected for the new millennium.

Keywords

New York • Manhattan • Urban planning • Urban history • Urban geography

Although, at least compared to the point of origin from which urban form took shape in America, planning the American metropolis is a rather recent phenomenon dating back only to the beginning of the past century, it includes an array of institutions, agencies, and interests. In many ways, managing urban systems in the United States differs significantly from European standards. By focussing on one of the most essential settings of American urban form, on New York City, this chapter wants to highlight some general difficulties of managing large metropolises in the United States. Then, some major arenas of

W. Gamerith (✉)
Fach Geographie der Universität Passau, Regionale Geographie, Innstraße 40, D-94032 Passau, Germany
e-mail: werner.gamerith@uni-passau.de

F. Kraas et al. (eds.), *Megacities: Our Global Urban Future*, International Year of Planet Earth,
DOI 10.1007/978-90-481-3417-5_15, © Springer Science+Business Media B.V. 2014

urban planning, once again in connection with America's biggest city, are depicted before concluding with a closer look at the prospective opportunities for dealing with New York's future problems in the new millennium. One should point out that due to its limited length, this chapter can only deal with a few exemplary aspects of New York's planning agencies and procedures.

15.1 Managing the Metropolis: New York Peculiarities

15.1.1 Overlapping Administrative Competencies

In most American metropolitan areas, the sheer number of administrative units increases the necessity for coordinating planning and management services. No matter whether you consider Chicago, Denver, Los Angeles, or New York, several hundreds of special-purpose districts cover entire metropolitan regions. While these entities sometimes compete with each other, their territories usually overlap, leading to additional potential conflict. Contrary to the norms in many European countries, administrative divisions may crisscross other units even at a superior level. School districts – even if they normally do not intersect state borders – do not necessarily coincide with other administrative divisions. This explains why American metropolitan areas appear as an intricate web of judicial, tax, fire, school, waste disposal, sewage, sanitation, and hospital districts. The New York City region is no exception to this, apart from the fact that metropolitan New York stretches across 24 counties in three states (New York, New Jersey, and Connecticut), which makes intra-regional planning an all the more complicated and sophisticated process.

Propositions to delineate the New York urbanized area have subsequently been manifold. As inconclusive as they are, they leave no doubt that New York still constitutes *the* all-American city leading the United States' urban system (Savitch 1988), even though Los Angeles is coming close to displacing New York. What has also become obvious is the city's shrinking population and its declining share of the total population of the metropolitan area. New York was particularly affected by this demographic trend from the 1960s to the 1980s (Drennan 1991). Only in recent years, New York City is once again increasing in population.

That managing an urban leviathan like New York can be a difficult task is well illustrated by the various attempts at grasping the complete size of the metropolis. Where does New York, defined as a functional not as a political term, end? And which areas on the city's fringes should be excluded from the management and planning processes? Answers to questions like these will be hard to obtain, and New York planning officials have decided to define their city according to the needs of the specific planning organs. This is why the Port Authority of New York and New Jersey, for instance, operates with an interpretation of the city quite different from that of the Regional Plan Association (RPA). According to the Port Authority, New York covers an area of 3,900 square miles in 17 counties, while the Regional Plan Association considers more than 12,700 square miles in 31 counties as the New York urban region which extends into three states (Fainstein and Harloe 1992).

15.1.2 Physical Shape and Consolidation

Contrary to Houston, Atlanta, or even Chicago, New York's physical terrain is far from being homogeneous, flat, and monotonous. Shaped by Pleistocene glaciation, the Atlantic coastline in the New York area shows a complex variety of islands, peninsulas, bays, and rivers. Only one of New York City's five boroughs is naturally part of the mainland (the Bronx); the other four boroughs form islands of their own or are part of a larger island. Branding New York as "America's Venice" is uncommon, yet this would actually not be too far-fetched, particularly since the city was water-bound until the 1950s. Today's city planning only rarely incorporates this aspect once characteristic for New York.

Abstaining from New York's water-bound and maritime history appears even more unsatisfying considering the advantages the city was bestowed by the region's physical shape from the very beginning. New York would never have managed to dominate the American urban system if topography had not provided an ideal location, especially in regard to nautical needs. "Fertile soil, rich waters, forests abounding with animals – all at the edge of a harbour vast and protected – combined to give seventeenth-century Europeans visions of a new Eden" (White 1987, xviii).

As early as the eighteenth century, travellers had recognized New York's benefits from the settlement's pristine position at the southern tip of an island that would later be named Manhattan, on the shores of a sizeable river connecting the Atlantic with a vast interior. In 1748, a Swedish visitor commented, "The situation of [New York] is extremely advantageous for trade: for the town stands upon a point which is formed by two bays; into one of which the river Hudson discharges itself, not far from the town; New York is therefore on three sides surrounded with water. [...] The port is a good one: ships of the greatest burthen can lie in it, quite close up to the bridge: but its water is very salty, as the sea continually comes in upon it; and therefore it is never frozen, except in extraordinary cold weather. This is of great advantage to the city and its commerce; for many ships either come in or go out of the port at any time of the year" (Pollara 1997, 155). Indeed, Upper Bay in the Hudson estuary is the northernmost natural port on North America's Atlantic coast that normally does not freeze in winter; additionally, low tides there facilitate navigation (Stern 1989; Moss 1980).

While this argument does not intend to support blatant geodeterminism, it appears obvious that physical shape helped to configure what would later become a political entity named "Greater New York". Initial settlements that gave birth to the future boroughs were centred on specific islands and peninsulas within the tidal marshes of the Hudson and East River systems. Though shipping as the principal mode of transportation until the first decades of the twentieth century made communication between the different towns on Manhattan, Long Island, and Staten Island possible, isolation was more or less common. Remoteness and tranquillity, intensified by a jagged coastline with secretive coves and inlets, helped to promote localism and neighbourhood pride among residents. This in turn fostered local identities and affiliations which are so prominent in New York City politics until the present day and which also (threaten to) impede urban planning and regional management in many ways.

Although ideas of a "Greater New York" already came up in the late eighteenth century (Hammack 1987) and the merging of Brooklyn and Manhattan had already been proposed in 1833, 1850, 1851, and 1856 (Schoener 1998), it took the political representatives until 1898 to complete "consolidation" in the New York region. Arguing for consolidation had purely economic reasons and was driven by a rationale for planning. Efforts to manage and improve traffic coordination formed the basis for the initiative to join New York proper with Brooklyn and other cities in the region. Also, the completion of the Brooklyn Bridge in 1883 and the unification of the school system served as an impetus for fusing New York with Brooklyn (Berrol 1997). No wonder the Chamber of Commerce played a vital role in disseminating the idea of consolidation on a most comprehensive level which was intended to include all New York State areas along the mouth of the Hudson, including Staten Island and the west of Queens (Hammack 1987).

Even though the New York metropolitan region had consisted of more than 40 local governments prior to consolidation, attempts at merging New York with Brooklyn and some surrounding areas were met with remarkable resistance, particularly on the Brooklyn side of the East River. However, linking the two cities seemed inevitable, especially due to Brooklyn's precarious financial situation which made seeking shelter in New York's substantial budget an attractive option (Hammack 1995). Yet, opposition to consolidation was not over after the merging in 1898. Sceptical Brooklynites had founded the "League of Loyal Citizens" which had the goal of achieving a new separation from New York and even upheld the possibility of a "Greater Brooklyn" instead of a "Greater New York"

(Hammack 1987). In Brooklyn, advocates were nearly on a par with opponents of the consolidation process. 64,744 votes for merging with New York opposed 64,467 votes against this (Kaplan 1979).

However narrow the triumph of the consolidation petitioners had been, consequences were significant. With a stroke of the pen, New York City nearly doubled from two million to almost three and a half million people. Consolidation not only meant that the city had grown bigger than Paris, Berlin, Tokyo, or Vienna but also could now compete with London as the most important metropolis in the world; only a few decades later, New York City had exceeded the British capital. Consolidation also stood for the improvement in planning and management processes on a basis that now encompassed most of the urbanized area, with the still notable exception of New Jersey. If it had not been so far, urban planning would now become a political necessity and was endowed with public funds. However, some parts of the city decided to conserve their own identities either on the neighbourhood or on the borough level. Still today, Queens is a matter of case: people there often refer to Manhattan as solely "New York", while they do not want to get their neighbourhood too closely associated with "New York". Even the way letters are addressed reflects this perception: it's not "Queens" one finds on the letter, but "Astoria", "Corona", or "Jackson Heights". Any such highly symbolic relationship with the neighbourhood has of course a somehow deleterious effect on a citywide process of urban planning and management.

15.1.3 Public and Private Planning Agents

During the nineteenth century, New York practically lacked any efficient instrument for planning land use and residential building (Buck and Fainstein 1992). Many New Yorkers felt that restricting specific forms of land use as well as enforcing building codes and regulations represented an attack on their constitutional right to private property. Thus, it took considerable time to convince the local elite that "laissez-faire" in building and architecture would be harmful to the city and that even developers might be disadvantaged by a completely unrestricted urban expansion. Although the city issued public stipulations in regard to hygiene, building standards, and the location of industries as early as the mid-nineteenth century (Gilmartin 1995), the influence of private people and organizations on New York's growth and planning processes is far from insignificant. In general, New York planning has profited from personal engagement by concerned citizens and associations alike. Yet, this combination of public and private initiatives makes the city's planning history and practice a rather complicated one in which numerous fractions compete with each other. Unravelling the web of personal and institutional planning relations is a tricky task that goes far beyond the limits and the aims of this chapter. Thus, a more or less cursory view must suffice here.

Facing an exploding and ever disorderly city, it was primarily the *Municipal Art Society* (MAS), founded in 1892, that efficiently launched projects and proposals for a better, cleaner, and more beautiful New York. In supporting the construction of monuments and statues, the MAS was one of the main forces to embellish the city. Not only in a proactive but also in a prohibitive manner, the MAS engaged in a battle against billboards threatening to flood Manhattan's public spaces, including trams and stations. "Billboards, placards, and posters sprang up wherever space could be rented" (Gilmartin 1995, 139). More often than not, the Municipal Art Society had to surrender to judicial verdicts favouring individual liberty and not the aesthetic quality of the city. "The Municipal Art Society tried to limit billboards by banning sky signs (unconstitutional), by controlling their height (which didn't work out), by getting them taxed (which the city wouldn't do), and by shaming advertisers (who proved to be fairly shameless)" (Gilmartin 1995, 148). Another controversy propelled by the MAS focussed on the delimitation of public space; here the MAS fought for an exact definition of the building line, between private houses and public streets, but – once again – failed in establishing a consistent solution. Still today, some privately used porches and yards remain on what is technically city property. Legally, these "grey" areas belong to the city even though they are privately occupied.

While the MAS often employed a fervent moralism when arguing for the embellishment of the city, the likewise private and non-profit *Regional Plan Association* (RPA), founded in 1922, emphasizes quality of life and economic competitiveness, covering 31 counties which stretch as far as New Jersey, Upstate New York, and Connecticut. With such spatial composition in mind, the RPA's suggestions for managing urban challenges are more encompassing than the MAS's concerns mainly for Manhattan and the city of New York. Until today, the RPA's aims comprise a multitude of issues ranging from mass transport and the improvement of metropolitan transportation to natural preservation and control of suburban expansion (Buttenwieser 2005). The Second Regional Plan, published in 1968, focussed on inner-city revitalization and provided a blueprint for the later accomplishment of structures such as the World Trade Center, numerous new arcades and plazas in Manhattan, and the redevelopment of downtown Brooklyn and of the region's waterfronts. The Third Regional Plan (1996) envisaged the stimulation of the region's economy. Recently, RPA has also contributed solutions to the question of how to rebuild parts of Lower Manhattan in the aftermath of 9/11.

The RPA's First Plan in 1929 proved to be of particular significance. Not only was it one of the most expensive schemes for urban planning ever drafted in the United States (Johnson 1996), but it also legitimized the agenda of Robert Moses who as "America's greatest builder" (Hall 1988, 229) transformed the New York metropolitan area more profoundly than any other planner before or after him. Convinced with unmatched providence, Moses gratefully picked up the ideas of the RPA in regard to parks and highways. Until World War II, the RPA's First Plan helped Robert Moses and the acceptance of his intentions by authorities and firms (Johnson 1985).

The case of Robert Moses serves to illustrate the strong connections between public planning and private figures in the New York arena. Even though Moses acted as a public official, it was his own personality that endowed him with prominence across the nation. "Beginning in the 1920s and extending into the 1970s, his name, or fingerprints, could be found in the planning and implementation of hundreds of public works projects […]" all over the New York metropolitan region (McDonald 2005, 1210). Only in the post-World War II decades, "[…] the mood in New York City turned against Moses's highways […]" (Schwartz 2005, 1015). Neither before nor after the era of Moses has an urban planner ever been so influential nor has the process of planning ever been so tightly connected with an individual and his or her own, sometimes colloquial expectations. Politically, Robert Moses survived eight mayors and five governors (Savitch 1988) – a fact which once again underlines his outstanding role in politics facing little opposition. Even in controversial issues like bulldozing lower-income housing projects and replacing them by highways, Moses was confronted with only minor objections. He was, in fact, on the same level as the mayor, with even more duties and competencies (Gilmartin 1995).

15.2 Arenas of Urban Planning: New York Experiences

15.2.1 Architecture

Not solely in the case of New York architecture can be interpreted as the predecessor of urban planning. First efforts to improve the physical shape of cities were launched by journalists, politicians, doctors, and most often by architects and landscape designers. The "City Beautiful" Movement is a case in point: its proponents struggled for urban ornamentation and thereby accentuated decorative elements in the city's scenes. Urban planning came to replicate some of the architects' aesthetic claims. Yet, with the refinement of building techniques and ever-increasing building heights, mere artistic and visual scepticism transformed into stern opposition to apparently limitless construction endeavours. New York is being honoured as the first

American city to have introduced building regulations (Cullingworth 1997; Weiss 1992). Based on the recommendations by the *Commission on the Heights of Buildings* in 1913, America's first zoning law came into existence in 1916 (Richman 1985).

The zoning law primarily aimed at solving conflicts between commercial and industrial property owners in Manhattan's central business district, not at improving the situation of New York residents. Different land use categories (residential, business, and "unrestricted") were established in order to avoid future disputes. In addition, height and volume of edifices to be constructed were regulated. The law also would thwart the darkening of complete streets by excessive buildings (Moseley 1986). Consequently, a new architectural style, "jumbles of cubic setbacks and towers", resulted from the 1916 provision: "One simply couldn't make sense of these shapes with the old vocabulary of columns, pedestals, and cornices. So the Zoning Resolution ushered in, quite by accident, the great experiment of Art Deco architecture" (Gilmartin 1995, 201).

New Yorkers doggedly clung to the first zoning law in the nation. As late as the 1950s, the zoning resolution of the World War I era was still practiced, though held together by about 2,500 amendments. After long public debates, this often revised regulation was followed by a new building law in 1961 (Moseley 1986). Its most important innovation turned out to be the "floor area ratio" (FAR), a kind of incentive zoning by which the city would offer developers a bonus for additional floors, if buildings reserved open or public areas for parks or plazas. However, nobody seemed to be aware that this deal would inevitably lead into unparalleled multiplication of dull and repetitive spaces which remained empty for most of the day (Willis 1995). The lack of systematic and widely accepted architectural preservation became a preoccupation of urban planning in the 1960s – even though there had been sporadic attempts at preserving New York landmarks as early as the late nineteenth century (Mason 2004). Protests against plans to demolish venerable Pennsylvania Station in the early 1960s reverberated without success, and in 1963 demolition began. Destruction was completed 2 years later.

Yet, the Penn Station incidence supported the formation of a *Landmarks Preservation Commission* in 1965 (Pouzoulet 2000) which gained further momentum when the Singer Building was torn down. Knocking down this skyscraper – with exception of the Twin Towers, the Singer Tower was the tallest building in New York to ever be destroyed – caused a public outcry as well. Soon, preservation received significant attention as a major field of architectural urban planning. Designations as historic landmarks for single structures and historic districts for conservation areas began to thrive. In the 1990s, 856 buildings, 79 interior designs, and 9 parks had been declared special status, while 52 neighbourhoods had been designated as historic districts (Zukin 1993; Deák 2000). That declaring buildings as landmarks could save from plans of demolition through lengthy legal battles was shown in the 1970s when Grand Central Station became the target of a new development scheme. "No More Bites Out of the Big Apple", was the preservationists' motto, and public opinion helped it gain momentum. Grand Central Station was rescued in 1977, following a Supreme Court judgement forbidding skyscraper to be built towering above or located in front of the station (Gilmartin 1995).

15.2.2 Traffic

In the automobile era, a city that is made up mainly of islands and peninsulas faces substantial traffic problems. To be aware of the resulting troubles, just consider that the most important concentration of office space in the Western Hemisphere is located on a small part of a narrow island close to the estuary of one of the major rivers in the Northeastern United States. To connect this island with the mainland other than by ship proved to be an extraordinary challenge generations of technicians have tried to overcome. In the context of New York's coastline, it is no wonder that early stages of urban planning were primarily preoccupied with connecting the different parts of an ever-growing metropolitan area. The construction and administration of bridges and tunnels (Solis 2005), together with planning a system of mass transport, kept New York's urban policy busy from the 1850s far into the twentieth century. Today, urban planning is dedicated primarily to the challenge of automobile traffic.

While in the immediate post-World War II years, the policy of urban transport mainly connected to the provision of fast traffic arteries for private motor vehicles, recent decades have seen a more sensitive approach, also stressing the needs of pedestrians and public transport. It is remarkable that in terms of traffic the island of Manhattan has been turned upside down in the course of the last one and a half centuries. In the formative years, from the colonial beginnings until the advent of the private car in the 1910s, the city was oriented towards and adjusted to the ocean. Many travellers noted the busy port activities that could be observed along all of Manhattan's shores. Former South Street Seaport, located on a 2-mile stretch of the East River, is a case in point: "By the 1820s, every available space along the west side of South Street had been taken over by a bewildering parade of counting houses, warehouses, marine insurers, importers, commission merchants, jobbers, chandleries, coopers, sail makers, and taverns [...]" (Burns and Sanders 1999, 47).

After 1880, South Street lost its premier naval position for New York, partly because the East River became too shallow for ocean-going vessels. Manhattan's West Side began to proliferate itself as the main location of port facilities. As late as 1960, a fifth of all imports to and exports from the United States were handled in via New York's ports; by 1971, this share had fallen to a mere 3 % (Pries 1998). Obviously, New York had lost the leading position in port activities, partly to New Jersey and states in the South but mainly to ports elsewhere in the world. Above all, dropping figures were most visible in the transatlantic traffic: while in 1955 700,000 passengers crossing the Atlantic from Europe were counted, in 1978 only 42,000 persons arrived by ship (Moss 1980). The decline of the ports resulted in a relentless economic downturn which culminated in the 1970s. By the end of the 1970s, only three of former 42 piers along the Hudson River in Manhattan were still used as passenger terminals, ten piers served as docks or parking lots, and 17 piers were not in use or to some extent dilapidated (Seltz-Petrash 1979).

This change in transportation mode and traffic was accompanied by transformation of planning objectives. New York's water-bound character gradually changed, turning into a view of the city with a stronger orientation towards the mainland and with only minor attention to the past maritime activities on Hudson and East rivers. Thus, urban planning focussed mainly on developments linked to automobile and railway traffic. Also, along with the deterioration of the ports came the problem of how to revitalize the former port areas – another aspect urban planning began to be obsessed with from the 1970s onwards. The Chelsea Piers Complex, used for cargo until 1968, provides a good example. Finally, these piers were used as parking lots, garages for garbage trucks, and as warehouses for goods confiscated by US customs. The remaining piers were derelict, demolished, or closed (Pries 1998). In general, many plots of land along the shoreline were carved out of former port areas and were intended to be converted into streets, highways, or parking lots. Correspondingly, New York's shift from a water-bound to an inland-oriented city propelled critics of an all-too-permissive use of private vehicles.

Particularly, the plans for the Westway project on Manhattan's West Side generated ire and fury among both residents and other New Yorkers. For the city, subsidizing new highway construction through the federal highway law seemed to be more lucrative than redeveloping existing highways. "Big projects bring in big bucks" (Savitch 1988, 82). In the mid-1970s, the new Westway consisted of four lanes in each direction, making it the most expensive highway in the world. When it came out that highway entrances and exits would be cutting through residential areas, opposition to the project swelled. Local media dubbed Westway "Wasteway" (Savitch 1988). In the long run, resistance to Westway was so successful that proponents of the project had to content themselves with only a fraction of the original highway – some 6 km from 42nd Street to Battery Park (Hochstein 1982). Interestingly, it was environmental concerns that finally limited Westway plans to such a reduced length. As preservationists claimed, landfills necessary for the highway construction would have endangered the spawning area of a native fish species, the striped bass. Not quite accidentally, money already promised for Westway by the federal government went into public transport instead (Wise et al. 1997).

15.3 Planning New York in the New Millennium

Architecture and traffic are but two fields of demand for coordination and regulation since urban planning became a subject for early metropolitan New York in the late nineteenth century. Of course, there are other aspects that need to be regulated and controlled in a metropolitan setting, such as social policy, public transport, urban redevelopment, environmental planning, zoning, and land use management, among others. The limitations of this chapter do not allow an all-embracing evaluation of these fields with a decisive influence on New York's shape in the new millennium.

There are at least two aspects that will guide and probably dominate New York urban planning and management in the twenty-first century, one of which affects the whole city, while the other has more local relevance. What will be crucial to the city and the metropolitan area in general in the next couple of decades is New York's reliance on technological equipment and public facilities. In this respect, New York does not differ from any other metropolis, yet the city is especially fragile and dependent upon technological devices. First and foremost, bridges and tunnels linking the boroughs sustain public life.

With the possible exception of San Francisco and Boston, hardly any other American city is as susceptible to damage, decay, or corrosion as New York City. As soon as one or two bridges or tunnels have to be closed, something similar to a state of emergency, at least for Manhattan, becomes the norm. One should also be aware of the fact that New York's means of public communication and transportation were constructed at least a century ago. The first subway began service in 1904 (Hood 1992); the first bridge crossing the East River, Brooklyn Bridge, opened more than two decades earlier, in 1883; and the first freshwater lines were constructed during the 1840s (Gandy 1997; Koeppel 2000). The sewerage system was not completed until the late nineteenth century by crowds of cheap labourers, mainly of immigrant background. "Waterpower, steam-power, Irish-power" (Burns and Sanders 1999, 86) was one typical slogan. Thus, a huge portion of New York's facilities is no less than 100 years old, and several facilities will soon be reaching the end of their life cycle, which means that there is dire need to repair many of these tracks, buildings, sewers, pipes, and cables. Certainly, this affects the whole city as it will be an important consideration for urban planning and management in the years and decades to come.

The other dominant aspect of future urban planning that might be of more relevance to the local level (even though it has of course symbolic meaning for global capitalism, too) is situated in Lower Manhattan. It is the reconstruction of the inner-city area that came to be known as "Ground Zero" after the terrorist attacks of September 11, 2001. However awful damage and destruction done to a series of skyscrapers, the twin towers of the World Trade Center among others, might have been, not to mention the tragic loss of thousands of lives, 9/11 also opens up unexpected opportunities for new urban planning and architectural design in a part of Manhattan which is among the oldest and most crowded in the city (Sorkin and Zukin 2002; Gamerith 2002). Yet, it would not be New York had the planning process for the reconstruction of "Ground Zero" proceeded smoothly and without conflict (Marcuse 2002; Ross 2002; Sorkin 2003). An ongoing debate has evolved on how the different panels deciding on the reconstruction of Lower Manhattan should be staffed. Also, the question of how a future "World Trade Center" complex should look like in architectural and functional terms has not been settled entirely. Scores of propositions have been made for numerous new plans to improve public transport in the Lower Manhattan area and to connect the former World Trade Center site with the suburban rail system, as well as to link it to the JFK Express connection for JFK airport (Paaswell 2002).

Recreating the Twin Towers area will almost certainly monopolize New York urban planning for the next couple of years; at least it will attract any international attention in regard to New York's planning problems and processes. In view of "Ground Zero", questions not only of architecture but also of zoning, public transport, retail, and cultural politics must be debated anew. Hardly any other city in the Western Hemisphere is confronted with an urban restoration project so colossal and oversized. With New York's endeavours principally absorbed by the aftermath of 9/11, "normal" urban planning and management runs the risk of being severely neglected in the first decades of the new millennium. New York is more than just the new Freedom Tower (One World Trade Center) and some of its adjacent plots.

References

Berrol S (1997) The empire city: New York and its people. Praeger, Westport//London, pp 1624–1996

Buck N, Fainstein N (1992) A comparative history. In: Fainstein SS, Gordon I, Harloe M (eds) Divided cities: New York & London in the contemporary world. studies in urban and social change. Blackwell, Oxford/ Cambridge, pp 1880–1973

Burns RJ, Sanders J (1999) New York: an illustrated history. Knopf, New York

Buttenwieser AL (2005) Regional plan association (RPA). In: Eisenstadt P, Moss LE (eds) The encyclopedia of New York State. Syracuse University Press, Syracuse

Cullingworth B (1997) Planning in the USA: policies, issues, and processes. Routledge, London/New York

Deák G (2000) Picturing New York: the city from its beginnings to the present. Columbia University Press, New York

Drennan MP (1991) The decline and rise of the New York economy. In: Mollenkopf JH, Castells M (eds) Dual city: restructuring New York. Russell Sage Foundation, New York

Fainstein SI, Harloe M (1992) Introduction: London and New York in the contemporary world. In: Fainstein SS, Gordon I, Harloe M (eds) Divided cities: New York & London in the contemporary world, Studies in urban and social change. Blackwell, Oxford/Cambridge

Gamerith W (2002) Die Vulnerabilität von Metropolen – Versuch einer Bilanz und Prognose für Manhattan nach dem 11.9.2001. Petermanns Geogr Mitt 146(1):16–21

Gandy M (1997) The making of a regulatory crisis: restructuring New York City's water supply. Trans Inst Br Geogr 22:338–358

Gilmartin GF (1995) Shaping the city: New York and the municipal art society. Clarkson Potter, New York

Hall P (1988) Cities of tomorrow: an intellectual history of urban planning and design in the twentieth century. Blackwell, Oxford/Cambridge

Hammack DC (1987) Power and society: greater New York at the turn of the century, Columbia history of urban life. Columbia University Press, New York

Hammack DC (1995) Consolidation. In: Jackson KT (ed) The encyclopedia of New York city. Yale University Press, New Haven/London

Hochstein S (1982) The New York City Westway – non-progress through public participation. ITE J 52(5):14–18

Hood C (1992) Subways, transit politics, and metropolitan spatial expansion. In: Ward D, Zunz O (eds) The landscape of modernity: essays on New York City. Russell Sage, New York, pp 1900–1940

Johnson DA (1985) Metropolitan planning in the region: a retrospective assessment. N Y Aff 9(2):95–112

Johnson DA (1996) Planning the great metropolis: the 1929 regional plan of New York and its environs, vol 18, Studies in history, planning and the environment. E & FN Spon, London/New York

Kaplan BJ (1979) Andrew H. Green and the creation of a planning rationale: the formation of greater New York City, 1865–1890. Urban Past Present 8:32–41

Koeppel GT (2000) Water for Gotham: a history. Princeton University Press, Princeton

Marcuse P (2002) What kind of planning after September 11? The market, the stakeholders, consensus – or…? In: Sorkin M, Zukin S (eds) After the World Trade Center: rethinking New York City. Routledge, New York/London

Mason R (2004) Historic preservation, public memory, and the making of modern New York City. In: Page M, Mason R (eds) Giving preservation a history: histories of historic preservation in the United States. Routledge, New York/London

McDonald DB (2005) Planning and regional planning associations. In: Eisenstadt P, Moss LE (eds) The encyclopedia of New York State. Syracuse University Press, Syracuse

Moseley H (1986) Development control in New York City. Planner 72(12):14–18

Moss ML (1980) Staging a renaissance on the waterfront. New York Affairs 6(2):3–19

Paaswell R (2002) A time for transportation strategy. In: Sorkin M, Zukin S (eds) After the World Trade Center: rethinking New York City. Routledge, New York/ London

Pollara G (1997) Transforming the edge: overview of selected plans and projects. In: Bone K (ed) The New York waterfront: evolution and building of the port and harbor. Monacelli Press, New York

Pouzoulet C (2000) New York City as capital of the world? Or the contested fortunes of the city of capital. In: Sohn A, Weber H (eds) Hauptstädte und Global Cities an der Schwelle zum 21. Jahrhundert, vol 9, Herausforderungen: Historisch-politische Analysen. Winkler, Bochum
Pries M (1998) Die Chelsea Piers in New York. Beispiel einer erfolgreichen Waterfront Revitalisierung. Arbeitskreis "USA" der Deutschen Gesellschaft für Geographie. Mitteilungsblatt Nr 24:14–23
Richman J (1985) New York City zoning: a resolution with no resolve. N Y Aff 8(4):48–61
Ross A (2002) The odor of publicity. In: Sorkin M, Zukin S (eds) After the World Trade Center: rethinking New York City. Routledge, New York/London
Savitch HV (1988) Post-industrial cities: politics and planning in New York, Paris, and London. Princeton University Press, Princeton
Schoener A (1998) New York: an illustrated history of the people. W.W Norton & Company, New York/London
Schwartz J (2005) Moses, Robert. In: Eisenstadt P, Moss LE (eds) The encyclopedia of New York State. Syracuse University Press, Syracuse
Seltz-Petrash A (1979) New York – water city. Civ Eng 49(6):83–87
Solis J (2005) New York underground: the anatomy of a city. Routledge, New York/London
Sorkin M (2003) Starting from zero: reconstructing Downtown New York. Routledge, New York/London
Sorkin M, Zukin S (eds) (2002) After the World Trade Center: rethinking New York City. Routledge, New York/London
Stern RAM (1989) Die Erbauung der Welthauptstadt. In: Klotz H (ed) New York Architektur 1970–1990. Prestel, München
Weiss MA (1992) Density and intervention: New York's planning traditions. In: Ward D, Zunz O (eds) The landscape of modernity: essays on New York City, 1900–1940. Russell Sage, New York
White N (1987) New York: a physical history. Atheneum, New York
Willis C (1995) Skyscrapers: after 1916. In: Jackson KT (ed) The encyclopedia of New York City. Yale University Press, New Haven/London
Wise MZ, Woods W, Bone E (1997) Evolving purposes: the case of the Hudson River waterfront. In: Bone K (ed) The New York waterfront: evolution and building of the port and harbor. Monacelli Press, New York
Zukin S (1993) Hochkultur und "wilder" Kommerz: Wie New York City wieder zu einem kulturellen Zentrum werden soll. In: Häussermann H, Siebel W (eds) Strukturen einer Metropole, New York, Edition suhrkamp, Neue Folge Band 798. Suhrkamp, Frankfurt am Main

E-Governance Initiatives in India: Case Studies from Hyderabad and Bangalore

16

Christoph Dittrich

Abstract

In India, efforts to achieve smaller government through efficiency, re-engineering service systems, government transparency and an emphasis on citizen services and satisfaction have received particular attention by the wider public. The recent implementation of various e-government projects can be considered part of this wider developmental goal. This chapter focuses primarily on recent challenges in governing India's emerging megacities and is followed by an overview of e-governance initiatives nationwide. The third part examines the perspectives, successes and shortfalls of different e-governance projects in the emerging megacities of Hyderabad and Bangalore.

Keywords

Urbanisation • Decentralisation • E-governance • India • Hyderabad • Bangalore

16.1 Challenges for Governing India's Emerging Megacities

In India the percentage of people living in urban areas is still relatively low, amounting to 30%. However, this figure equals more than 330 million people and on its own would constitute the third largest nation in the world. Despite the fact that the majority of the population still lives in rural environments, India is one of the major regions affected by mega-urbanisation. Presently, India has 40 large cities/urban areas with a population of more than one million people. A total of some 130 million Indians, or 12% of the national population, live in these cities. The Indian ranking of cities by size is headed by Mumbai, Delhi and Kolkata, each of which has more than 15 million residents. According to UN estimates, Mumbai crossed the 20 million mark in 2010, Delhi about 5 years later. The 'big three' are followed by the emerging megacities Chennai, Bangalore, Hyderabad and Ahmedabad which are rapidly moving towards the 10 million line with annual growth rates of 3–4% (UN 2006).

While the Indian economy is opening up to the world market, these large cities are becoming hubs

C. Dittrich (✉)
Institute of Geography, University of Goettingen,
Department of Human Geography,
Goldschmidtstrasse 5, D-37077 Goettingen
e-mail: Christoph.dittrich@geo.uni-goettingen.de

F. Kraas et al. (eds.), *Megacities: Our Global Urban Future*, International Year of Planet Earth,
DOI 10.1007/978-90-481-3417-5_16,

of industrial and economic activities and attractive destinations for foreign direct investment (FDI). They have become the prime engines of India's booming economy and the main generators of national wealth. Projections state that their contribution to the national economy has increased from 20% in 1951 to approximately 50% in 2001. In the context of globalisation and economic liberalisation processes, India's large cities are also exposed to profound urban restructuring. They are growing much faster than their infrastructure, and their uncontrolled expansion is leading to large traffic volumes, high pollution levels, ecological overload, unregulated and disparate land and property markets, insufficient housing development and income disparities, as well as a stark contrast between extreme poverty and great wealth in close proximity to each other.

In present-day urban India, extravagant skyscrapers and super-malls next to slums reflect an increasing socio-economic polarisation and spatial fragmentation (Dittrich 2007, 2008; Heitzman 2004; Mistelbacher 2005; Sabir 2006). Due to the integration of large cities into the world market as well as the newly emerging inter-city competition for FDIs, it has become crucial for city governments to provide internationally competitive infrastructure to citizens in general and to investors in particular. The increasing flow of investments into cities requires the availability of adequate infrastructure in the form of transportation, basic amenities and housing. Investors are particularly demanding regarding the performance of city governments. Combined with the latent crisis of urban planning, this has led to a need to devolve power and authority to lower tiers of government, the Urban Local Bodies (ULBs), which are largely responsible for urban planning activities and the provision of infrastructure facilities within the city limits. Consequently, decentralisation as one of the main issues of good urban governance has become a prerequisite for making ULBs more accountable to the stakeholders.

India's decentralisation initiative – distributing power, functional responsibilities and authority to the ULBs – started with the 74th Constitution Amendment Act of 1992 that came into effect on June 1, 1993. This was the first time in the history of urban governance that municipal bodies were given the constitutional status of the third tier of government. The amendment aimed at creating a democratic governance structure at the local level. Its objective was to redefine the relationship between the states and the municipal bodies in order to firmly establish elected ULBs as institutions of self-government. The rationale of the amendment lay in the simple fact that planning and execution is most efficient when it is closest to the stakeholders and its area of impact. It also provided the Twelfth Schedule (Article 242W) which listed the functions of local urban units along with their planning, regulation and development powers. It made provisions for ward committees in areas exceeding 300,000 inhabitants, and it specified the powers and responsibilities of municipal units and the ward committees.

Elections have been held under the new dispensation in most states, in many more than once. The enthusiasm among urban residents for the ULBs has been demonstrated by the fact that the voter turnout in elections was 65–70%, much higher than the participation in elections to parliament and state legislature (GOI, Planning Commission 2005). However, experience shows that most state governments did not fully translate the required amendments into municipal laws. The distribution of competences to ULBs is still incomplete, and their financial resources are uncertain and inadequate (Bagchi 2005; Ramanathan 2005). Another important reason for the insufficient implementation of urban decentralisation policy is that most of India's large cities have special city development authorities that are in charge of all spatial issues: development of the cities' long-term plans, responsibilities for zoning and land use and approval of layouts.

For example, Bangalore alone has six core agencies responsible for a variety of civic functions leading to overlapping and competing administrative jurisdiction and to confusion among politicians, bureaucrats and citizens. As a result, the ULBs usually have little strategic power and a narrow financial margin, limiting their capacity to determine the destiny of the city. Ramanathan (2005) discusses the confusion and apparent lack of coordination in his article 'Too many cooks in the urban services kitchen'.

In December 2005, the federal government announced the National Urban Renewal Mission which has several objectives (GOI, Ministry of Urban Development 2005). Apart from the implementation of decentralisation measures as envisaged in the 74th Constitutional Amendment, it comprises a community participation law aiming to institutionalise citizen participation and urban infrastructure governance by introducing and strengthening e-governance initiatives.

16.2 E-Governance in India

While 'e-government' refers to the use of information and communication technology (ICT) to promote more efficient and cost-effective government and to facilitate more convenient government services, it is also a way of displaying the links between government and its broader political, social and administrative environment. E-governance refers to the processes and structures which encompass all forms of electronic interaction between the government, the corporate sector and the public that have the aim of increasing accountability and transparency of government processes. E-governance can also help to create new spaces for citizen participation (Kettl 2002; Schuppert 2006; Sheridan and Riley 2006; Traunmüller 2003).

In recent years, India has taken great strides in promoting e-government applications. Spending has increased by around 25% annually, and it crossed the 600 million US dollar line in 2007. According to Madon (2004), the country's experience in e-governance can be divided into two main phases: the first from the early 1970s to the late 1990s and the second from the late 1990s onwards. During the first phase, the focus was on in-house government applications in the areas of defence, economic monitoring, planning and the deployment of ICT to manage data-intensive functions related to elections, census and tax administration. During the second phase, the implementation of the National IT Task Force and state government IT policies indicated a paradigmatic shift in e-governance policies towards using IT for a wider range of sectoral applications reaching out to a large number of people in urban as well as rural areas.

Moreover, there has been a movement towards a greater involvement by NGOs and private sector organisations in providing services to the public. These projects have been increasingly influenced by international agencies such as the UNDP and the World Bank under the slogan 'E-governance for Development' (Heeks 2001). In 2002, the Government of India approved the National E-Governance Action Plan to be implemented between 2003 and 2007. The plan aims to create citizen- and business-centric environments for good governance through appropriate governance and institutional mechanisms, core infrastructure and policy. Other key objectives include provision of single-window services to citizens on an 'any time, anywhere' basis, the enhancement of efficiency and productivity of ULBs and the provision of timely and reliable management information relating to municipal administration for effective decision making (3iNetwork 2006).

Presently, many different types of e-governance projects are being implemented simultaneously in different Indian states (a list of various projects is available at www.dqindia.com/content/top_stories/103101501.asp, Accessed 10 Feb 2009). Some projects like the MCA-21 initiative, recently introduced by the Ministry of Company Affairs (MCA), aims to reduce the bureaucratic burden of the corporate sector (www.mca.gov.in, Accessed 10 Feb 2009). Other projects attempt to improve the transparency and accountability within the government by introducing electronic file handling and public grievance systems. One of the prominent e-governance projects in rural areas is Gyandoot in the state of Madhya Pradesh (www.gyandoot.nic.in, Accessed 10 Feb 2009). It is an Internet-based service delivery portal which is set up in one rural district and aims to create a cost-effective, replicable, economically self-reliant model for taking the benefits of ICT to rural citizens.

The services offered encompass a wide range of government departments and can be accessed from a Gyandoot kiosk by any citizen, after paying of a nominal transaction fee. An independent evaluation study reports the success of the project due to the following factors: a single-window facility makes it possible to benefit from the information, the empowerment of citizens, and, finally, the facilitation of entrepreneurship among the rural population through the private ownership of the kiosks. The citizens also perceive a reduction in corruption levels and a decrease in incidents of harassment by government officials. However, the study also highlights major shortfalls: besides a lack of reliable basic infrastructure such as power supply and connectivity, the study notes that Gyandoot does not pay adequate attention to information and application services related to schemes for the rural poor and the most vulnerable segments of society (Centre for Electronic Governance, IIM, Ahmedabad 2003).

Bhoomi ('land' in Hindi) is another promising e-governance project, launched by the Government of Karnataka. It has been selected as the role model for the entire country. Central to this project is a computerised administration of land records. This gives farmers a more direct access to their Record of Rights, Tenancy and Crop (RTC), an important document needed for many official tasks, such as obtaining bank loans, settling land disputes and even as collateral for bail. It is no less important than the identity card. Previously, farmers had to seek out the village accountant to get a copy of the RTC. There were frequent delays and harassment, and bribes had to be paid.

Small farmers, mostly illiterate, could do little to change this unfair system. It was almost impossible for them to obtain the RTC, either because they lacked information on its existence, because they did not know how or because they could not afford the bribe.

Today, for a small fee, a printed copy of the RTC can be obtained online at computerised land record kiosks (Bhoomi centres) in more than 170 subdistrict town offices. Despite the fact that Bhoomi has reduced harassment and corruption substantially, some serious criticism also has to be pointed out (Acharya 2003). While Bhoomi aims to help the poor, it is the poor who appear to be struggling most with the system. The project fails to address illiteracy and gender inequality, as the illiterate do not receive any help in filling the application for the RTC and the large majority of rural women, even though they own about 15% of the land, are not even aware of the new system.

16.3 Case Studies from Metropolitan India

While in rural India the concept of e-governance is being popularised through the tele-kiosk model, e-governance in urban India is being used primarily to provide public services, such as online payment of bills and taxes and the issuing of certificates. However, only very few cities and towns have portals which facilitate citizens to use government interfaces.

The emerging megacity of Hyderabad, the state capital of Andhra Pradesh, has been faster in adopting measures of e-governance than any other large Indian city. The main responsibility for this lies with the former Chief Minister N. Chandrababu Naidu, who governed the state from 1995 to 2004. Reform-oriented and strongly committed to information and communication technologies, he has contributed to the success of many e-governance initiatives. During the late 1990s, his government developed *Vision 2020*, a comprehensive forward-looking document that included the concept of a simple, moral, accountable, responsive and transparent (SMART) government. The realisation of this citizen-centred concept of government was to be achieved by e-governance initiatives. Hence, the state initiated a number of projects, for example, CARD (Computer Aided Administration of Registration Department), FAST (Fully Automated System for Transport), RAJiv (an Internet Village programme) and eSeva (Electronic Citizen Services, 'seva' in Sanskrit means 'service').

eSeva is the most prominent programme in urban areas (www.esevaonline.com). It is built on the success of the TWINS (Twin Cities Network Services) pilot project, which was launched in 1999 in the twin cities Hyderabad and Secunderabad. It is designed to deliver government information and services to citizens online. eSeva as a citizen service system provides a wide spectrum of services in a one-stop mode through state-of-the-art electronic technology. Over 60 government-to-consumer (G2C) and business-to-consumer (B2C) services are available online, ranging from the payment of bills and taxes, over e-registrations and the issuing of certificates, passport applications and licence renewals and the filing of income tax forms to the payment of cell phone bills and courier services and the reservation of cinema tickets. Technically, eSeva is based on a three-tier network architecture. Transactions are conducted on a real-time basis. Departmental servers are connected to the data centre, which in turn is connected through 2-Mbps lines to 46 eSeva centres with 400 service counters that can be found in different locations distributed all over the twin cities and the neighbouring Ranga Reddy District.

The centres operate from 8.00 a.m. to 8.00 p.m. on working days and from 9.00 a.m. to 3.00 p.m. on holidays, making the service convenient for citizens. All service counters are equipped with an electronic queuing system. Citizens are not charged for using the service, but the utilities cost Rs. 5 (about 0.1 euro) per transaction, regardless of the transaction amount. Payment is accepted by cheque, cashier's cheque, cash or credit card. In order to pay via Internet, eSeva has partnered with banks for direct debit transactions.

From an organisational point of view, eSeva is a government organisation based on a public-private partnership model. Infrastructure and buildings are provided by the government of Andhra Pradesh. The staff is employed by the government as well as by the private sector. Two private companies, RAM Informatics Ltd. from Hyderabad and the Mumbai-based CMS Computers Ltd., provide the network and the software. Apart from the technological equipment, maintenance, telephone bills, electricity costs and security are also the responsibility of the private partners.

In terms of transactions recorded, eSeva has shown excellent growth rates. Since it was initiated in 2001, more than 10 million transactions have been carried out, the majority through utility bill payment. Currently, the services are used by about 1,500 citizens a day. Studies on the public opinion on eSeva point out that users consistently cite time-saving, zero corruption, an improved redressing of grievances and the potential to pay their bills under one roof at any location as the main advantages of the system (Pardhasaradhi 2004; Somayajulu et al. 2003; Suthrum and Phillips 2003). The same studies reveal that the large majority of eSeva users are well-educated with a middle-class background, male and open-minded about technical advance. This leads to the assumption that eSeva is helping male middle-class citizens the most and that it is mainly this part of the population that is benefitting from the improvements in quality and speed of government services.

By contrast, the less educated, more or less marginalised urban dwellers, representing about half of the urban population in the twin cities, are rarely seen in eSeva centres. There are three major reasons for this: firstly, the centres are predominantly located in middle-class and upper-class neighbourhoods, some distance away from the places where lower middle-class people and the poor live; secondly, illiterates, many of whom are women, do not receive any support in understanding the logic and functioning of the new system; and thirdly, many of the marginalised urban dwellers are not aware of the existence of eSeva. For them, the e-governance initiative does not bring any positive effects. In order to pay their bills or to get a certificate, they still have to enter government offices where they are exposed to delays, widespread bribery and harassment. In contrast to the top-down approach pursued in Hyderabad, two recently launched e-governance initiatives in Bangalore – commonly known as India's hi-tech capital and one of Hyderabad's major competitors in the fields of ICT and biotechnology – attempt to reduce direct contact between government and the citizenry from a bottom-up perspective. One of these initiatives is called the Ward Works campaign and is organised by the local NGO Janaagraha with the objective of enabling citizens residing in a particular ward to discuss different aspects related to the implementation of works carried out by the city corporation in that ward, such as quality, timeliness and utility (www.janaagraha.org, Accessed 10 Feb 2009).

The second initiative is called the Public Record of Operations and Finance (PROOF) and was launched by five NGOs in 2002. The campaign is based on the premise that in a well-functioning democracy, it is important for citizens to keep a check on government performance, so that it will be motivated to deliver better services. PROOF aims at initiating a process of transparency, accountability and participation in the functioning of the city's municipal corporation. The PROOF campaign focuses on three aspects: obtaining quarterly financial statements from the government, developing performance indicators to evaluate work done by the municipal corporation including primary education and involving people in the functioning of the local government body. The campaign serves three purposes: it empowers local people by involving them in an analysis of government-released accounts, it directly increases the efficiency of the municipal corporation by introducing responsiveness and accountability, and it offers an opportunity for government and local citizens to join hands and ensure that public money is used for the public good.

Since its inception, various PROOF sessions have been held in the city. The initial result is that disclosure of important financial information so far restricted only to the authorities has made the corporation more accountable to its citizens. The initiative thus serves as an effective vigilance mechanism and makes it possible for citizens to keep a close watch on the functioning of the municipal corporation. Nevertheless, there is one major concern regarding the extent to which PROOF meetings are really 'public' and 'inclusive' (Madon 2004). The meetings usually seem very selective and attended only by people from the city corporation and other state and non-profit sector institutions.

16.4 Summary and Conclusion

In India, 'good urban governance' has received much attention along with other keywords such as decentralisation and local autonomy. Various federal and state government initiatives have undertaken considerable efforts to incorporate the principles of good urban governance in achieving sustainability in cities and to meet the challenges of rapid urbanisation. The promotion of e-government applications can be considered part of this wider developmental goal. Various studies report on the successes of these sophisticated ICT-based systems and highlight improved transparency and accountability in the governance of ULBs. Their revenue base has improved, and the interaction between citizens and government departments has been enhanced through quick and convenient access to automated services. However, the digital divide could be reduced only to a certain extent. Access to the ICT-based systems remains deterred by educational, linguistic and technological barriers. Critics also emphasise that the top-down perspective of e-government initiatives followed by federal and state government agencies is reinforcing rather than counteracting polarisation and inequality: men are benefiting more than women; the well-off are benefiting more than the poor. The challenge is to create favourable conditions for reversing the polarities – but this task can be tackled more appropriately by social movements rather than by computers.

References

3iNetwork (2006)

Acharya K (2003) India's move to e-governance exposes ancient system flaws. Panos London Online, July 2003. http://www.panos.org.uk/newsfeatures/featursdetails.asp?id=1127. Accessed 10 Feb 2009

Bagchi AK (2005) Governance – national, corporate and international: the Indian context. J South Asian Stud 28(2):265–281

Centre for Electronic Governance, IIM, Ahmedabad (2003) An evaluation of Gyandoot. http://www.iimahd.ernet.in/egov/documents/gyandoot-evaluation.pdf. Accessed 10 June 2007

Dittrich C (2007) Bangalore: globalisation and fragmentation in India's hightech-capital. Asien 103(April):45–58

Dittrich C (2008) Bedroht, verdrängt und doch gebraucht: Der mobile Kleinhandel mit Nahrungsmitteln in der indischen Metropole Hyderabad. Geogr Rundsch 60(4):22–29

GOI, Ministry of Urban Development (2005) Jawaharlal Nehru National Urban Renewal Mission. http://www.urbanindia.nic.in/moud/programme/ud/jnnurm.htm. Accessed 10 Feb 2009

GOI, Planning Commission (2005) Part II, Chapter 11, Urban infrastructure. http://planningcommission.nic.in:80/midterm/english-pdf/chapter-11.pdf. Accessed 29 Dec 2006

Heeks R (2001) Understanding e-Governance for development. iGovernment Working Paper Series 11. http://www.unpan1.un.org/intradoc/groups/public/documents/NISPAcee/UNPAN015484.pdf. Accessed 10 Feb 2009

Heitzman J (2004) Network city: planning the information society in Bangalore. Oxford University Press, New Delhi

Kettl DK (2002) The transformation of governance. John Hopkins University Press, Baltimore

Madon S (2004) Studying the developmental impact of e-governance initiatives: an exploratory framework. Electron J Inf Syst Dev Countries 20.5:1–13. http://http://unpan1.un.org/intradoc/groups/public/documents/APCITY/UNPAN021286.pdf. Accessed 20 June 2007

Mistelbacher J (2005) Urbanisierungsdynamik in Indien: das Beispiel Delhi. Geogr Rundsch 57(10):20–29

Pardhasaradhi Y (2004) Improving citizen – government interface through e-governance: a study of India. http://www.unpan1.un.org/intradoc/groups/public/documents/APCITY/UNPAN021281.pdf. Accessed 10 Feb 2009

Ramanathan R (2005) Too many cooks in urban services kitchen. Financial Express 17, Jan 2005

Sabir A (ed) (2006) Dimensions of urban poverty. Rawat Publications, Jaipur

Schuppert GF (ed) (2006) Governance-Forschung: Vergewisserung über Stand und Entwicklungslinien. Nomos, Baden-Baden

Sheridan W, Riley TB (2006) Comparing e-government vs. e-governance. http://www.electronicgov.net/pubs/research_papers/SheridanRileyComparEgov.shtml. Accessed 20 June 2007

Somayajulu G, Vanka S, Vedulla V, Phani K (2003) Demand driven and customer-oriented government initiatives in India: the eSeva model of Andhra Pradesh. Hyderabad

Suthrum P, Phillips J (2003) Citizen centricity: e-governance in Andhra Pradesh. University of Michigan Business School, Ann Arbor. http://www.nextbillion.net/node/1446. Accessed 20 June 2007

Traunmüller R (ed) (2003) Electronic Government, Second international conference EGOV 2003, vol 2739, LNCS. Springer, Heidelberg

UN Population Division (2006) World urbanization prospects: the 2005 revision. United Nations, New York

www.dqindia.com/content/top_stories/103101501.asp

www.gyandoot.nic.in

www.janaagraha.org

www.mca.gov.in

Index

F. Kraas et al. (eds.), *Megacities: Our Global Urban Future*, International Year of Planet Earth, DOI 10.1007/978-90-481-3417-5, © Springer Science+Business Media B.V. 2014

MIX
Papier aus verantwortungsvollen Quellen
Paper from responsible sources
FSC® C105338

If you have any concerns about our products,
you can contact us on
ProductSafety@springernature.com

In case Publisher is established outside the EU,
the EU authorized representative is:
Springer Nature Customer Service Center GmbH
Europaplatz 3, 69115 Heidelberg, Germany

Printed by Libri Plureos GmbH
in Hamburg, Germany